SCIENCE AS METAPHOR

SCIENCE AS METAPHOR

The Historical Role of Scientific Theories in Forming Western Culture

edited by

RICHARD OLSON

University of California,
Santa Cruz

Wadsworth Publishing Company, Inc.
Belmont, California

L.C. Cat. Card No.: 71-152550
ISBN-0-534-00039-8
Printed in the United States of America

1 2 3 4 5 6 7 8 9 10—75 74 73 72 71

Preface

Few informed citizens of the Western world are unaware of the dramatic changes in the material conditions of life which have resulted from the application of modern scientific knowledge. Even fewer are aware of the extent to which scientific theories have influenced the mental framework from within which we view the world. Scientific methods and concepts have often been transferred and transformed to provide the basis of some of our deepest social, aesthetic, and religious beliefs; it is the central concern in the following readings to understand the mechanism and impact of this transfer.

The idea for this set of readings arose initially out of an interdisciplinary course, "Science, Philosophy, and Religion," offered to freshmen entering Crown College at the University of California at Santa Cruz. The course — taught by a group which included a biologist, a psychologist, two philosophers, a political theorist, and an historian of science — sought to investigate the impact of the works of Copernicus, Newton, Darwin, and Freud on broad aspects of our contemporary world-view. It was conceived in the dual conviction that nothing is as important for the quality of men's lives as their fundamental beliefs and assumptions about themselves and their relations with nature, men, and God, and that the scientific tradition in Western culture has done more than most of us realize to influence these fundamental beliefs. The course was offered, as is this book, in the hope that an increased awareness of the scientific element within our culture will help men in their ability to control and to benefit from it rather than to be controlled and intimidated.

My thanks go to Ralph Berger, Lawrence Blinks, J. Peter Euben, Josiah Gould, and Paul Lee, with whom I shared responsibility for "Science, Philosophy and Religion," and with whom I gladly share the credit for whatever value this book might have. I can no longer remember who introduced me to what piece of literature or who presented a particular point of view that may appear in my comments. In addition I wish to thank Miss Toni Klimberg who acted as my research assistant in tracking down materials, to the members of the Crown College steno pool who patiently deciphered my handwriting to produce a manuscript, and to Michael Helm who urged me to turn some of my concerns with the cultural impact of science into a manuscript. In addition, thanks go to reviewers John Heilbron, University of California, Berkeley, Roger Hahn, University of California, Berkeley, Allen Dirrim, San Fernando Valley State College, J. Peter Euben, University of California, Santa Cruz, and Carroll Pursell, University of California, Santa Barbara, who read sections of my manuscript and made valuable criticisms. Some of these I adopted and many I pigheadedly and perhaps foolishly ignored. Of course, I alone am responsible for my intransigence in maintaining those errors of fact or interpretation which remain.

Contents

Images are not arguments, (they) rarely even lead to proof, but the mind craves them.

Henry Adams

1 Scientific Thought: Key to the Western World-View

One of the central requirements of a flourishing civilization is the existence of a widely shared set of beliefs about man's relationship to the natural world, to his society of fellow men, and to the Divine. Such a set of beliefs, a so-called world-view, defines what is important in the civilization. It provides the foundation for predominant moral and aesthetic values and for common patterns of behavior. Of course, not all individuals or groups living at a given time and place will assent to the predominant world-view, and in any given culture, not all of the elements of the predominant world-view will be equally shared. But the notion of prevailing world-views as crucial elements of civilization is nonetheless useful, for it provides a way to understand those important and often hidden assumptions which underlie men's feelings, thoughts, and actions.

The following readings are concerned with basic changes which the world-view of Western man has undergone since the Renaissance. The theme which unites them is that our world-view has become dominated by one of its aspects—the scientific understanding of our natural world—to an unprecedented extent seldom understood by students of Western civilization. In most advanced cultures—in the Far East until very recent times, in Islam, and even in medieval Europe—man's relationship to Divinity, or his religion, has formed the core of his world-view, and social and natural knowledge have played subservient roles. In nearly all other cultures—in classical Greece and Rome, for example—man's relationship with his fellows, or his politics, has been dominant. But a dramatic and unique change began to occur in Western Europe as a result of the scientific revolution of the sixteenth and seventeenth centuries; modern Western man began to think, feel, and perceive himself and the world around him in ways fundamentally unlike those of any other people in any prior age.

It is my contention that an overriding interest in scientific method and in successive scientific theories was instrumental in leading Western man away from the pervasive late medieval view of a harmonious, animistic cosmos filled with spirits and sympathies, in which the notions of community and coherence pervaded social thought and in which the awesome power and glory of God was continuously manifest. The new astronomy, physiology, and mechanics of the seventeenth century challenged deeply rooted medieval beliefs and gradually replaced them with a new, mechanistic world-view. In the new world, God began to resemble an engineer or geometer rather than a father, and atomistic individualism replaced the strong traditional emphasis on communal belonging as the basis for social concerns.

During the eighteenth century, scientific theories, dominated by the masterful works of Newton, stimulated men to seek a behavioristic social physic to replace the traditional moral understanding of man and society and led to a valuing of rational over mystical religious beliefs and of democratic over hierarchical social structures. The nineteenth century saw evolutionary ideas drawn from biology used to justify—almost deify—misery and struggle as the necessary conditions for human progress; it saw in thermodynamics the basis for a profound pessimism; and it saw in the new science of psychology an almost total denial of the uniqueness of man. Finally, the new physics of the twentieth century has been central in questioning anew the possibility of the kind of mechanically deterministic world whose existence has been assumed or feared for the past three centuries.

The complex interactions among ideas, events, institutions, and material conditions which have gone into forming Western culture are vastly oversimplified in this scenario and in many of the readings collected here. Of course I realize that scientific ideas have not been the sole driving force of our culture. Scientific theory itself arises only out of and under the influence of its social and intellectual milieu; that is, it is a product as well as a determinant of culture. Furthermore, science influences the development of civilization in many ways, often modifying material conditions more dramatically and directly than intellectual ones. But in spite of the acknowledged limitations of my approach, a twofold conviction has led me to emphasize the interaction between scientific ideas (not their technological manifestations) and other ideas (social, aesthetic, and religious) and to concentrate on the flow of influence from science toward other aspects of culture. First, I feel that nothing is as important to the quality of human life as a man's fundamental beliefs about himself and his relationships with nature, men, and God; and second, I feel that the great influence which the scientific tradition has had on these fundamental beliefs is too often underestimated or ignored.

Sigmund Freud has provided a good introduction to the impact which scientific ideas have had on Western man's world-view, arguing that three of the greatest blows to human pride and dignity have arisen out of the works of scientists. He writes:

> Humanity has, in the course of time, had to endure from the hands of science two great outrages upon its naive self-love. The first was when it realized

> that our earth was not the center of the universe, but only a tiny speck in a world-system of a magnitude hardly conceivable; this is associated in our minds with the name of Copernicus, although Alexandrian doctrines taught something very similar. The second was when biological research robbed man of his peculiar privilege of having been specially created, and relegated him to a descent from the animal world, implying an ineradicable animal nature in him: this transvaluation has been accomplished in our own time upon the investigation of Charles Darwin, Wallace, and their predecessors, and not without the most violent opposition from their contemporaries. But man's craving for grandiosity is now suffering the third and most bitter blow from present-day psychological research which is endeavoring to show the "ego" of each of us that he is not even master in his own house, but that he must remain content with the veriest scraps of information about what is going on unconsciously in his own mind.[1]

Freud's contentions about the implications of Copernicus', Darwin's, and his own theories for man's self-esteem and for his relation with God will be dealt with in detail later in this book, but there is one aspect of what he implied that I must discuss here in order to establish the framework within which the readings have been chosen.

By his use of the words "realized," "robbed," and "show," Freud implied a mechanism by which scientific statements about the natural world come to have social and religious power. He believed that scientific knowledge divested men of their moral and religious illusions, leaving only the core truths of human existence, which were scientific and naturalistic. It was as if truth were like the meat of a walnut from which layers of outer covering had to be removed before it could be eaten. According to this point of view, adopted by many positivistic scientists, by Marx and his followers and by modern political scientists, science has its power to move men because all *true* understanding is fundamentally scientific. Science operates to de-mythify and objectify the world. That is, it demonstrates the fallacies inherent in religious, mystical, and other intensely personal ways of understanding the world and develops an alternative detached and unprejudiced means of viewing phenomena—so that one sees what is *really* there. On the surface, this interpretation of the impact of science is seductive. But there is an important sense in which the de-mythifying image of science is far too limited to account for the variety of ways in which scientific theories have interacted with each other and with other realms of knowledge. In fact, in many ways science has merely justified the successive substitutions of more modern myths for obsolete ones as the basis for our understanding of the world.

In order to understand how science replaces old myths with new, we must consider briefly what the term "explanation" means. Suppose I say I

---

[1]Sigmund Freud, *A General Introduction to Psychoanalysis*, authorized English translation of the revised edition by Joan Riviere (New York: Simon and Schuster, 1935), p. 252.

am going to *explain* something. This almost always implies that I am going to demonstrate its similarity to some objects or events or experiences which are more familiar and are assumed to be better understood. For example, consider the following discussion of the function of rain given by the Greek playwright Aeschylus:

> Holy sky passionately longs to penetrate the earth, and desire takes hold of the earth to achieve this union. Rain from her bedfellow, sky, falls and impregnates the earth, and she brings forth for mortals pasturage for flocks and Demeter's livelihood.

Here Aeschylus "explains"—*i.e.*, makes intelligible—a set of natural phenomena in terms of the sexual desires and activities which take place on the human level and which are presumably more familiar to him. This explanation is clearly anthropomorphic and nonobjective; but consider the following scientific explanation of the function of the values in the veins written by Fabricius ab Aquependante about 1578:

> The mechanism which Nature has devised is strangely like that which artificial means have produced in the machinery of mills. Here millwrights put certain hindrances in the water's way so that a large quantity of it may be kept back and accumulated for the use of the milling machinery. These hindrances are called . . . sluices and dams. . . . Behind them collects in a suitable hollow a large head of water and finally all that is required. So nature works in just the same way in veins (which are just like the channels of rivers) by means of floodgates, either singly or in pairs.

The structure of Fabricius' scientific argument is very much like that of Aeschylus' mythopoeic one. Fabricius uses an analogy between valves and floodgates and between veins and rivers to show that the operations of the human body are just like those more familiar operations of mills; Aeschylus uses a metaphor—or implicit analogy between rain and seminal fluid and between the fertility of woman and the fertility of earth—to show that a natural phenomenon is like the more familiar act of sexual intercourse.

Not all scholars agree that explanations are necessarily analogical in form. In fact, since the last third of the nineteenth century, most students of scientific thought have subscribed to an altogether different theory of explanation. They hold that a scientific explanation is accomplished when a phenomenon to be explained has been subsumed under a general law which is subject to some kind of verification or falsification procedure. There is a great deal of controversy over precisely how general a covering law must be and over the details of what constitutes verification of a scientific law; but most contemporary philosophers of science agree that theoretical explanations can be purely formal deductive systems, some of whose terms are connected with experienced phenomena so that statements from the theory can be tested.

The arguments in this book do not depend upon a denial of this so-called "logical-positivist" interpretation of scientific explanation, though a strong case can be made for the statement that historically, all scientific theories have depended on the use of analogies of one form or another.[2] Even the logical positivists acknowledge that analogies provide powerful heuristic devices to suggest the fundamental laws used in building scientific theories. Thus, while it may be true that not all explanations depend on the use of analogies or metaphors, it is clear that many of the most striking and fruitful modes of generating knowledge or understanding do so. And it should become very clear through the readings collected in this book that much of the influence of science upon the Western world-view has arisen out of the models, analogies, metaphors, and symbols which science has provided to modern man—hence my statement that scientific theories may often merely provide a way of supplanting old, no longer convincing, myths with new models as a basis for social and humanistic thought.

Theories from natural science have been the prime sources for metaphorical understanding, even in social, moral, and religious realms in the Western world precisely because they have that property of simplicity which a good model should have and because, since the seventeenth century, they have been so widely disseminated that they have become familiar to all men. Scientific theories have also provided popular analogies because they have been so impressively successful in explaining natural phenomena.

Analogies are crucial to understanding, but it is important to be aware of their limitations. No two sets of phenomena are exactly alike, and the use of analogies, whether explicit or implicit in the form of metaphor or symbol, may severely limit our modes of understanding by causing us to lose sight of differences as a result of our enthusiasm for the similarities. A striking example of this power of analogy occurs in the work of René Descartes. He began in 1637 to compare the human body to a machine, writing that the way in which the organs of the body are arranged "will not seem strange to those who know how many different automata or moving machines can be made by the industry of man. . . . From this aspect," Descartes says, "the body is to be regarded as a machine." In 1637, Descartes did not believe that the living human body was merely a mechanism; he simply thought that certain aspects of living beings could best be illuminated by considering mechanistic models. But as he continued to use the analogy, comparing a sick man to a poorly made watch (1641) and interpreting the human nervous system as a hydraulic system in which fluid flow transmits information (1649), he began to forget the original differences between man and machine and to identify the two. In *L'Homme*, published posthumously in 1662, Descartes dropped all pretense of analogical thinking as he wrote of the function of the circulatory system:

---

[2]Mary B. Hesse, in her *Models and Analogies in Science* (Notre Dame, Ind.: University of Notre Dame Press, 1966), makes a strong case for this.

> These functions follow in a completely natural way from the disposition of the organs alone *neither more nor less than the movements of a clock or other automaton follow from their counter weights and their wheels*; in such a way that there is no occasion to conceive in them any other vegetative or sensitive soul, no other principle of movement or of life, than its blood and its spirits agitated by the heat of the fire which burns continually in its heart, which has no other nature than those fires which are in inanimate bodies.

Now, there can be no doubt about the fruitfulness of Descartes' vision of the similarities between organisms and machines. His analogy has been and continues to be the basis for the great bulk of work done in biology and psychology; but too often men lose sight of the fact that men and machines may not be identical and they are blinded to alternative images which may be equally valuable.

The temptation to ignore negative analogies in our pleasure with seeing positive ones is understandable and compelling. But we must try constantly to recall the basic analogical nature of our arguments if we are to remain masters of our own belief, capable of discriminating between fruitful insights into the world and gross distortions of it.

Analogies can also be misleading when their users do not truly understand the models which are presumed to be beyond need of explanation. During the nineteenth century, for example, many social philosophers saw in the evolutionary "survival of the fittest" a normative implication—a justification for the status quo—which simply was not present in the biological theory. And in the early twentieth century, historians, literary critics, and art critics were misled by the term "relativity" into believing that Einstein's work had removed all absolute standards and bases for judgment. In fact, Einstein's principle might just as well have been called the principle of absolutivity, since it is based on the absolute invariability of the velocity of light, no matter what the observer's position or condition might be; and since it provides an exact way of determining what an event will look like in any physical frame of reference once it is described in one arbitrary frame. Of course, perversions of analogies like those by the social Darwinists and relativistic humanists do not lessen the social and cultural influence of scientific metaphors, but they do provide important distortions in our beliefs and in our subsequent behavior.

In recent years, scholars have become increasingly aware of the often unconscious and uncritical application of analogies from natural science to other domains and there has been a tendency to condemn such uses because they often seem to lead to misunderstandings. I do not know if the use of scientific theories as models or analogs has led on the whole to more penetrating insights or to more deep misapprehensions about the human, aesthetic, moral, and religious aspects of Western life, but I do know that their use has had a major impact.

As Marshall McLuhan said about his study of moveable type technology, *The Gutenberg Galaxy*, "the theme of this book is not that there is anything good or bad about this development ... but that the unconsciousness of the effect of any force is a disaster, especially a force that we have made ourselves." The readings in this book have been selected to make the reader conscious of the detailed and dramatic forces which the use of scientific models and metaphors have unleashed on Western culture. They deal with a wide variety of fields—literature, politics, social theory, history, law, and theology—in order to demonstrate the pervasive influence of these forces. These readings have also been arranged chronologically, but the reader should not be misled into believing that the influence of older theories is gone. The crisis felt by seventeenth century men on discovering the vastness of the universe is relived by every generation; the behavioristic social thought initiated by followers of Newton is a basic feature of modern political science; and the rugged individualism reinforced by social Darwinism continues to inform much of American society today, for better or for worse. We may applaud, abhor, or try to ignore our past, but we cannot escape it.

I have made no attempt at all to achieve a balanced assessment of the overall importance of scientific belief in forming Western culture in my comments and in the following readings. Huge and influential works have been written completely ignoring or denying any such importance; so I feel wholly justified in presenting a one-dimensional case. Some of the articles selected are guarded in their imputation of influence to science, but even *they* demonstrate the undeniable truth that our modern understanding of the human condition is, as Robin Collingwood has stated, "based on the analogy between the processes of the natural world as studied by natural scientists and the vicissitudes of human affairs as studied by historians."

2

The "New Philosophy" of the Scientific Revolution: A Crumbling of the Old Cosmos and a Reassessment of the Extent and Nature of Human Knowledge

One of the major effects of the new science of the sixteenth and seventeenth centuries was to reinforce and even extend the crisis of authority, order, and certainty arising out of the Protestant Reformation and civil wars. The line from John Donne's "The First Anniversary"—"And new philosophy calls all in doubt"—which introduces the selection of Marjorie Nicolson in this section seems to properly symbolize one of its principal roles.

The late medieval and early Renaissance world-view that was challenged and eventually supplanted by the new science was dominated by the notion of Cosmos, an overwhelming harmony, a unified whole, in which the universe, the earth, society, and man were all images of one another. All were living entities pervaded by God's presence and spirit, and all were ordered in such a way that the parts of one corresponded directly with the parts of each of the others. Thus the organs of the human body corresponded with the parts of the body politic and with the signs of the zodiac in the macrocosm, or body of the universe, as well as with the different elements out of which the earth's body was composed. Not only were all realms of being integrated into one harmonious whole, but within each realm a perfect hierarchical and immutable order was established. Sir Thomas Elyot (1490-1546) demonstrates this belief in his political treatise, the *Governor*:

> Hath God not set degrees and estates in all his glorious works? First in his heavenly ministers, whom he hath constituted in divers degrees called hierarchies. Behold the four elements, whereof the body of man is compact, how they be set in their places called spheres, higher or lower according to the sovereignty of their natures. Behold also the order that God hath put generally in his creatures, being at the most inferior or base and ascending upward. He made not only herbs to garnish the earth but also trees of a more eminent stature than herbs. Semblably in birds, beasts, and fishes some be good for the sustenance of man, some bear things profitable to sundry uses, others be apt to occupation and labor. Every kind of trees, herbs, birds, beasts, and fishes have a peculiar disposition appropered unto them by God, their Creator; so that in everything is order, and without order may be nothing stable or permanent.

And it may not be called order except it do contain degrees, high and base, according to the merit or estimation of the thing that is ordered.[1]

The important thing to be aware of about this world-view is its amazingly interlocking nature, the way in which the hierarchies of orders—of angels, of heavenly bodies, of beasts, and of elements—paralleled one another; so that when the new science called into question the immutability—even the very existence—of one ordering, then it called into question the entire structure.

For this reason, the new astronomy of Copernicus, Kepler, and Galileo drastically undermined the world view of their contemporaries. The old Aristotelian ordering of heavenly bodies was destroyed, and the sun took the place of the earth at the center of the universe. The appearance of the spectacular new star of 1604 as well as the telescopic discoveries of Galileo—mountains on the supposedly spherical moon, and spots coming and going on the supposedly immutable surface of the sun—showed that the heavens were both imperfect and in a constant state of flux. The speculations of many new astronomers even seemed to make the universe indefinite if not infinite in size.

Some people found it easy to live with the new heavenly structure, but to most seventeenth century thinkers the vastness and uncertainty of the new universe was terrifying. John Milton called it "a wide, gaping void" that threatened even the angels with utter loss of being; and Blaise Pascal, French scientist and religious mystic, presented a graphic picture of man's position in an infinite universe that seemed to characterize the despair of many:

What is man in nature? A nothing in comparison with the infinite, an all in comparison with the nothing, a mean between nothing and everything. Since he is infinitely removed from comprehending the extremes, the ends of things and the beginnings are hopelessly hidden from him in an impenetrable secret; he is equally incapable of seeing the nothing of which he is made, and the infinite in which he is swallowed up.

What will he do then, but perceive the appearance of the middle of things in an eternal despair of knowing either their beginning or their end?[2]

Pascal's cry may not give rise to great emotion today; for his keen awareness of human limitation and insignificance in the universe has become a part of the accepted world-view. But when it was first uttered as a statement about the consequences of accepting the new science, it signaled a great change—the downfall of a comfortable and coherent conception of the

---

[1]Quoted from E. M. W. Tillyard, *The Elizabethan World Picture* (London: Chatto and Windus, 1960), pp. 24-25.

[2]*Pensees* (London: J. M. Dent & Sons, Ltd., 1908) Fragment 72, pp. 17-18.

world created *by* God and largely *for* man who was the focus of Divine concern. In "The 'New Astronomy' and English Imagination," Marjorie Nicolson discusses the widespread literary response to the new science, illustrating how it was first used largely for its topical imagery and how its immensely corrosive implications were only gradually recognized and articulated in the works of John Donne.

Like the Reformation, which it paralleled in time, the scientific revolution did not merely question old authorities and beliefs, it offered its own new alternatives. S. F. Mason, in "The Scientific Revolution and the Protestation Reformation," talks of the parallels between the reforms in science and religion, emphasizing their mutually reinforcing rejections of old hierarchies in favor of more democratic principles of order, and showing how the content of both the new astronomy and the new anatomical works of William Harvey began to provide powerful metaphors for religious thought.

Barbara Shapiro's "Law and Science in Seventeenth-Century England" deals with a totally different side of the scientific impact on seventeenth century culture. Perhaps even more important than the content of the new philosophy was its great emphasis on method. It was a new way of philosophizing, not merely a new system of philosophy; and in many cases the method rather than the content of the new science provided important models for social thought. Shapiro quickly outlines the new scientific methodology, discusses the widespread dissemination of scientific ideas, and then shows how principles of method deriving from the new philosophy influenced two important aspects of seventeenth century law—the systematization of statutes, and far more important, the rules of evidence.

The "New Astronomy" and English Imagination

Marjorie Nicolson

"And new Philosophy calls all in doubt," wrote John Donne in 1611, in "The First Anniversarie," a poem marked by cosmic reflection of a sort Donne had not shown before. His "new Philosophy" was more than "new astronomy." As *Ignatius his Conclave* had implied a year earlier, accepted ideas of the Renaissance were everywhere challenged by those of the "Counter Renaissance," other political concepts cutting across conventional belief in "order" and "degree," new theories of medicine attacking the old, Coperni-

Source: "The 'New Astronomy' and English Imagination," in Marjorie Nicolson, *Science and Imagination* (Ithaca, N.Y.: Cornell University Press, 1956) pp. 30-57. Originally published in *Studies in Philology* 32 (1935), 428-462. Reprinted by permission of the University of North Carolina Press and Cornell University Press.

canism challenging accepted ideas of astronomy and astrology. Donne was right in feeling that the "new astronomy" was an important element in the melancholy of the day, leading him to feel, " 'Tis all in peeces, all cohaerence gone."

The "new astronomy" had a dual origin in England. Imported from Italy, it also had a native English background, since England had produced important works of the *nova* of 1572, as well as telescopic observers who might have been better remembered had Galileo's *Sidereus Nuncius* been less dramatic than it was. When the star of 1572 appeared, John Dee published one work and wrote another, which remained in manuscript, to prove that the star was in the celestial region and therefore a *nova*. One of the most important treatments was the *Alae seu Scalae Mathematicae* of Dee's pupil, Thomas Digges, published in 1573. So significant was the treatise that Tycho Brahe in his analysis of various papers written on the phenomenon devoted the longest section of his work to an analysis of Digges' opinions. While the star of 1604 was visible in England, no work of the first importance seems to have been written there. Later English scientists were content to accept the conclusions of Galileo, Kepler and other continental writers.

So far as the telescope is concerned, there is evidence of the development of the instrument in England earlier than on the continent. John Dee worked with "perspective glasses," the advantages of which, he felt, would be their service to a commander in time of war—a belief similar to Galileo's when he developed his early telescopes. Between 1580-1590 William Bourne wrote a *Treatise on the Properties and Qualities of Glasses for Optical Purposes*. Thomas Digges carried research and invention farther. It was said that he "not onely discovered things farre off, read letters, numbred peeces of money . . . but also seven myles of declared what hath been doon at that instante in private places." American scholars have suggested that Digges' interest in the question he raises in his *Perfit Description of the Caelestial Orbes*, "whether the world have his boundes or bee in deed infinite and without boundes," may have been the result of the fact that Digges used his telescope for celestial observation.

If we must still conjecture about the celestial observations of Dee and Digges, we are on firmer ground in the case of Thomas Hariot, though final evaluation of his place in the history of astronomy must wait for the publication of his manuscripts. Hariot is of particular importance in any study of the influence of astronomy on English literary imagination because of his place as scientist-royal to the "School of Night." Hariot had long been interested in perspective glasses. In his *Brief and True Report of the New Found Land of Virginia*, published in 1588, he commented on the astonishment of the natives over equipment Europeans took for granted, "as Mathematicall instruments, sea compasses, the vertue of the loadstone in drawing yron, a perspective glasse whereby was shewed many strange sights. . . ."

Hariot was corresponding with Kepler at least as early as 1606, having apparently come in contact with him through their friend John Erikson. Kepler's earliest inquiries were about Hariot's theories of color and light, to which Hariot replied with a discourse on the rainbow. For two years they seem to have written largely on such problems, interchanging news of books on various phases of natural philosophy. In 1609 they were chiefly concerned with optical problems on which Kepler was engaged in connection with his *Dioptrice*. Evidently Hariot urged the study of Kepler upon his disciples, since Sir William Lower wrote him from Wales on February 6, 1610: "Kepler I read dilligentlie, but therein I find what it is to be so far from you. For as himself, he hath almost put me out of my wits." A later statement in the letter that Kepler "overthrowes the circular Astronomie" indicates that Lower was grappling with Kepler's theory that the planets move not in circles but in ellipses.

Whether Hariot's first interest in the telescope came from his foreign correspondence, or was a natural development from his own early "perspective-glasses" his papers may show. Certainly in 1610, before he had heard about Galileo's discoveries, Hariot was making important astronomical observations by means of his own instrument. Early in 1610, Hariot sent Lower a "perspective Cylinder" with instructions that Lower observe "the Mone in all his changes." Lower's comments suggest that Hariot had already made many of the same observations as Galileo on the nature of the moon. Lower indeed continued to feel that it was Hariot who had been the real pioneer in the new astronomy, and urged him more than once to make public his observations and theories. Hariot, however, remained to the end little interested in persuading others; his observations were made largely for his own interest, and either through carelessness or lack of any desire for public fame, he did little to make others aware of his own importance in the history of astronomy.

No matter how interesting and important, then, were these early observations of Digges and Hariot, there is no question that the real stimulation of imagination in England among poets, philosophers, churchmen, came rather from the Italian than from the native astronomy. Whether because, as Lower had suggested, Hariot made little effort to popularize his discoveries among any except his immediate associates, or, as is more probable, because his theories as yet lacked the cosmic significance of Galileo's, it is clear enough to one who reads the popular literature of the next few years in England that the implications which affected imagination came from Italy, and were much of a sort with the stimulation that immediately followed Galileo's pronouncements in his own country. The mind of the poet was stirred then as now by implications rather than by such careful observations as those of Hariot and Lower, and the seventeenth-century English writers were quick to read such implications into Galileo, as Campanella had done. It is the "Italian's moon" and the "Florentine's new world" we find in English poetry, not the same moon as it had appeared to Hariot in London and to Lower in Wales. To many more than Milton, the telescope remained that

> optic glass the Tuscan artist views
> At evening from the top of Fesole,
> Or in Valdarno.

Indeed, English poets often read implications into Galileo that would have surprised him, as did Lovelace in his "Advice to my best Brother":

> Nor be too confident, fix'd on the shore,
> For even that too borrows from the store
> Of her rich Neighbour, since now wisest know,
> (And this to Galileo's judgement ow)
> The palsie Earth it self is every jot
> As frail, inconstant, waveing as that blot
> We lay upon the Deep; That sometimes lies
> Chang'd, you would think with's botoms properties,
> But this eternal strange Ixion's wheel
> Of giddy earth, n'er whirling leaves to reel
> Till all things are inverted, till they are
> Turn'd to that Antick confus'd state they were.

Two centuries before Browning English poets felt, if they did not say:

> Lo! the moon's self,
> Here in London, yonder late in Florence,
> Shall we find her face, the thrice-transfigured.
> Curving on a sky imbrued with color,
> Drifted over Fiesole by twilight.

News of Galileo's discoveries travelled quickly to England, undoubtedly from many sources. Sir Henry Wotton, Ambassador to Venice, had arrived at his post on September 23, 1604, the week before the *nova* was observed by Kepler's teacher, Michael Maestlin of Tubingen, and only a little more than two weeks before Galileo first observed it. With his great interest in matters of contemporary science and the close association he maintained throughout his embassy with professors and students at Padua, he must have heard much at first hand of the popular interest in Galileo's lectures, though the letters written by him during the first year of his ambassadorship are no longer extant. Fortunately, his first letter on the telescope is still in existence, a letter written on the day the *Sidereus Nuncius* appeared in Venice, March 13, 1610. Wotton wrote to the Earl of Salisbury:

> Now touching the occurents of the present, I send herewith unto his Majesty the strangest piece of news (as I may justly call it) that he hath ever yet received from any part of the world; which is the annexed book (come abroad this very day) of the Mathematical Professor at Padua, who by the help of an

> optical instrument (which both enlargeth and approximateth the object) invented first in Flanders, and bettered by himself, hath discovered four new planets rolling about the sphere of Jupiter, besides many other unknown fixed stars; likewise, the true cause of the *Via Lactae*, so long searched; and lastly, that the moon is not spherical, but endued with many prominences, and, which is of all the strangest, illuminated with the solar light by reflection from the body of the earth, as he seemeth to say. So as upon the whole subject he hath first overthrown all former astronomy—for we must have a new sphere to save the appearances—and next all astrology. For the virtue of these new planets must needs vary the judicial part, and why may there not yet be more? These things I have been bold thus to discourse unto your Lordship, whereof here all corners are full. And the author runneth a fortune to be either exceeding famous or exceeding ridiculous. By the next ship your Lordship shall receive from me one of the above instruments, as it is bettered by this man.

Among others who may have written back to England were a group of British students at Padua, some of them resident pupils of Galileo. Three at least were closely associated with him during the period of his spectacular discoveries, one the "Mr. Willoughby" who was the guide of that curious traveller Thomas Coryat, another the Scottish Thomas Seggett, whose epigrams on Galileo have already been mentioned. Seggett was the first student at Padua whom Wotton knew, since the ambassador's earliest official act was to secure Seggett's release from a Venetian prison into which he had been thrown for speaking too freely of a Venetian nobleman. To another Scottish student Galileo entrusted his defense against the attack of the German, Martin Horky, who, shortly after the publication of the *Sidereus Nuncius*, wrote a pamphlet against the possibility of the new planets Galileo believed he had discovered.

It is entirely possible that Kepler himself sent word of the telescopic discoveries to Hariot. Evidently Hariot had not been startled "to see every day some of your inventions taken from you." With characteristic generosity, Hariot was among the first to praise the *Sidereus Nuncius*. The letter he wrote to Lower on the subject is not yet available, if it is extant, but Lower's reply reflects the early Italian attitude:

> Me thinkes my diligent Galileus hath done more in his three fold discoverie than Magellane in opening the streightes to the South sea or the dutch men that weare eaten by beares in Nova Zembla. . . . I am so affected with this newes as I wish sommer were past that I mighte observe these phenomenes also. . . . We both with wonder and delighte fell a consideringe your letter, we are here so on fire with thes thinges that I must renew my request and your promise to send mee of all sortes of thes Cylinders. . . . Send me so manie as you thinke needful unto thes observations, and in requitall, I will send you store of observations.

As Lower and his friends in Wales were "affected with this newes," so were many readers of the *Sidereus Nuncius* in England. The widespread literary response of poets, dramatists, satirists, essayists to every aspect of the "new astronomy" is the best evidence of lay interest in science during the early years of the seventeenth century.

"New stars" shone in English literature throughout the century. Drummond of Hawthornden, in his "Shadow of Judgement," saw in them an indication of the end of the world:

> New stars above the eighth heaven sparkle clear,
> Mars chops with Saturn; Jove claims Mars's sphere.

"The Heavens are not only fruitful in new and unheard-of stars," wrote Sir Thomas Browne in the *Religio Medici*, "the Earth in plants and animals, but men's minds also in villany and vices." *The Anatomy of Melancholy*, particularly the "Digression of Air," is rich in conjectures of the significance of "new motions of the heavens, new stars, *palantia sidera*, comets, clouds, call them what you will." Phineas Fletcher suggested in his idyllic description of Virginia:

> There every starre sheds his sweet influence
> And radiant beams; great, little, old, and new
> Their glittering rayes, and frequent confluence
> The milky way to God's high palace strew.

Cowley in his "Ode to the Royal Society" mentioned the fear inspired by the early *novae*:

> So when by various Turns of the celestial Dance,
> In many thousand Years,
> A Star, so long unknown, appears,
> Though Heaven it self more beauteous by it grow,
> It troubles and alarms the World below,
> Does to the Wise a Star, to Fools a Meteor show.

Samuel Butler, who lost no opportunity of poking fun at the new astronomy, described Sidrophel, gazing through his telescope at a boy flying a kite, and leaping to the conclusion that this remarkable phenomenon was

> A comet, and without a beard!
> Or star, that ne'r before appeared!

While new stars begin to shine in poetry immediately after 1604, they are much more common after the invention of the telescope, and figures drawn from them are customarily involved with more complex allusions to various of Galileo's discoveries.

References to the telescope are so frequent in seventeenth-century English literature that any attempt to list them would result only in a dull catalogue. The instrument became increasingly familiar to the general public. By the middle of the century it was a common spectacle in public parks, one of the holiday sports of the people. Its names are many: it is now the "Mathematicians perspicil" or the "perplexive glasse" of Ben Jonson; now the "optick magnifying Glasse" of Donne; again the "trunk-spectacle" or "trunk," the "perspective" or "the glass." Most often it is "the optick tube" or merely "tube." Its lenses are the "spectacles with which the stars" man reads "in smallest characters," as Butler said in *Hudibras*. Its appearance is frequently commented upon, as in Davenant's reference in *Gondibert* to "vast tubes, which like long cedars mounted lie." Innumerable figures were coined from it, compliments paid, insults hurled by its means. "Tell my Lady Elizabeth," the second Viscount Conway wrote of Elizabeth Cecil, "that to see hir is better than any sight I can see at land, or at sea, or that Galileo with his perspective can see in heaven." James Stephen in 1615 called his mistress "my perspective glasse, through which I view the world's vanity." Marvell wrote "To the King," playing upon the spots Galileo's tube had discovered in the sun:

> So his bold Tube, Man, to the Sun apply'd,
> And Spots unknown to the bright Stars descry'd;
> Show'd they obscure him, while too near they please,
> And seem his Courtiers, are but his disease.
> Through Optick Trunk the Planet seem'd to hear,
> And hurls them off, e're since, in his Career.

The telescope could readily be adapted to religious implications, as in John Vicars' "A Prospective Glasse to Looke into Heaven," Arthur Wilson's "Faith's optic," or Donne's lines written in 1614 in connection with the "Obsequies to the Lord Harrington":

> Though God be our true glasse, through which we see
> All, since the beeing of all things is hee,
> Yet are the trunkes which doe to us derive
> Things, in proportion fit, by perspective,
> Deeds of good men; for by their living here,
> Vertues, indeed remote, seeme to be neare.

Vaughan concluded his reflection, in "They Are All Gone,"

> Either disperse these mists, which blot and fill
> My perspective (still) as they pass,
> Or else remove me hence unto that hill
> Where I shall need no glass.

The literary adaptations of the telescope were so many and various that Kepler's rhetoric in his continuation of Galileo's work seems hardly exaggerated:

> What now, dear reader, shall we make out of our telescope? Shall we make a Mercury's magic-wand to cross the liquid ether with, and, like Lucian, lead a colony to the uninhabited evening star, allured by the sweetness of the place? or shall we make it a Cupid's arrow, which, entering by our eyes, has pierced our inmost mind, and fired us with a love of Venus? . . . O telescope, instrument of much knowledge, more precious than any sceptre! Is not he who holds thee in his hand made king and lord of the works of God?

Throughout the century we find charming figures of speech describing "the new galaxie," the Milky Way, none lovelier than Milton's

> broad and ample road, whose dust is gold,
> And pavement stars, as stars to thee appear
> Seen in the galaxy, that milky way
> Which nightly as a circling zone thou seest
> Powdered with stars.

The literature of Galileo's moon is so extensive that one might devote a long section to that "coast i' th' Noone (the Florentine's new World)" as Phineas Fletcher called it in "The Locusts." The new "planets" became involved with the new world in the moon, as half-curiously, half-fearfully man began to question whether moon and planets are inhabited. Drummond saw in the discovery of new planets as of new stars still another omen of the end of the world:

> New Worlds seen, shine
> With other suns and moons, false stars decline,
> And dive in seas; red comets warm the air,
> And blaze, as other worlds were judged there.

To Dryden, later in the century, the possibility of such worlds bespoke the goodness of parental Deity:

> Perhaps a thousand other worlds that lie
> Remote from us, and latent in the sky,
> Are lightened by his beams, and kindly nurs'd.

Optimism and pessimism, we shall find, were both involved in the "new astronomy." From the extended cosmos discovered by the telescope, one group might turn, as did Pope, declaring:

> Thro' worlds unnumber'd tho' the God be known,
> 'Tis ours to trace him only in our own.

Others believing that "boundless mind affects a boundless space," exulted with Young:

> The soul of man was made to walk the skies,
> Delightful outlet of her prison here!
> There, disencumber'd from her chains, the ties
> Of toys terrestrial, she can rove at large,
> There, freely can respire, dilate, extend,
> In full proportion let loose all her powers;
> And, undeluded, grasp at something great.

... Among English poets, none showed a more immediate response to the new discoveries than John Donne, nor is there a more remarkable example of the effect of the *Sidereus Nuncius*. Whether there was depth in the manifold interests of Donne some critics question; no one who observes his reaction to novelty can doubt the breadth of his interests. His curious mind—that "hydroptic, immoderate thirst of human languages and learning"—was always hungry for new fare; for a time he fastened upon the new astronomy as another source for figures of speech, another vehicle for his restless imagination. No matter what our ultimate conclusion may be as to the effect upon him of the Copernican or the Galilean hypotheses, no one who reads his poetry thoughtfully, with due attention to chronology, can fail to see the stimulation of his mind from those two great moments in the history of astronomy—the discovery of Kepler's new star of 1604 and Galileo's contagious enthusiasm in the *Sidereus Nuncius*.

The *Songs and Sonets*, the majority of which were written before the turn of the century, contain no significant astronomical figures of speech. References to the sun, moon, and stars appear, though not in the proportion we find later, but these are purely conventional:

> And yet no greater, but more eminent,
> Love by the Spring is growne;
> As, in the firmament,
> Starres by the Sunne are not inlarg'd, but showne.

The same conventions appear in the *Epigrams*, the *Elegies and Heroicall Epistles*, and the *Satyres*, in which the figures of speech drawn from the heavens are even fewer than those in the *Songs and Sonets*, and equally lacking in significance. Even in the first *Progresse of the Soule*, in which one might expect some stirring of the cosmic imagination, we do not find it. Not unjustly we may conclude that Donne unconsciously suggested his own early attitude in "Elegie XVIII":

Although we see Celestial bodies move
Above the earth, the earth we Till and love.

In the *Verse Letters to Severall Personages* we begin to feel the stirring of a different kind of imagination. In one of the poems to the Countess of Bedford, we find Donne's first poetic references to the Copernican theory and his association of the "new astronomy" with the "new Philosophy":

As new Philosophy arrests the Sunne
And bids the passive earth about it runne,
So we have dull'd our minde.

At this time new stars begin to shine in his poetry. The stars of 1572 and 1604 came to his mind when he sought a figure of speech to describe the geographical and astronomical expansion of his own period:

We have added to the world Virginia, and sent
Two new starres lately to the firmament.

Both the new star of 1604 and the comet of 1607 came to mind when he wrote to the Countess of Huntingdon:

Who vagrant transitory Comets sees,
Wonders, because they are rare: But a new starre
Whose motion with the firmament agrees,
Is miracle, for there no new things are.

The regret which Tycho and Kepler, in pre-telescopic days, may have felt as they observed the disappearance of new stars is caught by Donne in three passages:

But, as when heaven lookes on us with new eyes,
Those new starres every Artist exercise,
What place they should assigne to them they doubt,
Argue, and agree not, till those starres goe out.

At another time he looked back to the Golden Age when man's leisurely life was so indefinite that

if a slow pac'd starre had stolne away
From the observers marking, he might stay
Two or three hundred yeares to see't againe,
And then make up his observation plaine.

Again he wrote, unconsciously predicting his own later attitude toward such figures of speech:

> And in these Constellations then arise
> New starres, and old doe vanish from our eyes.

For a short time new stars eclipsed the old in Donne's mind, then vanished from his poetry, either because the possibility of *novae* became so widely accepted that the metaphor ceased to be novel or because Donne's imagination passed on to other stimuli. But the new stars did not really become important to Donne until they were interpreted in the light of Galileo's discoveries.

The *Sidereus Nuncius*, we remember, appeared in Venice on March 13, 1610. Donne's prose satire, the *Conclave Ignatii*, was entered in the *Stationer's Register* on January 24, 1611; on May 18 of the same year an English translation was entered under the title, *Ignatius his Conclave.* The work began in Donne's mind as a Lucianic "Dialogue in Hell." He introduced "innovators" of the Renaissance, in "a secret place, where there were not many, beside Lucifer himselfe; to which, onely they had title, which had so attempted any innovation in this life, that they gave affront to all antiquities, and induced doubts, and anxieties, and scruples, and after, a libertie of beleeving what they would; at length established opinions, directly contrary to all established before." Here he found Machiavelli, Paracelsus, Ignatius, and, most important for our purposes, Copernicus:

> As soone as the doore creekt, I spied a certaine Mathematitian, which till then had bene busied to finde, to deride, to detrude Ptolomey; and now with an erect countenance, and setled pace, came to the gates, and with hands and feet (scarce respecting Lucifer himselfe) beat the dores, and cried: "Are those shut against me, to whom all the Heavens were ever open, who was a Soule to the Earth, and gave it motion?"
>
> By this I knew it was Copernicus ... To whom Lucifer sayd: "Who are you?" ... "I am he, which pitying thee who wert thrust into the Center of the world, raysed both thee, and thy prison; the Earth, up into the Heavens; so as by my meanes God doth not enjoy his revenge upon thee. The Sunne, which was an officious spy, and a betrayer of faults, and so thine enemy, I have appointed to go into the lowest part of the world. Shall these gates be open to such as have innovated in small matters? and shall they be shut against me, who have turned the whole frame of the world, and am thereby almost a new Creator?"

Although *Ignatius his Conclave* took form in Donne's mind as a dialogue in Hell, as published it seems to begin as another literary form, also ultimately derived from Lucian, a cosmic voyage: "I was in an Extasie," wrote the author, "and

My little wandring sportfull Soule
Ghest, and Companion of my body,

had liberty to wander through all places, and to survey and reckon all the roomes and all the volumes of the heavens, and to comprehend the situation, the dimensions, the nature, the people, and the policy, both of the swimming Ilands, the Planets, and of all those which are fixed in the firmament." Only at the beginning and toward the end of the work does Donne do anything with his cosmic voyage, with the result that *Ignatius* is a curious mingling of two literary *genres*. In the original Dialogue of the Dead Copernicus is a major character; at the beginning and end of the work Donne is interested in Galileo and Kepler, who are mentioned immediately after the passage on "the swimming Ilands, the Planets, and of all that are fixed in the firmament." "Of which," Donne writes, "I thinke it an honester part as yet to be silent, than to do Galileo wrong by speaking of it, who of late hath summoned the other world, the Stars to come nearer to him, and give him an account of themselves. Or to Keppler, who (as himselfe testifies of himselfe) *ever since Tycho Braches death hath received it into his care, that no new thing should be done in heaven without his knowledge.*"

Later in the dialogue Donne returns to the theme of Galileo's discoveries, when Lucifer trying "earnestly to thinke, how he might leave Ignatius out" decided on a device:

> I will write to the Bishop of Rome: he shall call Galilaeo the Florentine to him; who by this time hath thoroughly instructed himselfe of all the hills, woods, and Cities in the new world, the Moone. And since he effected so much with his first Glasses, that he saw the Moone, in so neere a distance that hee gave himselfe satisfaction of all, and the least parts in her, when now being growne to more perfection in his Art, he shall have made new Glasses, and they received a hallowing from the Pope, he may draw the Moone, like a boate floating upon the water, as neere the earth as he will. And thither . . . shall all the Jesuites bee transferred. . . . And with the same ease as you passe from the earth to the Moone, you may passe from the Moone to the other starrs, which are also thought to be worlds, and so you may beget and propagate many Hells, and enlarge your Empire.

In *Ignatius* "Jack" Donne was smiling over a "new astronomy" that did not yet matter to him. Within a year he began to understand its implications more fully. Early in 1610 occurred the death of Elizabeth Drury, daughter of Sir Robert Drury, who was to be Donne's patron. At the time of the young girl's death, Donne wrote his *Funerall Elegie*, conventional enough in its references. For the anniversary of the death in 1611 he composed *The First Anniversarie*. Barely a year had passed, yet Donne who had laughed in *Ignatius* had come to realize that "new Philosophy" might indeed call all in doubt.

The themes of the *Anniversary Poems* were literary commonplaces. Donne himself had used most of them before, though never with such moroseness. Elizabeth Drury was not the real subject of Donne's poem. The death of a young girl was only his point of departure. When we turn from the *Funerall Elegie* to the poem written only a year later on the anniversary of Elizabeth Drury's death, we find a changed world and a man aware of change. All the former motifs are here; the "Decay of Nature" is the central theme; the new stars appear and fade again, but they have ceased to be mere figures of speech, and have taken on new meaning, as Donne sees the relation to cosmic philosophy. They are a symbol of the "Disproportion" and the "Mutability" in the universe of which Donne has become compellingly aware:

> It teares
> The Firmament in eight and forty sheires,
> And in these Constellations then arise
> New starres, and old doe vanish from our eyes.
> As though heav'n suffered earthquakes, peace or war,
> When new Towers rise, and old demolish't are.

It is man, he is forced to conclude, who has slain "Proportion":

> Man hath weav'd out a net, and this net throwne
> Upon the Heavens, and now they are his owne.
> Loth to goe up the hill, or labour thus
> To goe to heaven, we make heaven come to us.
> We spur, we reine the starres, and in their race
> They're diversly content t'obey our pace.

Donne's most quoted lines, in which he reflects the poignant regret of a generation which had inherited from the past centuries conceptions of *order, proportion, unity,* which had felt the assurance of the immutable heavens of Aristotle, take on new meaning when one reads them, remembering the revolution in thought that was occurring in 1610:

> And new Philosophy calls all in doubt,
> The Element of fire is quite put out;
> The Sun is lost, and th' earth, and no mans wit
> Can well direct him where to looke for it.
> And freely men confesse that this world's spent,
> When in the Planets, and the Firmament
> They seeke so many new; then see that this
> Is crumbled out againe to his Atomies.
> 'Tis all in peeces, all cohaerence gone;
> All just supply, and all Relation.

Here is Copernicanism to be sure, but it is less the position of this world than the awareness of new worlds that troubles the poet; less the disruption of this little world of man than the realization how slight a part that world plays in an enlarged and enlarging universe that leads Donne to his conclusion.

The *First Anniversarie* marks the climax of Donne's interest in the Galilean astronomy. Indeed, were Donne the main interest of this essay, it would be tempting to study in detail a highly significant trait of his imagination, admirably illustrated by his reaction to the "new stars" and Galileo's discoveries: his almost immediate response to new ideas, followed by cooling of his interest. The "new star" motif, as we have seen, is persistent from his first use of it in the *Verse Letters* through those letters—that is, it appears some time between 1604 and 1609 and persists until about 1614. After that it almost disappears. The *Sidereus Nuncius* obviously affected him greatly at the time; those themes, too, tend to disappear. This was, of course, partly the result of the change in Donne's own personal life: his visit to the continent with Sir Robert Drury, followed by his entrance into Holy Orders, the death of his wife—all these turned his thoughts in other directions. One must remember that immediately after the period dealt with here occurred the change from his writing on secular subjects to the period of his religious poetry and prose. While Donne continued to refer to astronomical ideas in his letters and in such personal works as his *Devotions upon Emergent Occasions*, references to the new astronomy—both Copernican and Galilean—are rare in his formal *Sermons* and *Divine Poems*. He occasionally mentioned the telescope, as when he wrote to Mr. Tilman:

> If then th'Astronomers, whereas they spie
> A new-found Starre, their Opticks magnifie,
> How brave are those, who with their Engine, can
> Bring man to heaven, and heaven againe to man?

or when he says, in a sermon:

> God's perspective glass, his spectacle, is the whole world . . . and through that spectacle the faults of princes, in God's eye, are multiplied far above those of private men.

In another sermon we find a reference to Galileo's discovery of the nature of the Milky Way, which recalls the earlier figure in *The Primrose:*

> In that glistering circle in the firmament, which we call the Galaxie, the milkie way, there is not one starre of any of the six great magnitudes, which Astronomers proceed upon, belonging to that circle: it is a glorious circle, and possesseth a great part of heaven, and yet it is all of so little starres, as have no name, no knowledge taken of them.

Only occasionally in the sermons does Donne venture upon more philosophical connotations of the Galilean discoveries. The idea of a plurality of worlds, for a churchman, was indeed a dangerous tenet, even, as it came to be called, the "new heresy." The condemnation of Bruno listed that belief as one of the chief charges against him; many orthodox Protestants, as well as Catholics, felt that such a conception struck at the roots of the Christian idea of the sacrifice of Christ. In his *Holy Sonnets*, written after the death of his wife in 1617, Donne does not hesitate to suggest the possible existence of other worlds, though without theological connotation:

> You which beyond that heaven which was most high
> Have found new sphears, and of new lands can write,
> Powre new seas in mine eyes, that so I might
> Drowne my world with my weeping earnestly.

In his *Devotions upon Emergent Occasions*, he plays with the idea:

> Men that inhere upon Nature only, are so far from thinking, that there is anything singular in this world, as that they will scarce thinke, that this world it selfe is singular, but that every Planet, and every Starre, is another World like this; They find reason to conceive, not onely a pluralitie in every Species in the world, but a pluralitie of worlds.

But he is cautious in his expression upon one of the few occasions in which he raised the problem in his sermons:

> And then that heaven, which spreads so farre, as that subtill men have, with some appearance of probabilities, imagined, that in that heaven, in those manifold Spheres of the Planets and the Starres, there are many earths, many worlds, as big as this world which we inhabit.

Only on one occasion does he approach the vexing question of orthodox theology, and then, so guarded is his statement that it is difficult to tell whether he is really referring to the "new heresy." He writes in one of his more dramatic sermons of

> the merit and passion of Christ Jesus, sufficient to save millions of worlds, and yet, many millions in this world (all the heathen excluded from any interest therein) when God hath a kingdom so large, as that nothing limits it . . . .

Whether the churchman found it expedient in his sermons to keep away from those disputed matters, or whether the poet had ceased to feel the appeal of figures of speech that once had led him to new reaches of poetry is a matter that cannot be determined.

Ideas that were startling and revolutionary in 1610 were becoming more familiar. But the "new Philosophy" did not cease to be "new" in Donne's lifetime, did not settle into a literary convention that was almost a commonplace as it became a half-century later. The unrest Donne had felt in 1611 at the disruption of his universe was still felt by Pascal with his terror of the silence of infinite spaces. Donne apparently ceased to ponder the new hypotheses. Perhaps his attitude was that of Milton's Angel, who recognized the appeal of the new ideas to man's curiosity, but warned Adam:

> Solicit not thy thoughts with matters hid;
> Leave them to God above; him serve and fear.

"Paradox and Probleme" Donne remains to his modern critics, who will probably never agree about the "conversion" that transformed Jack Donne into Dr. John Donne, Dean of St. Paul's. Read against the scientific background of his time, against the inruption of new stars and the dramatic discoveries of Galileo's telescope, his experience seems an epitome of the experience of his generation. I shall continue to believe that the discoveries of the new astronomy, coinciding with a troubled period in his own personal life and in his age, proved the straw that broke the back of his youthful scepticism and led John Donne "from the mistresse of my youth, Poesy, to the wife of mine age, Divinity."

The Scientific Revolution and the Protestant Reformation

S. F. Mason

From the inception of the scientific revolution and the Protestant Reformation during the sixteenth century, it has been noted by various authors that there were some similarities between the new science and the new religion, and that Protestant beliefs have been more conducive than the Catholic faith to the promotion of scientific activity. The sixteenth-century medical writer, Richard Bostocke, held the view that the reform of religion had been indispensible to the reform of medicine, and that Copernicus and Paracelsus had restored the sciences just as Luther and Calvin had restored

Source: S. F. Mason, "The Scientific Revolution and the Protestant Reformation—I: Calvin and Servetus in Relation to the New Astronomy and the Theory of the Circulation of the Blood," *Annals of Science* 9 (1953), 64-81. Reprinted by permission of Messrs. Taylor and Francis, Ltd.

religion. Referring to the new iatro-chemistry, Bostocke wrote, that Paracelsus

> was not the author and inventor of this arte as the followers of the Ethnicks physicke doe imagine, . . . no more than Wicklife, Luther, Occolampadius, Swinglius, Calvin, etc., were the authors and inventors of the Gospell and religion in Christes Church when they restored it to his puritie, according to God's word . . . And no more than Nicolaus Copernicus, which lived at the time of this Paracelsus, and restored to us the place of the starres according to the trueth . . . is to be called the author and inventor of the motions of the starres.

During the following century, Thomas Sprat, who was himself an Anglican Churchman and a Fellow of the newly formed Royal Society, noted "the agreement that is between the present design of the Royal Society, and that of our Church in its beginning. They both may lay equal claim to the word Reformation; the one having compassed it in Religion, the other purposing it in Philosophy . . . ."

The impulse which the religious ethos gave to scientific activity was perhaps the most important single element integrating science with religion in seventeenth-century England, but it cannot be said that the men of the time fully separated their natural philosophy from their theology. Moreover the medieval view of the world had been composed of a theology and a natural philosophy which were closely integrated and its overthrow was accomplished simultaneously, though in a piece-meal fashion, on the one hand by the Protestant Reformers who criticized the theological aspects, and on the other by the scientists who criticized the cosmological features. In the development of the new sciences and the new theologies it is possible to discern that the criticism of the Calvinists and of the astronomers proceeded along lines which bore some similarity one to the other, and that both prepared the way for a new mechanical-theological world view, which enjoyed considerable popularity during the late seventeenth and eighteenth centuries.

The *leit-motif* of the medieval view of the universe to which both the Protestant Reformers and the early modern scientists took exception was the concept of hierarchy. The concept was rooted in the idea that the universe was made up of a graded chain of beings, stretching down from the Deity in the empyrean Heaven at the periphery of the world, through the hierarchies of angelic beings inhabiting the celestial spheres, to the ranks of mankind, animals, plants, and minerals of the lowly terrestrial sphere at the centre of the cosmic system. In such a view of the world even the material elements were regarded as graded in perfection. In particular the four terrestrial elements were thought to be susceptible naturally only to linear motions, which had a beginning and an end like all terrestrial phenomena, whilst the fifth element of the heavenly bodies moved naturally with the eternality and perfection of uniform circular motion. According to the generally received theory of mechanics a body in motion required the constant action of a mover, and an important integration of ancient natural philosophy with early Christian theology had occurred through the identification of the movers of the heavenly bodies with the angelic beings mentioned in the Scriptures. The pseudo-Dionysius the Areopagite in the fifth century had arranged these angelic beings into a hierarchy of nine orders, and by the time of the medieval Schoolmen it had come to be generally accepted that these orders were responsible for the motions of the nine celestial spheres. Because of their hierarchical arrangement, the superior celestial entities governed the inferior, and in general the government of the universe was such that a given being had domination over those below it in the scale of creatures and served those above it in the scale.

It was against such a conception of a hierarchically ordered universe that the Protestant Reformers, particularly Calvin, and the early modern scientists rebelled. The pseudo-Dionysius the Areopagite, by means of his celestial hierarchy of angelic beings, had justified the setting up of the ecclesiastical hierarchy of Church government on earth, an organization which Calvin strongly opposed. [Calvin wrote:]

> To the government thus constituted, some gave the name of Hierarchy—a name in my opinion improper, certainly one not used by Scripture. For the Holy Spirit designed to provide that no one should dream of primacy or domination in regard to the government of the Church.

Replying to the justifications offered by the pseudo-Dionysius the Areopagite, Calvin affirmed that there was no "ground for subtle philosophical comparisons between the celestial and earthly hierarchy", and he averred that mankind could not know whether the angelic beings were ordered by rank or not. [Calvin wrote:]

> None can deny that Dionysius (whoever he may have been) has many shrewd and subtle disquisitions in his Celestial Hierarchy: but only looking at them more closely, everyone must see that they are merely idle talk ... in regard both to the number and rank of angels, let us class them among those mysterious subjects, the full revelation of which is deferred to the last day, and accordingly refrain from inquiring too curiously or talking presumptuously.

In formulating a positive point of view on these matters, Calvin tended to minimize the role of the angelic beings in the government of the universe, and to assign to the Deity a more absolute and direct control over His creatures. Speaking of the relation between God and the angelic beings, Calvin affirmed:

> Whenever He pleases, He passes them by and performs His own work by a single nod; so far are they from relieving Him of any difficulty.

Not only did the Deity govern the universe directly, but also, according to Calvin, He had predetermined all events from the beginning. [Calvin wrote:]

> We hold that God is the disposer and ruler of all things—that from the remotest eternity according to His own wisdom, He decreed what He was to do, and now by His power executes what He has decreed. Hence we maintain that, by His Providence, not heaven and earth and inanimate creatures only, but also the counsels and wills of men are so governed as to move exactly in the course which He has destined.

Thus the workings of the Calvinist universe were orderly, and were fully predeterminate. Miraculous happenings contravening the laws of Nature were no longer to be expected: "God alters no law of Nature", as it was put by John Preston, 1587-1628, the Puritan Master of Emmanuel College, Cambridge. The angelic beings lost their power, and ultimately their place in the cosmic scheme, and by the end of the seventeenth century they no longer played an important part in Calvinist theology. Such a development was regretted by some divines, but they admitted that it was the case. The Presbyterian theologian, Richard Baxter, 1615-91, remarked in a work published during the last year of his life:

> It is a doleful instance of the effect of a perverse kind of opposition to Popery, and a running from one extream to another, to note how little sense most Protestants shew of the great benefits that we receive by Angels: How seldom we hear them in publick or private give thanks to God for their ministry and helps, and more seldom, pray for it?

Baxter himself confessed that he had not come across many instances of the ministry of angelic beings: most of the stories he related were concerned with the activities of evil spirits. However it seems that even the evil spiritual beings had disappeared from educated English opinion by this time, for John Aubrey (1626-97) tells us:

> When I was a child, and so before the civill warres, the fashion was for old women and maydes to tell fabulous stories night-times, and of sprights and walking of ghosts. This was derived down from mother to daughter from the monkish balance, which upheld the holy Church; for the divines say, 'Deny spirits and you are an atheist'. When the warres came, and with them liberty of conscience and liberty of inquisition, the phantomes vanish.

Some doubts concerning the existence of spiritual beings appear to have been raised before the Civil Wars, as Thomas Browne (1605-82) writing about 1635, remarked that it was a riddle to him, "how so many learned heads should so far forget their metaphysics, and destroy the ladder and scale of creatures, as to question the existence of Spirits".

The removal of the angelic beings from the government of the universe in Calvinist theology was indeed a criticism of the idea that the universe was peopled by a graded scale of creatures, or rather it was a criticism of the concept of hierarchy which was the kernel of the idea in the medieval world-picture. The Deity no longer ruled the universe by delegating His authority to a hierarchy of spiritual beings, each with a degree of power which decreased as the scale was descended, but now He governed directly as an Absolute Power by means of decrees decided upon at the beginning. These decrees were nothing other than the laws of Nature, the theological doctrine of predestination thus preparing the way for the philosophy of mechanical determinism. Indeed it seems that both the term and the concept of 'laws of Nature' were first used consistently by the primary exponent of the mechanical philosophy, notably in the *Discourse on Method*, where Descartes spoke of the "laws established in Nature by God". The historian of the term, 'the laws of Nature', has ascribed the usage of the phrase to the hypostatization into the cosmic realm of the earthly rule through statute law developed by the absolute monarchs of the sixteenth and seventeenth centuries. "It is not mere chance", wrote Zilsel, "that the Cartesian idea of God as the legislator of the universe, developed only forty years after Jean Bodin's theory of sovereignty". Perhaps it is also not a matter of chance that, some forty years before Bodin, Calvin was working towards the conception of God as the law-giver of the universe, an Absolute Ruler who exercised His power directly, and not through the mediacy of subordinate beings.

Whilst the Calvinists in theology were moving away from the hierarchical conception of the government of the universe towards an absolutist theory of cosmic rule, the early modern scientists were effecting a not dissimilar transformation in natural philosophy. Copernicus, whose heliocentric system of the world was published in 1543, rejected, implicitly at least, the gradation of the elements, for he assigned to the earth that circularity of motion which hitherto had been the prerogative of celestial matter. Furthermore he invested the heavenly bodies with the property of gravitation, which previously had been considered to be peculiar to the earth, implying once more that the earth was similar to the other planets, and was not inferior. Again, according to his pupil, Rheticus, he rejected the hierarchical view that the higher celestial spheres influenced the motions of the lower. [Rheticus wrote:]

> In the hypothesis of my teacher, the sphere of each planet advances uniformly with the motion assigned to it by nature, and completes its period without being forced into any inequality by the power of a higher sphere.

In the stead of the traditional hierarchical view, Copernicus advanced the conception that the sun had an absolute rule over the bodies of the solar system:

> In the middle of all sits the Sun enthroned. In this most beautiful temple, could we place this luminary in any better position from which he can illuminate the whole at once? He is rightly called the Lamp, the Mind, the Ruler of the Universe: Hermes Trismegistus names him the Visible God, Sophocles' Electra calls him the All-Seeing. So the Sun sits as upon a royal throne, ruling his children, the planets which circle round him.

In most of the new systems of the world put forward during the sixteenth and seventeenth centuries the sun assumed a position of particular importance in the cosmic order. Such was the case even in the conservative system of Tycho Brahe, who following the dominant tradition, placed the earth immobile at the centre of the universe. [Brahe wrote:]

> I think that the celestial motions are disposed in such a manner that only the sun and the moon, together with the eighth sphere, the most distant of all, have the centres of their motions in the earth: the five other planets turn about the Sun as round their Chief and King, the Sun being always at the middle of their orbs, which accompany it in its annual motion. Thus the Sun is the Rule and the End of all the revolutions, and, like Apollo in the midst of the Muses, he rules alone the entire celestial harmony of the motions which surround him.

William Gilbert of Colchester adopted yet another system of the world, which was similar to the Tychonian scheme save that the earth performed a diurnal spin upon its axis, but his cosmic values were identical with those of Copernicus. The sun, Gilbert thought, was the noblest body in the universe, "as he causes the planets to advance on their courses", and he served as "the chief inciter of action in nature". The earth and the planets were equal in status, according to Gilbert, for "The earth's motion is performed with as little labour as the motion of the other heavenly bodies, . . . neither is it inferior in dignity to some of these". In rejecting the concept of hierarchy, Gilbert in fact appears to have sensed some connexion between theology and natural philosophy.

> The sun is not swept round by Mars' sphere (if sphere he have) and its motion, nor Mars by Jupiter's sphere, nor Jupiter by Saturn's. The higher do not tyranise over the lower, for the heaven of both the philosopher and the divine must be gentle, happy, tranquil, and not subject to changes.

With Kepler the connexion became more explicit, for he located the domicile of the theological Ruler of the universe upon the central power of the solar system:

> The sun alone appears by virtue of his dignity and power, suited for this motive duty [of moving the planets] and worthy to become the home of God himself. . . . For if the Germans elect him as Caesar who has the most power in the whole empire, who would hesitate to confer the votes of the celestial motions on him who already has been administering all other movements and changes by the benefit of the light which is entirely his possession?

The physician-philosophers, Robert Fludd (1574-1637), and John Baptist van Helmont (*c.* 1577-1644), similarly placed the dwelling place of the Deity upon the sun.

A transition from the hierarchical to the absolutist conception of cosmic government was effected not only in the theory of the macrocosm, or the universe at large, but also in the theory of the microcosm, or the little world of the human body. Here the Protestant Reformation and the scientific revolution found perhaps their most intimate and direct connexion, for in early modern times the theory of the lesser circulation of the blood first appeared in a work which was primarily a theological treatise, the *Restitutio Christianismi* (1553) of the Reformer, Michael Servetus, whose particular theology enabled him to overcome some of the difficulties which had stood in the way of the circulation theory. The main intellectual obstacles to the development of the theory of the circulation of the blood were the Aristotelian notion that only celestial matter was susceptible naturally to circular motion, natural terrestrial motions having a beginning and an end, and the conception, which had come down from Galen, that the human body was governed physiologically by a hierarchy of three organ systems, each with its own separate function. The vegetative function of nourishment and growth had its seat in the liver, and it was mediated by the ebb and flow of the dark red blood in the veins under the impetus of the natural spirit, which, like the venous blood, was prepared in the liver. The vital function of animal heat and muscular activity had its seat in the heart, which prepared the vital spirit, the mover of the bright red arterial blood. The nervous function of irritability and sensitivity had its seat in the brain, and it was exercised by a fluid in the nerves, moved by the animal spirit. Galen's analysis of the physiology of the human body in terms of a trichotomous hierarchy of functions was a particularized expression of the general tendency in late antiquity and the middle ages to classify entities into triadic ranks at all levels of gradation—classes, orders, genera, and species. According to the late medieval theologian, Raymond de Sebonde, for example, the world was peopled by three general kinds of beings: those that were entirely material, like minerals, plants, and animals; those that were entirely spiritual, such as the angelic beings; and those that were mixed, namely, human beings. Each kind was subject to a threefold division, and each group in turn subdivided triadically. Thus the vegetable kingdom was made up of herbs, plants, and trees; the animal kingdom of birds, fishes, and land animals; mankind consisted of labourers, burghers, and nobles, together with a separate triadic hierarchy of ecclesiastics; and above man there were three triadic orders of angelic beings; whilst at the head of the scale of beings inhabiting the world stood the supreme Trinity of the Godhead.

The most important single element in the theology of Servetus was his rejection of the doctrine of the Trinity. He denied that the Son was co-eternal with the Father, and he held that the Holy Spirit was nothing other than the breath of God, or the all-pervading pneuma of the world. Just as he denied the supreme Trinity, so Servetus questioned the general principle of triadic classification. In particular, having had a medical training, Servetus denied that the human body was governed physiologically by the threefold hierarchy of the natural, vital, and animal spirits, claiming that "in all of these there is the energy of the one spirit and of the light of God". Thus there were not two kinds of blood, differentiated by the natural and the vital spirits, but only one blood, and only one kind of spirit in the blood, as "the vital is that which is communicated by the joins from the arteries to the veins, in which it is called the natural". Servetus suggested that the spirit of the blood was the soul of man, or rather than, "the soul itself is the blood", a view which he supported with texts from the Old Testament. Such a view implied that the soul perished with the body, and that man was wholly mortal, an implication of his theology which was charged against Servetus at his trial by Calvin.

The traditional doctrine of the three physiological fluids, the two bloods and the nervous fluid, had been a considerable obstacle to the development of the circulation theory, for any large-scale movement of the blood from the arteries to the veins and from the veins to the arteries, which the circulation theory required, would have involved the complete mixing of what were regarded as quite different fluids, each with its own distinct function. Having denied that there was an important difference between the venous and arterial bloods, the way to the circulation theory was clear for Servetus, though he put forward positively only the theory of the lesser circulation of the blood from the right to the left chamber of the heart through the lungs, as he was primarily concerned with the relations between the blood and the atmosphere, that is, the soul and the Holy Spirit. He held that "the Divine breath is in the air", and that in the lungs the inspired air mixed with the blood circulating through them. Thereby the blood was purified, for "by air God makes ruddy the blood", while the soul participated in the divine pneuma.

It is probable that the *Restitutio Christianismi* had little influence in the sixteenth and seventeenth centuries, as most of the copies of the work were burnt in the same year as its author (1553). However, Servetus's point of view lived on, and in some cases where his theology was sympathetically received, the scientific implications of that view were discussed. The contemporary reformer who was the most interested in the views of Servetus, and to whom Servetus addressed a considerable portion of his *Restitutio Christian-*

ismi, was the German, Philip Melanchthon (1497-1560). Melanchthon had criticized the doctrine of the Trinity in 1521, and in 1552 he published a work in which he controverted the Aristotelian view that there were three ventricles in the heart. The view that there were three ventricles in the heart was not a live issue during the sixteenth century, but the criticism of Melanchthon, like the work of Servetus, illustrates the tendency of those who had doubts concerning the theological doctrine of the Trinity to deny the existence of trichotomous divisions in Nature. Of more importance perhaps, was the influence which the pantheistic and anti-Trinitarian attitude of Servetus exerted in Italy. The founders of the Unitarian movement, Laelius Sozini (1525-62), and his nephew, Fausto Sozini (1539-1604), were Italian, and so too were the main theorists of the circulation of the blood intermediate between Servetus and Harvey, namely, Colombo, Cesalpino, and Bruno. Cesalpino and Bruno, with whom the theory was but an abstract speculation, were inclined towards a pantheistic theology, like Servetus. Pantheism carried the criticism of the concept of hierarchy to an extreme within the framework of theology, for it implied that all beings were equally divine, and, in particular, that terrestrial entities, as well as the celestial, possessed the attributes of nobility and perfection, notably circularity of motion. The same conclusion could be derived from the notion that man was a microcosm, or a complete epitome of the universe in miniature, an idea to which Harvey subscribed and which was central to Bruno's system, as, suitably interpreted, the notion indicated that circulation should be as much a feature of the microcosm as of the macrocosm.

With Cesalpino, Bruno, and Harvey, an absolutist theory of the government of the microcosm was evolved, the heart being accorded the same primacy of place in the human body that the sun had assumed in the new systems of the world. For Servetus the blood as vector of the soul was the most important constituent of the body, and with the blood was associated the heart as the prime organ: "The heart is the first to live, the source of heat in the middle of the body." Such a conception found favour with Cesalpino, Bruno, and particularly Harvey, who identified his cosmic evaluations with those of the astronomers, ascribing to the heart the same attributes as Copernicus had accorded to the sun. [Harvey wrote:]

> The heart is the beginning of life; the sun of the microcosm, even as the sun in his turn might well be designated the heart of the world; for it is the heart by whose virtue and pulse the blood is moved, perfected, made apt to nourish, and is preserved from corruption and coagulation; it is the household divinity which, discharging its function, nourishes, cherishes, quickens the whole body, and is indeed the foundation of life, the source of all action. . . . The heart, like the prince in a kingdom, in whose hands lie the chief and highest authority, rules over all; it is the original and foundation from which all power is derived, on which all power depends in the animal body.

Such a view was supported by the works of Aristotle, who had assigned the central control of the body to the heart, in contrast to Galen who had assumed that there was a more decentralized control through a triad of organs, the liver, the heart, and the brain. But the early modern development went beyond Aristotle; and Cesalpino, Bruno, and Harvey accorded to the blood that circularity of motion which had been considered, in general, as unique to the celestial bodies. For such a purpose these men drew upon the one example that Aristotle had given of natural circular motion within the terrestrial sphere, the example, Harvey indicated, which had suggested to him the idea of the circular motion of the blood:

> I began to think whether there might not be a motion as it were in a circle . . . in the same way as Aristotle says that the air and the rain emulate the circular motion of the superior bodies: for the moist earth warmed by the sun evaporates; the vapours drawn upwards are condensed, and descending in the form of rain moisten the earth again. . . . And so in all likelihood does it come to pass in the body, through the motion of the blood.

Harvey indeed appears to have searched constantly for examples of circular motion in terrestrial bodies in order to give them parity of status with the celestial bodies and to remove the hierarchical gradations between them. In his work on the generation of animals, Harvey suggested that the development of the chick embryo in the early stages was a kind of circular expansion, and that the succession of individuals constituting a species was a form of circular motion emulating the movements of the heavenly bodies:

> This is the round which makes the race of common fowl eternal; now pullet, now egg, the series is continued in perpetuity; from frail and perishing individuals an immortal species is engendered. By these, and means like to these, do we see many inferior or terrestrial things brought to emulate the perpetuity of superior or celestial things.

Such changes in cosmic evaluations brought about by the early modern scientists appear to have had an influence upon the metaphors and similes of majesty used during the sixteenth and seventeenth centuries. It had been customary to compare a monarch in his realm to the mind in the body, or to the primum mobile at the periphery of the universe, governing the world from above, but now the sun and the heart at the centre came into fashion as the images and analogies of majesty. John Norden in his *Christian Familiar Comfort* (1600) described Elizabeth I as the primum mobile of England, and Francis Bacon too adopted the analogy in his *Essay on Seditions*. But when William Harvey published his theory of the circulation of the blood in 1628 he dedicated his book to Charles I as "the sun of the world around him, the heart of the republic", and when Louis XIV came of age in 1660 he was hailed, not as the primum mobile of France, but as *le Roi Soleil*.

The new cosmic values of the early modern scientists also had an influence upon the theology of one of the Protestant sects, the Mortalists, so called because they held that the soul of man perished with his body. The most notable member of this sect was the poet, John Milton, while the chief exponent of the Mortalist theology was Richard Overton, one of the leaders of the Leveller movement during the English Civil Wars. Overton held that God must reside in the most noble part of the universe, and since the scientists considered the sun to be the most exalted of the heavenly bodies, the Deity must dwell in the sun. The Deity, Overton argued,

> ascended upward from the Earth into some part of the coelestiall bodies above, therefore, without doubt he must be in the most excellent, glorious, and heavenly part thereof, which is the SUN, the most excellent piece of the whole Creation, the Epitome of God's power, conveyour of life, groweth, strength, and being to everie Creature under Heaven . . . and according to the famous Copernicus and Tycho Braheus, it is highest in station to the whole Creation: And it is called by the Learned, Cor Coeli, Anima et Oculus Mundi, Planetarum et Fixarum Choragus, Author Generationis. . . . As for the Coelum Empyreum which the Astronomers have invented for his residence, I know no better ground they have for it than such as Dromodotus the Philosopher in Pedantius had to prove there was devils: Sunt Antipodes, Ergo Daemones. Sunt Coeli, Ergo Coelum Empyreum.

The Mortalists' central doctrine that the soul of man perished with his body was supported by the view of Servetus, Harvey, and others, that the soul was itself the blood, a view which Overton quoted, but did not draw upon extensively, discussing at greater length the idea that the soul was the sum total of man's faculties, those faculties perishing with the body. The doctrine that the soul of man perished with his body served to carry further the criticism of the concept of hierarchy which had been initiated by the Calvinists and the early modern scientists. Calvin and the astronomers had questioned the existence of the celestial hierarchies, but the Mortalists now came to doubt the reality of the terrestrial hierarchies of plants, animals, and mankind. If man were wholly mortal, then he did not differ essentially from the animals, Overton argued, for animals as well as man would ultimately rise again:

> For all other Creatures as well as man shall be raised and delivered from Death at the Resurrection. . . . That which befalleth the sonnes of man, befalleth beasts, even one thing befalleth them all: as one dyeth, so dyeth the other; yea they have all one breath, so that man hath no preheminence above a beast.

The Mortalists seem to have disappeared from England after the Restoration, but it is probable that they have had a continuous history in America, as they returned from thence to England during the nineteenth century as the Christadelphian sect. By this time, with the advance of astronomy, the domicile of the Deity was placed at the unknown centre of the universe, but the other doctrines of the earlier Mortalists remained substantially unchanged in the Christadelphian theology. Thomas Wright of Durham (1711-86), a pioneer of sidereal astronomy, appears to have been the first to suggest that the dwelling place of the Deity lay at the unknown centre of the universe. In 1750 he put forward the hypothesis that the sun and the stars of the Milky Way moved round a common centre, thus forming a giant sidereal system. At this centre, Wright suggested, "the Divine Presence, or some corporeal agent full of all virtues and perfections, more immediately presides over His creation".

The Mortalists were only a minor sect, and their synthesis of Protestant theology with early modern natural philosophy was not an important one. By equating the status of man with that of the beasts they carried the early modern criticism of the concept of hierarchy to lengths which were not acceptable to most of the Protestant theologians and natural philosophers who were influential during the seventeenth century, and in tacitly accepting the view that there was but one world, by placing the sun at its centre, they by-passed the important doctrine that there was a plurality or an infinity of worlds in the universe, a doctrine which helped to bring together the main streams of Calvinist theology and early modern natural philosophy in seventeenth-century England. The doctrine of the plurality of worlds even in its most restricted form, in which it was supposed that the planets of the solar system, or the moon alone, were inhabited by creatures similar to those on earth, carried implications which ran counter to the traditional concept of hierarchy, for it was derived from, and gave added weight to, the notion that the heavenly bodies were of the same qualitative nature as terrestrial bodies. "Through arrogance", wrote Montaigne, "man dareth imaginarily place himselfe above the circle of the Moone, and reduce heaven under his feet. It is through the vanity of the same imagination that he dare equall himselfe to God, that he ascribeth divine conditions unto himselfe, that he selecteth and separateth himselfe from out the ranke of other creatures." In its extended form, wherein it was supposed that each star was the centre of a planetary system and that there was an infinity of inhabited worlds in the universe, the doctrine tended to assign the same status to all finite existents. This was one of the aspects of the infinite which led to Pascal's inquietude:

> In comparison with all these Infinites all finites are equal, and I see no reason for fixing our imagination on one more than on another. The only comparison which we make of ourselves to the finite is painful to us.

Giordano Bruno, with whom the doctrine of the plurality of worlds came into vogue in modern times, made the same point, though he indicated that, while the various entities in the universe had equal powers, they might be graded according to the perfection of their natures. [Bruno wrote:]

> I declare to you that there is in truth one prime and principal motive power; but not prime and principal in the sense that there is a second, a third, and other motive powers descending down a certain scale to the midmost and last, since such motive powers neither do nor can exist. For where there is infinite number, there can be neither rank nor numerical order, although there is rank and order according to the nature and worth, either of diverse species and kinds, or of diverse grades of the same kinds and species. . . . Thus there is no *primum mobile*, no order from it of second and other mobile bodies either to a last body or yet to infinity. But all mobile bodies are equally near to and equally far from the prime and universal motive power.

Thus with Bruno the integument of the traditional scale of beings was dissolved. The orders, genera, species, and the individuals within those species, retained their gradations of perfection, but there were no relations of domination and servitude between the creatures composing the scale of beings, and there was no flow of power down the scale from the higher to the lower, for the creatures were now autonomous, each deriving its motion from an immanent source.

Law and Science in Seventeenth-Century England

Barbara J. Shapiro

. . . Seventeenth-century England underwent what has been called a scientific revolution. This revolution was not confined to a narrow circle of professional scientists. The scientific ideas associated with the names Copernicus, Galileo, Newton, and Boyle became the common property, and changed the basic modes of thought, of the entire literate community of England and certainly of the community of gentlemen to which the barristers and judges belonged. The change was not only in beliefs about the nature of the physical world, but more fundamentally in beliefs about what methods were best for finding the truth, how certain men could be about the truths they found, and how they might best communicate those truths to one another. Science came to shape men's views of what was and was not "common sense," of what was and was not well argued, and of what was and was not assumed to be true. Thus even if English law were based exclusively on a taught tradition of commonsense reasoning, much of that tradition would necessarily be dictated by the scientific modes of thought that by the end of the century were becoming common to all literate Englishmen.

We need not, however, content ourselves with linking science to lawyers by way of the general category of gentlemen. Instead it can be demonstrated

---

Source: Barbara J. Shapiro, "Law and Science in Seventeenth-Century England," *Stanford Law Review* 21 (1969), 727-763.

that several of the major legal scholars and leaders of the bar of that day were immersed in the new science, and that these men not only viewed the two activities as compatible but frequently drew on the same central core of ideas for both their legal and scientific pursuits.

Perhaps most important it is possible to demonstrate that two major intellectual developments of the 17th century occurred almost simultaneously in law and science. The first was the drive for systematic arrangement and presentation of existing knowledge into scientifically organized categories. This concern for systematization is not only a characteristic of 17th-century English science, but is also reflected in the first comprehensive and systematic treatises on English law, Sir Matthew Hale's *Analysis of Law* and *History and Analysis oj the Common Law.* Indeed, the original inspiration for this Article was my reaction to the way Hale's treatises are usually treated as the first systematic work on English law and then blithely linked to Bracton on one side and Blackstone on the other as if legal treatises were somehow independently fated to move from the obscurities of Littleton to the latest elegance from West or Little Brown. Hale's work is part and parcel of the distinctly 17th-century concern for organized and simplified presentation in which he participated as both a lawyer and a scientist. A similar point could be made about casebooks and court reports, which are usually treated as purely independent developments. They began to flourish only after Bacon, again as both a scientist and a lawyer, emphasized the need for the careful and accurate collection and correlation of data from which generalizations might be drawn.

The second major movement of the century shared by law and science was the concern with degrees of certainty, or, in more modern terminology, probability. There was a new emphasis on the grading of evidence on scales of reliability and probable truth. In science, statements about the real world became probabilistic hypotheses. In law, an examination of the credibility of witnesses and a concern for truth beyond a reasonable doubt replaced the search for absolute truth. Here again there are striking overlaps between the vocabularies and methods found in law and science as well as an overlap in the actual persons employing these notions.

Accordingly, the first part of this Article sketches the scientific revolution and its effects on general intellectual life, the second concerns the involvement of lawyers in scientific activities, the third describes the movement toward systematization in science and law and the fourth the development of degrees of certainty.

Science in Seventeenth-Century England

The Nature of the Revolution

In the late 16th and 17th centuries, Europe experienced a scientific revolution. Although the ancient and medieval precursors of the developments of this period can be traced, beginning about 1550 there was a very rapid acceleration in mathematical learning, the elaboration of the scientific method, and the accumulation of empirical results derived from mathematical and scientific inquiry. Nor were these developments the product of a few isolated scientists, for during this period an increasing proportion of the intellectual community enlisted itself in scientific pursuits, either as active investigators or amateurs of the new learning. Ideas nurtured in the scientific milieu became the common coin of intellectual discourse, even the discourse of those who at first glance might seem far removed from mathematics and experimentation.

While in biology and botany the revolution took the form of new findings and classification, something a good deal more startling occurred in the realm of astronomy and mechanics. For in these areas the medieval conceptions were rejected and new ways of thinking established that dominated inquiry until the 20th century. Not only were fundamentally new explanations of the workings of the natural world offered, but a whole new canon of scientific investigation as well. This new scientific method was adopted or at least aspired to in fields of knowledge far beyond the boundaries of astronomy and mechanics.

The revolution is most clearly seen in astronomy, where a centuries-old conception of the cosmos was overthrown. The traditional Ptolemaic cosmos, which placed man firmly at the center of the universe, conformed well with theocentric and Christian notions and neatly fitted the medieval urge toward hierarchy. For all the emotional and theological satisfactions it produced, however, the system created incredible difficulties for astronomers who sought to describe it mathematically. In the mid-16th century, Copernicus offered his hypothesis of the central position of the sun as a solution to many of these mathematical complexities, but not until astronomical observations in the 17th century seemed to verify the Copernican hypothesis was its impact widely felt. It then became accepted as a description of reality rather than simply a mathematical convenience. Once the earth was removed from its central position, it was no longer as easy to view the cosmos in terms of God's purpose for man. Thus the revolution in astronomy was more than simply a rejection of the authority of the Ptolemaic system; it resulted in a major adjustment in man's view of his place and purpose in the universe.

Although advancements in the area of mechanics did not affect the layman as dramatically as the verification of the Copernican hypothesis, Galileo's mathematical formulations of the movement of terrestrial bodies were also important. Isaac Newton, in the latter portion of the 17th century, combined the new celestial physics and the new terrestrial physics of Galileo into a single system that again provided a coherent view of the cosmos. This view, however, could be understood only by mathematical reasoning and scientific observations. The Newtonian system became the unquestioned basis of European assumptions about the nature and operation of the cosmos.

The important shift in intellectual outlook and the enormous accomplishments of the individual sciences that marked the scientific revolution were made possible at least partially by changes in attitudes. For science to develop and to gain some kind of popular acceptance it was necessary to eliminate the imputation of superior and final knowledge to ancient authorities and to substitute the notion that the acceptance or rejection of statements concerning natural phenomena must depend on contemporary reason and observation. A notion of the possibility of an increase or progress in knowledge was thus requisite to, as well as an outgrowth of, scientific inquiry.

Concurrent with the attack on traditional authority was an attack on traditional methods of verifying statements or obtaining truth. The virtuosi rejected not only the earlier emphasis on metaphysics, but also the deductive method by which the scholastics had approached all subject matters. Although the virtuosi reached no consensus on the proper approach to natural phenomena, the concern with method was an overriding one in the 17th century. The question was constantly discussed and refined until at the end of the century something like the modern scientific method had emerged.

One of the initially most popular as well as powerful attacks on the traditional scholastic method was that of Francis Bacon. Rejecting the deductive approach, Bacon asserted that once the proper scientific method was adopted knowledge could be harnessed for the use of society. This visionary side of Bacon had perhaps greater impact than did the radically inductive approach that he advocated. Bacon argued that by collecting numerous instances of the particular, generalizations would emerge. This radical empiricism had its limitations and was not the method finally adopted by science; yet it was one of the streams of thought that contributed to the new method.

The other side of the new 17th-century approach to truth was provided by René Descartes, one of the greatest mathematicians of the century, whose mathematization of scientific inquiry made an enormous impact on the development of specific sciences as well as on the development of scientific method. It is too simple to say that the "scientific method" adopted by the virtuosi was a union of Bacon's empiricism and Descartes' essentially logical and deductive approach to the problem of knowledge, but clearly both the observation and collection aspects of the former and the hypothesizing and theorizing aspects of the latter were essential to the new method and were often practiced in combination by 17th-century investigators.

By the end of the 17th century then, traditional views of the cosmos and its functioning had been upset and new methods of determining truth and investigating the natural world had replaced those that had been accepted for centuries. Even though there was no unanimous agreement on the newer methods and epistemologies, it was clear that new types of standards were accepted. This scientific revolution of the 17th century did more than alter men's view of natural phenomena: It resulted in a new set of philosophical propositions about the nature of man and his ability to know the world.

The Diffusion of Scientific Knowledge

Nor were these new views the exclusive possession of a small or isolated scientific community. Sixteenth- and 17th-century scientists did not think of themselves as a closed professional community. Most scientific publications were directed to the general reading public. Writers like Bacon and Galileo displayed an almost missionary zeal to spread the faith of the new science and to convey their vision of the brighter future to be gained through man's new understanding. Over 10 percent of the works published in England between 1475 and 1640 were on scientific subjects, and the majority of these were written in the vernacular rather than Latin. Some were original contributions to the development of scientific knowledge while others were designed to convince the ordinary man of the validity and utility of scientific knowledge and the legitimacy of scientific endeavor.

This spate of general publication was reinforced by the development of several widespread correspondence networks among scientists, a number of which resulted in the creation of scientific journals. The desire to reach both broader domestic and foreign audiences created a linguistic dilemma, for the vernacular was obviously better suited to one and Latin to the other. As a result there were several efforts to create a universal language with which to communicate scientific information.

Closely associated with the desire for a spread of scientific knowledge was the movement toward systematization and classification of knowledge. Bacon and John Wilkins, proponents of a universal language, were advocates of systematically collecting scientific information, and Wilkins was largely responsible for stimulating the creation of the great biological and botanical classifications of John Ray and Francis Willoughby. In every learned discipline, and we shall find law to be no exception, there was a strong movement toward arranging both concepts and data into some rational ordering that could be easily communicated and fitted into the materials of other fields so that a universal knowledge might emerge.

In the scientific, as in the literary world, men met together first informally and then in societies and academies to discuss their findings and experiments. The most important of these was the Royal Society of London which received its first charter in 1662 and numbered among its members not only men making important scientific contributions, but gentlemen, clergymen, businessmen, and politicians for whom scientific discussion was largely a diversion. In addition to reporting on research and undertaking new experiments, the Society propagated the "new philosophy" to the wider world.

The composition of the scientific community in England was extremely diverse—socially, economically, and religiously. Science in the 16th and 17th centuries was often an avocation rather than a profession. Widespread but frequently peripheral scientific interest had been made possible by an educational system in which more people were receiving instruction than at any earlier period or than would again until well into the 20th century. New chairs of mathematics and astronomy were established at Oxford. Most of the famed scientists of the century were educated in the universities and several held administrative or academic posts at some time during their careers. Science was also taught in nonuniversity settings, the most important being Gresham College in London. Thus scientific knowledge was available to most gentlemen and indeed to many who could not quite claim that title. The Royal Society boasted an "equal balance of all professions . . . ," and by the Restoration period, science was widely recognized to be part of that general culture that a gentleman was expected to possess.

Yet science involved more than a pleasant hobby, for its mode of operation, its methodological concerns, and its general approach to empirical problems affected all modes of thought. Even the most cursory examination of 17th-century religion, literature, philosophy, and social thought indicates how much the intellectual classes had absorbed the scientific ideology. Certainly philosophy was affected by the scientific revolution; in one sense the scientific revolution meant the victory of the "new philosophy." The names Bacon, Descartes, and Locke were as important for philosophy as they were for science. In fact, the distinction between philosophy and science was not clear—rationalist and empiricist theories were not only the basis of 17th-century epistemology but of the scientific method as well.

Social and political thought, too, came under the sway of the "new philosophy." Hobbes' discussion of politics in the *Leviathan* would have been virtually inconceivable prior to the 17th century. His aim was to create a scientifically accurate description and analysis of social and political behavior. Locke attempted to define the reality of political life and organization by rejecting divine and authoritative principles. Natural law, though hardly a novel conception in European thought, moved to the center of political analysis and was then gradually transformed from a religious to a more secular and scientific conceptual framework that sought to deal with the universal regularities of men in societies as others dealt with the regularities of physical nature.

A number of recent studies have persuasively demonstrated the impact of science upon 17th-century literature. The new science seems to have had a marked effect in simplifying English prose style and to have contributed to the ultimate victory of prose over poetry as the general vehicle for the presentation of serious discourse. New astronomical and geographical discoveries provided subject matter for literary speculation in such works as More's *Utopia* and Shakespeare's *The Tempest*, and such literary men as Bacon, Raleigh, Sprat, Pepys, Cowley, Glanvill, Evelyn, and Dryden were associated with scientific circles. John Wilkins, who is best known among literary historians for his contribution to the simplification of prose style, used that style to popularize Copernican astronomy.

Perhaps the most striking feature of 17th-century science is that, while representing a very fundamental alteration of our world view and permeating every aspect of intellectual life, it was so rapidly and generally accepted in a nation that was subject to grave religious and political factionalism. Neither Anglicans nor Puritans—the two major religious-political groupings—expressed real hostility to the "new philosophy." A view of science as the study of one of God's two great books—nature (the other being scripture)—was extremely important in making scientific pursuits acceptable to society at large. The virtuosi believed that God worked in orderly ways and that man might to some extent discern these ways. They were thus able to incorporate the traditional concepts of Providence and natural law into the scientific investigation of nature. As the century progressed, God was gradually turned into little more than a first cause, which set the original mechanism in motion. The secondary causes that subsequently moved the world could be the subject of strictly scientific inquiry without theological disturbance. The scientific approach to philosophical and natural problems was not only compatible with but also directly affected religious culture. The virtuosi had developed a canon of scientific disputation that stressed tentative, nondogmatic statement, the full exchange of all relevant theories and data, and the suspension of judgment where proofs were insufficient. They sought, with some success, to carry this canon over into the religious realm, and the work of the scientists was a major component of the liberalization of religion that came to be called latitudinarianism. . . .

The Concern for Systemization: Sir Matthew Hale and the Scientific Study of Law

We have already examined the urge toward popularization and the interest in a universal language among 17th-century scientists. Stated more broadly, a major thrust of intellectual life in this period was toward the systematic organization and presentation of the whole of human knowledge in such a way as to make it available to all literate men. It still seemed possible at this time to describe every branch of knowledge in a way understandable to laymen and then to relate every part to every other by some system of master concepts, so that the ideal of universal knowledge might be attained. Symptoms of this movement can be found in the constant stream of popularizing texts by even the greatest scientific minds, such as Galileo, and by the enormous energy spent on international scientific communication and the correlation of such correspondence, and in the efforts of the Comenians to create an international community of learned men who would share a basic fund of information organized according to a common set of concepts and categories. It is expressed by actions as small as the attempt of the virtuosi marooned at Oxford by the Civil War to construct a subject index for the books in the Bodleian Library and as grand as the universal classification of human knowledge attempted by John Wilkins and his associates of the Royal Society. In law the movement can be seen most clearly in the works of Sir Matthew Hale, after Bacon the most scientific jurist that England has seen.

Hale, the greatest lawyer of his day and the model 17th-century judge, was, like North, engrossed in the scientific discoveries of the period. Although he never became a member of the Royal Society, Hale was a close friend of its chief founder, John Wilkins, and numbered many of its members as intimates. While it is unclear whether Hale took advantage of the scientific opportunities available at the university, he may have become interested in the new philosophy at Magdalene Hall, Oxford, where he studied at about the same time Wilkins picked up his early scientific interest. During the course of his residence at Lincoln's Inn he became very interested in mathematical and scientific studies. Beginning with arithmetic he went on to "*Algebra* both *Speciosa* and *Numerosa* and through all the other Mathematical Sciences . . . ," becoming "very conversant in *Philosophical* Learning and in all the curious Experiments, and rare Discoveries of this Age . . . ." He collected scientific books and instruments and performed many experiments, to "recreate himself" when he tired of his legal studies. Hale also developed considerable interest and skill in anatomy and medicine, the latter to such an extent that a physician indicated that he had gone as far in the study of medicine as "Speculation without Practice could carry him."

Hale contributed several volumes to the growing body of scientific and semiscientific literature. In 1673 he published an *Essay touching the Gravitation of Fluid Bodies*, and the following year *Difficiles Nugae: or Observations touching the Torricellian Experiment.* When Henry More rejected the views presented in this latter work, Hale replied with *Observations touching the Principles of Natural Motions, and especially touching Rarefaction and Condensation.* Although these volumes show an awareness of the current scientific literature and controversies, they were somewhat old-fashioned and failed to make a serious contribution to the development of scientific thought. They do, however, exhibit a sensitivity to some major problems of scientific philosophy and method. For example, Hale distinguished two approaches to finding scientific truth. The first begins with observations of the senses, proceeds to experimentation, and ends by constructing theorems to explain the experimental results. The second was deductive. Its foundation lay in speculation and its followers manipulated natural phenomena in accordance with their hypotheses. Hale himself favored the inductive approach because he felt that practitioners of the deductive method tended to distort the data to fit their hypotheses. He was not, however, a naive empiricist and was particularly critical of empirics in the field of medicine. Although the distinction between these two approaches was not highly original, Hale did provide the first detailed attempt to describe the mental processes and procedures required for invention and discovery.

Hale was sensitive to the scientific community's demand for a clear, uncomplicated, unadorned style. Like the members of the Royal Society, he insisted that eloquence and wit be used sparingly if at all in the communication of serious matters. He therefore opposed eloquence and rhetoric at the bar or on the bench and insisted that such language would confuse and corrupt juries by "*bribing their Fancies, and biassing their Affections* ...." As a judge "he held those that Pleaded before him to . . . the main Hinge of the Business, and cut them short" when they strayed from the main point. He detested violent language not only in the courtroom but everywhere. Sounding like an echo of the credo of the Royal Society, he insisted that "you must not speak that as upon knowledge which you have by conjecture or opinion only," and that it was necessary to think before speaking and to present one's views in "significant, pertinent, and inoffensive" expression.

Hale's conduct on the bench, then, seems to be directly related to his scientific studies. But the linkage is far more extensive and important than that, for his considerable contributions to English jurisprudence are marked by an approach distinctly in accord with the best canons of theorizing and data collection then current in the scientific community. As might have been easily anticipated from his scientific publications, the basis of his legal scholarship is an inductive method that emphasized the collection of data and based the construction and reform of legal principles on cautious and tentative theorizing from past experience. Thus in Hale we find a combination of systematic presentation and the urge to reform and modernize on the one hand and, on the other, the rejection of radical changes in law based on abstract rational systems such as Hobbes'. While legal scholars have been accustomed to think of this combination as peculiar to the tradition of the common law, Hale's position is not only a part of that tradition, but typical of the approach followed in the most advanced scientific circles of his day—circles in which he himself played a conspicuous part. Hale is best remembered for his attempts to systematize the law. Legal scholarship, however, has not usually recognized that his interest in such systematization was in harmony with and perhaps even an outgrowth of the virtuosi's effort to classify natural knowledge methodically. . . .

Hale's work on the civil part of law, *The Analysis of the Law: Being a Scheme or Abstract of the Several Titles and Portions of the Law of England, Digested into Method*, was one of the first efforts to treat English law by "*Analytical Method*." Despite his admission that the complexity of the law would not permit him to "reduce it to an exact *Logical Method*," he thought his *Analysis* provided a good start. It sought to prove, first of all, that it was "not altogether impossible, by much Attention and Labour, to reduce the Laws of *England* at least into a tolerable *Method* or *Distribution*." Second, it would give the opportunity to himself and others "to rectify, and to reform what is amiss in this, . . . whereby, in Time, a more *Methodical System* or *Reduction* of the Titles of the Law, under Method, may be discovered." His last reason for attempting the work was that

> although, for the most Part, the most *Methodical Distributors* of any Science rarely appear subtle or acute in the Sciences themselves, because while they principally study the former, they are less studious and advertent of the latter; yet a Method, even in the *Common Law*, may be a good Means to help the Memory to find out *Media* of Probation, and to assist in the Method of Study. . . .

I do not attempt here a detailed analysis of Hale's approach, but a few major points should be noted. Contrary to most earlier practice, Hale rigorously built his schema on the basis of substantive interests, segregating procedural matters at the end. Thus we get the modern division of adjective from substantive law. Similarly the rigorous separation of "rights of persons" (political, economic, and civil rights) from "rights in things" (property law), while certainly not a new idea, provides a far more rigorous and intellectually satisfying sequence of topics than was typical for the period. The treatment of rights in personal relationships—husband and wife, master and servant, and landlord and tenant—fall neatly together. The separation of various wrongs also provides a clarity that alternative organizations in terms of trespass or other traditional categories would have obscured. Finally, the great care in handling remedies and the procedural aspects of trials as general categories is a considerable advance over treating each individual writ-remedy-procedure combination as a separate entity, as they necessarily were by practice-oriented manuals. This is not to say, however, that Hale had freed himself from or rejected the writ-oriented practice of his day. The point is that Hale realized that, whatever the demands of professional practice, some new method of organizing legal knowledge was necessary if law was to be treated as an integral part of man's body of knowledge.

As so frequently happens, however, the theorizing of one age became the practice of another. Hale's analytical outline was adopted by Blackstone and thus eventually became a basic part of the practical education of several generations of American lawyers. We have Blackstone's testimony that "[o]f all the schemes hitherto made public for digesting the Laws of England, the most natural and scientifical of any, as well as the most comprehensive, appeared to be that of Sir Matthew Hale, in his posthumous 'Analysis of the Law.'"

Yet Hale's classifications, though they were the most sophisticated that the 17th century produced, were not unique. John Wilkins, a friend of Hale's and one of the leading virtuosi of the day, attempted to organize and classify "judicial relations" in the course of his effort to organize all knowledge into a systematic and philosophically sound system. Wilkins subdivided his legal category into persons, actions, crimes, and punishment. These subcategories were to be all-inclusive; "persons," for instance, included lawyers, witnesses, mediators, arbitrators, and judges, as well as plaintiffs and defendants. "Actions" was arranged in temporal sequence proceeding from pretrial through trial to such posttrial matters as appeal, execution, and pardon. It is particularly interesting that as early as the 17th century the application of science to law yielded an approach bearing certain startling similarities to modern sociological and political jurisprudence. In particular, there developed an emphasis on process rather than doctrine. Wilkins' treatment was essentially "behavioral"; he took the stance of a scientist seeking to describe accurately and interrelate all facets of a complex social phenomenon labeled law rather than of one seeking to provide the traditional explanations, orderings, and rationalizations of written legal materials. Thus his work, at least at the outline stage, moves somewhat beyond that of Hale, which is more fully embedded in concern for the taught, doctrinal tradition. Wilkins may even have had the assistance of Hale in his efforts at legal classification, for he used the talents of many of his friends to complete his *Essay Towards a Real Character and a Philosophical Language*. Wilkins may also have seen manuscript copies of Hale's legal treatises and classificatory efforts. He almost certainly had the benefit of his good friend's conversation. In any event, portions of Wilkins' treatment of "judicial relations" bear a striking resemblance to Hale's work. Wilkins, however, was extremely brief and provided only note headings, while Hale frequently provided at least some description of material belonging under each heading.

Sir Geoffrey Gilbert, a respected judge, brilliant mathematician, and member of the Royal Society, also contributed to the systematization of legal concepts and materials. His numerous posthumously published writings were part of an effort to compose a "general History of the Courts of Justice," a work that "would have been of great utility to those who may have Occasion to study the *English* law on fundamental Principles." The completed portions have been described as "simple and concise," not "pompous and elaborate," brief, and sparing in their use of authority. A comment of the editor of his *Treatise of the Court of Exchequer* was equally appropriate to his other legal treatises:

> there is such a Consistence betwixt the Parts of the System, as makes it necessary to expatiate, where they occur in this work, on the Certainty of many Particulars, on Account of their being natural and unquestionable Deductions from others sufficiently proved; though taken in a separate Light, they might demand some Demonstration.

Gilbert evidently wishing not only to analyze the law systematically but also to describe the work of the courts, was responsible for separate treatises on the courts of exchequer, common pleas, and chancery as well as a series of topical works.

The Doctrine of Certainty

The systematizing work of Hale, so frequently viewed as part of the progress of an essentially autonomous legal discipline, was thus related to the scientific culture of his day; a concern for classification and systematic communication was a general feature of 17th-century intellectual life. The development of other areas of legal thought likewise paralleled advances in scientific thought during this period. Perhaps one of the most important areas was in theories of certainty.

Although the doctrine of relative certainty—the notion that we can be surer of some truths than others—did not fully develop until the 17th century, it can be traced ultimately to Aristotle and was used in an early form by several 16th-century, legally trained French humanists. Thus among the earliest proponents of the theory were men trained in law and concerned with the history of law. The reason is not hard to guess: Historians as well as lawyers and judges have always shared the problem of attempting to determine the truth of statements about past occurrences. They all reach decisions about "what happened," or probably happened, when direct empirical evidence is unavailable. They are forced to weigh evidence which turns largely on the testimony of those who do claim to have direct empirical experience. Probabilities rather than certainties are therefore the concern of both jurists and historians. In the 16th century, Melchior Cano, Francois Baudouin, and Jean Bodin, faced by Renaissance skepticism about the validity of historical knowledge, argued for reasonable doubt and evaluation of the probity of individual historians rather than the complete dismissal of historical writing. Indeed, Baudouin and Bodin used the analogy of courtroom practice.

Certainty and the Scientific Revolution

In the 17th century, the doctrine of relative certainty came to the fore as part of the great religious controversy of the period. It was employed against both atheists and persecuting zealots to demonstrate that there could be very low levels of certainty about specific points of religious doctrine, and almost absolute certainty about such basic tenets as the existence of God, his creation of the world, the immortality of the soul, and the existence of future rewards and punishments. A very considerable portion of those who used this approach in the mid- and later-17th century were men associated with the scientific movement who wished not only to uphold the basic principles of religion against disbelief, but to lower the temperature of religious dispute by showing that most disputed points were of a very low order of certainty. In fact, nearly all English proponents of this theory were religious latitudinarians.

William Chillingworth, a liberal Anglican theologian, introduced the certainty arguments into England after their earlier development by Castellio and Grotius in a religious context on the continent. Chillingworth described three levels of certainty. The highest was available to God alone. The second was based on evidence that virtually excluded the possibility of error. Moral certainty, the third, was the level a reasonable person might achieve after considering all the available evidence. Religious matters, with the exception of those principles clearly stated in Scripture, fell into the third category.

Henry More, a liberal theologian of the Neoplatonist school with a scientific bent, and Seth Ward, Savilian Professor of Astronomy and one of the founding members of the Royal Society, used the doctrine to prove the existence of God. Walter Charleton, a physician member of the Royal Society, used the theory to prove the immortality of the soul, not by "Demonstrations Geometrical," which were inappropriate, but by "Proofs sufficiently Persuasive, for all such, who come not to examine them with invincible Prejudice and resolution not to be convinced." This emphasis on the unprejudiced observer or the reasonable man was a common thread among writers using the doctrine, whether in the area of religion, science, or law. The "reasonable man" will not, Charleton argued, demand demonstration or proofs that "exclude all Dubiosity, and compel assent," but will accept moral and physical proofs that are the best that may be gained in metaphysical matters. Thus one can gain a "competent certitude where Demonstration is impossible. . . ."

John Tillotson, a leading latitudinarian churchman, also adopted this approach, first in his attack on atheism and later in a repudiation of the Roman Catholic view of faith.

> *Mathematical* things being of an abstracted nature are only capable of *clear Demonstration*; but Conclusions in *Natural Philosophy* are to be proved by a sufficient *Induction* of experiments; things of a *moral* nature, by *moral* Arguments; and *matters of Fact*, by *credible* Testimony; and though none of these be capable of *strict Demonstrations*, yet we have an undoubted assurance of them, when they are proved by the best Arguments that the nature and quality of the thing will bear.

He also insisted that "pure Negatives," that is, proving the nonexistence of a thing, are extremely difficult and often impossible to demonstrate, an observation not unknown in legal circles.

Some latitudinarians went beyond their associates to develop the concept of certainty in terms of a theory of knowledge appropriate to the natural sciences. Joseph Glanvill and John Wilkins are perhaps the two most important figures responsible for the transition. Glanvill wrote several tracts publicizing the scientific credo of the Royal Society as well as the more moderate and rationalistic approach of religion. Wilkins, with whom we are already familiar, demonstrated how this approach to certainty of knowledge could incorporate both religious and secular truth. For Wilkins there were two fundamental categories. Knowledge or certainty was the first and was derived from evidence that did not admit of any reasonable doubt. The second was opinion or probability. Knowledge had three subcategories: physical, mathematical, and moral. The first was derived from the sense data; the second included all matters capable of the same certainty as mathematics; the third was more complex because it was not dependent on evidence that "necessitated every man's assent." Nevertheless, it was so clear "that every man whose judgment is free from prejudice will consent to them. And though there be no natural necessity, that things be so, and that they cannot possibly be otherwise . . . yet may they be so certain as not to admit of any reasonable doubt concerning them." The first two resulted in "infallible" certainty, the third in "indubitable" certainty. Most things were capable of only the latter, a certainty that did not admit of any reasonable doubt. When evidence was unclear or reasonable doubt existed, probability or opinion rather than knowledge resulted. In such cases impartial observers were to "incline to the greater probabilities" or, if necessary, suspend judgment.

Henry Van Leeuwen has demonstrated that this theory of certainty was developed further by such active scientists as Robert Boyle and Isaac Newton, whose friendship and association with the latitudinarian theologians is well known. Boyle and Newton used this comprehensive theory to deal with problems of both religious and scientific beliefs. A generalized version of the theory that comprehended all knowledge was then formulated by John Locke in his *Essay Concerning the Human Understanding*. Locke's espousal might have been expected, for he was intimately connected with the virtuosi of the Royal Society, was a close friend of Tillotson, Boyle, and Newton, and was himself a latitudinarian.

Certainty and the Law

These theorists assumed quite naturally that legal evidence was subsumed under their theory of evidence and knowledge and thus did not attempt to deal with it separately. Boyle, for example, in describing the differing degrees of certainty and probability to be ascribed to mathematical, physical, and moral demonstration, noted that men's actions were in the realm of probability and used "the practice of our courts" as a vivid example.

> For though the testimony of a single witness shall not suffice to prove the accused party guilty of murder; yet the testimony of two witnesses though but of equal credit, that is, a second testimony added to the first though of itself never a whit more credible than the former, shall ordinarily suffice to prove a man guilty; because it is thought reasonable to suppose, that though each testimony singly be but probable, yet a concurrence of such probabilities (which ought in reason to be attributed to the truth of what they jointly tend to prove)

may well amount to a moral certainty, i.e., such a certainty as may warrant the judge to proceed to the sentence of death against the indicted party.

Boyle also indicated that this approach could be applied to witnesses.

> You may consider . . . that whereas it is as justly generally granted, that the better qualified a witness is in the capacity of a witness, the stronger assent his testimony deserves; . . . for the two grand requisites, of a witness [are] the knowledge he has of the things he delivers, and his faithfulness in truly delivering what he knows. . . .

Locke, too, in the course of his discussion of various kinds of evidence and the certainty they produce, dealt with the evaluation of testimony and, like Boyle, easily saw the applicability of this approach to the law. For example, both he and Boyle noted that an attested copy of a record is good evidence that an event occurred but that a copy of a copy is not as good. The testimony of a witness is good evidence that an event occurred but "a report of his report is not and will not be admitted in a court of law. The further from the source, the weaker the evidence becomes." Thus the scientific community felt that the rules for determining the truth in legal matters were the same as in other areas of investigation.

Judges and lawyers also found that the theory suited their needs. Lord Nottingham, for example, when Lord Keeper, used the language of certainty quite naturally when he defended the King's Declaration of Indulgence in 1673: "A Mathematical security we cannot have: a moral one we have from the King." John Selden, the jurist, also suggested the mutual borrowing of the theory from one field to another. When discussing the truths of history and methods of historical proof, he turned to the terminology of certainty and reasonable doubt.

More significant, however, in demonstrating the application of the theory of certainty to legal thinking is the full description and acceptance of the theory in the writings of Sir Matthew Hale. In *The Primitive Origination of Mankind*, Hale categorized knowledge in much the same way as his friends Tillotson and Wilkins. He was most concerned with the evidence for "matters of Fact." Although the evidence of the senses was the "best evidence" in these matters, it was obviously inapplicable to "things transacted before our time, and out of the immediate reach of our Sense." Here only "moral and not demonstrative or infallible" evidence was available. Yet a "variety of circumstances renders the credibility of such things more or less, according to the various ingredients and contributions of credibility that are concentered in such an evidence." To elicit assent, it was necessary to weigh

> the veracity of him that reports and relates it. And hence it is, that that which is reported by many Eyewitnesses hath greater motives of credibility

> than that which is reported by few; that which is reported by credible and authentic witnesses, than that which is reported by light and inconsiderable witnesses; that which is reported by a person disinterested, than that which is reported by persons whose interest it is to have the thing true, or believed to be true; . . . and finally, that which is reported by credible person of their own view, than that which they receive by hear-say from those that report upon their own view. . . .

Such evidence might be "of high credibility, and such as no reasonable man can without any just reason deny. . . ." While Hale developed these ideas principally in the context of history and general knowledge, he readily transferred them to law. "That evidence," for example, "at Law which taken singly or apart makes but an imperfect proof, *semiplena probatio*, yet in conjunction with others grow to a full proof, like *Silurus* his twigs, that were easily broken apart, but in conjunction or union were not to be broken."

The first treatise devoted entirely to the problems of legal evidence, that of Sir Geoffrey Gilbert, followed the same approach as the rational theologians, the scientists, and Hale. His *Law of Evidence*, which viewed exclusively in the context of legal scholarship has often been treated as quite revolutionary, in fact represents an advance only in explicitly employing the 17th-century doctrine of certainty as the central basis for a systematic treatment of legal evidence. Moreover, its purpose was not to reform but to describe the state of the law at the time the treatise was written, and it was not actually published until nearly three decades after Gilbert's death. Thus the work, which might casually be taken as introducing the doctrine of certainty into law in the middle of the 18th century, actually reveals the earlier widespread judicial adoption of that doctrine.

The work begins with a discussion of the "rules of probability," by which evidence offered to the jury "ought to be weighed and considered." Citing the observations of that "very learned man," Locke, Gilbert notes that

> there are several degrees from perfect Certainty and Demonstration, quite down to Improbability and Unlikeness, even to the Confines of Impossibility; and there are several Acts of the Mind proportioned to these Degrees of Evidence, which may be called the Degrees of Assent, from full Assurance and Confidence, quite down to Conjecture, Doubt, Distrust and Disbelief.
>
> Now what is to be done in all Trials of Right, is to range all Matters in the Scale of Probability, so as to lay most Weight where the Cause ought to preponderate, and thereby to make the most exact Discernment that can be, in Relation to the Right.
>
> Now to come to the true Knowledge of the Nature of Probability, it is necessary to look a little higher, and see what Certainty is, and whence it arises.

He proceeds to discuss certainty in terms of sense perceptions and necessary inferences or demonstrations from fixed data. Since most litigation depends on transient data "retrieved by Memory and Recollection" rather than demonstration, "the Rights of Men must be determined by Probability." Probability is then considered in terms of degrees of credibility of witnesses and the ability to accept their statements of facts beyond "any more reason to be doubted than if we ourselves had heard and seen it. . . ." This line of thought was subsequently adopted by Blackstone and incorporated into the 19th- and 20th-century texts on evidence, where, of course, it is still to be found.

We have seen that Hale's and Gilbert's treatment of questions of evidence and matters of fact is very much in accord with the most advanced thinking of the period and that both were aware that developments in epistemology and scientific method had an impact on law. Moreover, other legal intellectuals could not have immunized themselves from the developments in the theory of relative certainty that occurred in the scientific, philosophical, and religious fields during the 17th century. Yet legal historians have suggested that a sophisticated view of evidence did not develop in England until the 18th century. Their failure to acknowledge the 17th century contribution can be explained in several ways. There is a major gap in legal history between the medieval period and the mid-18th century and a resulting tendency to assume that medieval conceptions continued to rule until they were suddenly replaced at the point where legal history picks up again. In the realm of evidence part of the difficulty no doubt arises from excessive scholarly concentration on the rules of procedure and admissibility of evidence rather than the principles of proof, the ratiocinative process of continuous persuasion that Wigmore thought was of far more importance than rules of admissibility. Concern with this latter process has been very limited, and Wigmore himself noted that he was the first scholar since Bentham to call attention to the principles of proof as distinct from admissibility. It is this area that "bring[s] into play those reasoning processes which are already the possession of intelligent and educated persons." Yet it is also precisely the area of persuasion and belief that changed so substantially in the course of the 17th century. Because such matters of evaluation lie largely in the habitual patterns of thought of judges and juries rather than in the formality of procedure, they leave few skeletal remains in the form of changes in rules of admissibility. Yet there can be no doubt that the major shift in intellectual climate created by the introduction of notions of relative certainty in theological and scientific discourse played an important role in shaping English legal practice long before Gilbert recorded them in the middle of the 18th century.

Still another reason that these 17th-century developments have been obscured is the notion that a sophisticated and consistent treatment of the law of evidence could not develop until the jury had ceased to be witnesses as well as judges of matters of fact. That argument may be analytically useful and logically satisfying, but it plays hob with the real dynamics of history. New ideas and processes can develop alongside and gradually surround an anachronism long before it finally disappears. And this is particularly true within a profession as conservatively wedded to old forms as the law.

Indeed, there has probably been too great an effort to link the law of evidence to the evolution of the jury. Rules of evidence and techniques for classifying types of evidence and witnesses on the basis of credibility may have had their origin, and certainly enjoyed much of their development, in chancery and ecclesiastical courts, which did not use juries. Sir Matthew Hale, whose highly developed views on certainty of evidence we have already examined, and who insisted on the impartiality of juries, announced these ideas at a time when he was still permitting jurymen to know and present information somewhat in the manner of witnesses. He sought to harmonize his two positions by arguing that additional information contributed by jurymen could be used to improve the court's assessment of the credibility of ordinary witnesses. Thus the new ideas on evidence had come in long before the old jury practices went out.

It has also been suggested that the medieval practice of treating all evidence given under oath as of equal weight continued into the 18th century. Some residues of the older notions undoubtedly did survive, but 17th-century judges clearly made judgments as to credibility. Locke's six criteria for evaluating testimony—"the number of witnesses, their integrity, their skill at presenting the evidence, their purpose, the internal consistency of the evidence and its agreement with the circumstances, and lastly the presence or absence of contrary testimony"—and John Wilkins' statement—"and as for the evidence for Testimony which depends upon the credit and authority of the Witnesses, these may be so qualified as to their ability and fidelity. . . ."—are echoed in greater or lesser degree in the legal literature and in several important cases of the period. In *The History and Analysis of the Common Law*, which attempted to describe the existing legal system, Hale made several comments that suggest how well established the notion of credible as opposed to merely lawfully sworn witnesses was during the Restoration period. At one point he noted that the testimony of legal witnesses can be attacked "either as to competency of the evidence, or the competency or credit of the witnesses. . . ." At another he indicated that if the jury has

> just cause to disbelieve what a witness swears, they are NOT bound to give their verdict according to the evidence, or testimony of THAT witness. And they may sometimes give credit to ONE witness, though opposed by more than one. And indeed it is one of the excellencies of this trial [the jury trial], above the trial by witnesses, that although the jury ought to give a great regard to witnesses and their testimony, yet THEY ARE NOT ALWAYS BOUND BY IT; but may either upon reasonable circumstances, inducing a blemish upon their credibility, though otherwise in themselves in strictness of law they are to be heard, pronounce a verdict CONTRARY to such testimonies; the truth whereof they have JUST cause to suspect, and may and DO OFTEN, pronounce their verdict upon one single testimony; . . . they are to weigh the credibility of witnesses, and the force and efficacy of their testimonies. . . .

Hale also noted the advantage "for the true and clear discovery of the truth" of observing the contradiction of witnesses sometimes of the same side." His *The History of the Pleas of the Crown* similarly notes the distinction between legal and credible witnesses, indicating that the jury is to judge the "probability or improbability, credibility or incredibility of the witness and his testimony. . . ."

This distinction between credible and lawful witnesses was also made in a few contemporary cases. Judge Hale, summing up evidence for the jury, noted that a witness was "a person, I think, of no great Credit. . . ." In a 1681 case of assault and battery, the defense counsel indicated that "we shall prove (by substantial and credible men) that not one blow was given. . . ." In 1679 Lord Chief Justice North also distinguished between lawful and credible witnesses, and in a 1696 conspiracy trial the judge instructed the jury to consider the "Fairness and Credibility" of the evidence that was given.

In cases where certain defense witnesses were not permitted to testify on oath, the notion of credibility also appeared. The Solicitor General, summing up the evidence in the case of Lord Mohun before the House of Lords, noted that the peers were to believe the defendant's witnesses though not under oath, "so far, as your Lordships shall Judge was said Credible, about Consideration of all that you have heard." One of the numerous Popish Plot trials turned on the question of the credibility of the witnesses. When the defendant, Langhorn, himself a lawyer, indicated that his "Whole Defense must run to disable the witnesses . . ." and that he could "have no defense unless it be by lessening their Credit. . ." Lord Chief Justice North advised him: "Do lessen it if you can." In the process of his defense Langhorn further noted that "[i]f I can Disprove a Witness in any one material thing that he says then it will take off from his Credit in every thing he says." In summing up, North instructed the jury that they must judge the credit of the witnesses on both sides, those who had testified under oath as well as those who had not been so permitted.

In a related development, the employment of multiple witnesses testifying as to the same event lost its oath-helper quality and became instead a means of improving the scientific certainty of judicial factfinding.

> [I]f to any one *quantum* of fact there be many but probable evidences, which taken singly have not perchance any full evidence, yet when many of those evidences concur and concenter in the evidence of the same thing, their very multiplicity and consent makes the evidence the stronger; as the concurrent testimonies of many Witnesses make an evidence more concludent.

Isaac Barrow and Robert Boyle, like Hale, indicated that the preference for a larger rather than a smaller number of witnesses was based on considerations of probability. Thus although the rhetoric of oaths and the multiplication of witnesses may not have been substantially altered between the 13th and 17th centuries, their meaning and significance as modes of legal proof had changed considerably.

The newer views are particularly evident in several late 17th-century court decisions in which judges sought to distinguish levels of proof needed for various kinds of cases. The recorder in a 1681 case insisted that in assassination cases "exact and positive proof" was unattainable so that the court "must not expect it should be so clear as in a Matter of Right between Man and Man. . . ." In the trial of Carr for publication of a libelous book, this position was even more clearly elaborated. The presiding judge argued that "you very well know, that Evidences of Fact, are to be expected according to the Nature of the Thing." Forgery could not be proved in the same way as the sealing of a document because witnesses were not ordinarily present; "in things of that nature, we are fain to retreat to such probable and conjectural Evidence as the matter will bear." In cases involving murder, juries should not expect "a direct Proof of the Act or the actual Killing; but yet, you [have] such Evidence by Presumption as seems reasonable to conscience." The judge therefore advised the jury:

> You must take Evidence in this case, as you do all the Year long; that is, in other Cases, where you know there is an absolute certainty, that the thing is so: for human frailty must be allowed: that is, you may be mistaken. For, you do not Swear, nor, are you bound to Swear here, that he was the Publisher of this Book: but, if you find him guilty, you only Swear, you believe it so.

The jury was therefore instructed to reach their verdict "according to reason and the probable Evidence of Things."

Statutes, particularly those concerned with treason, also indicated the growing concern with problems of credibility and standards of proof. While the 16th-century statutes confined themselves to demanding the testimony of "lawful witnesses," the revised treason statute of 1661 required "two lawful and credible witnesses." The statute of 1696 was even more rigorous in its demand of proof and in the opportunities it afforded the accused for his defense; it has been characterized as embodying "almost the difference between medieval and modern." In 1697 the Blasphemy Act provided that conviction be based on the testimony of two or more credible witnesses.

Closely brigaded with this more sophisticated approach to evidence was the increasing concern for the impartiality of judges to be found after the Restoration. Judicial practice might still have been far from ideal, but the judicial model in the minds of the literate public shifted more and more from the prosecuting servant of the government toward a detached seeker of truth. By the end of the century impartiality was expected of judges as much as of scientists. One measure of the change in public attitude is the contrast between the acceptance of the highhanded judicial behavior of Coke and the indignation at the outrageous behavior of Jeffreys and Scroggs. It should thus not be totally unexpected that Sir Matthew Hale, one of the most scientifically inclined judges of the period, was so widely acclaimed as an ideal judge. It was Hale's impartiality that made him famous in his own day.

I do not wish to suggest that Hale's impartiality, or even the ideal of judicial impartiality, was a creation of the 17th century or of the scientific movement, but only to indicate that those most affected by the new views concerning the search for truth were more sensitive to the ideal of impartiality. Isaac Barrow, a theologian and scientist, who was not himself connected with the legal profession, provides a good example of what the laymen of the 17th century expected of judges.

> A Judge should never pronounce final Sentence, but . . . upon good grounds, after certain proof, and upon full conviction. Not any slight conjecture, or thin surmise; any idle report, or weak pretence is sufficient to ground a Condemnation upon: the Case should be irrefragably clear and sure before we determine on the worse side. . . . Every Accusation should be deemed null, until both as to matter of fact, and in point of right, it be firmly proved true; it sufficeth not to presume it may be so; to say, It seemeth thus, doth not sound like the voice of a judge. . . .
>
> Moreover, a Judge is obliged to conform all his determinations to the settled rules of Judgment, so as never to condemn any man for acting that which is enjoyned, or approved, or permitted by them; he must not pronounce according to his private fancy, or particular affection, but according to the standing Laws; . . . he that proceedeth otherwise, is an arbitrary and a slippery Judge; . . . a Judge should be a person of good knowledge and ability; well versed and skilful in the Laws concerning matters under debate; endowed with good measure of reason, enabling him as to sift and canvas matters of Fact, so to compare them accurately with the rules of right. . . .
>
> Lastly: It is the property of a good Judge to proceed with great moderation, equity, candour and mildness. . . .

3 The Mechanical Philosophy of the Seventeenth Century: God, the Divine Engineer, and a Political Theory of Atomic Individualism

The dominant trend of the scientific movement of the seventeenth century was certainly toward the interpretation of all natural phenomena in terms of the motion of mechanical artifacts. Christian Huygens presented the view of his scientific contemporaries in speaking of "the true Philosophy, in which one conceives the causes of all natural effects in terms of mechanical motions." And he added this nearly universal opinion, "this we must necessarily do, or else renounce all hopes of ever comprehending anything in Physics."

But the mechanical philosophy popularized by the scientists Huygens, Robert Boyle, René Descartes, Johannes Kepler, Galileo, and Pierre Gassendi was by no means limited to physics. It spread from astronomy into biology, then into psychological theory, metaphysics, political theory, and even into theology with dramatic consequences for men's fundamental beliefs.

One of the fields in which the use of mechanical analogs was highly developed very early was astronomical theory. In 1370, struck by the beauty, intricacy, harmony, and purposiveness of weight-driven clockworks that were being made by artisans to simulate the heavenly motions, Nicole Oresme wrote of the angelic powers that God placed in the heavens to move the planets in their orbits:

> And these powers are so moderated, tempered, and ordered against their resistances that the movements are made without violence. And except for the lack of violence it is like the situation when a man has made a clock and lets it go and be moved by itself. Thus it was that God let the heavens be moved continually according to the proportions that the moving powers have to their resistances and according to the established order. . . .[1]

---

[1]Nicole Oresme, "Maistre Nicole Oresme; Le Livre du Ciel et du Monde," edited by A. Deverny and A. Menut, *Medieval Studies* 4 (1942), 170. The translation of this passage is my own and differs materially from that by A. C. Crombie in *Medieval and Early Modern Science, Vol II* (Garden City: Doubleday & Co., 1959), p. 76.

By the beginning of the seventeenth century, Oresme's mild suggestion that God's universe could somehow be understood as similar to the mechanical productions of man had become immensely suggestive, and we find Kepler writing to a friend in 1605:

> I am now much engaged in investigating physical causes; my goal is to show that the celestial machine is not the likeness of the divine being, but is the likeness of a clock (he who believes that the clock is animate ascribes the glory of the maker to the thing made). In this machine nearly all the variety of movements flows from one very simple magnetic force just as in a clock, all the motions flow from a simple weight.[2]

This attitude had an immense impact on the religious beliefs of the seventeenth century, for it envisioned God as a kind of grand engineer and focused men's attention on the so-called argument from design.

The argument from design appeared in various garbs but it usually went like this: complicated machinery is always produced by some intelligent artisan for some specific end. The world is obviously like a very complicated piece of machinery, parts of which have been produced with ends in mind. Hence there must have been some extremely intelligent artisan who created the whole thing—and this creator we call God. There is, of course, an extreme danger for traditional theology connected with the emphasis on God as creator; and that danger became obvious during the seventeenth century. If an artisan God made the world mechanism perfect and self-perpetuating, there is no reason for God to be immanent in the world, and as some deists began to feel, God might have created the world and then retired. This specter of a world without God was presented very clearly by Robert Boyle in 1686. Boyle wrote:

> The world is like a rare clock such as that at Strasburg, where all things are so skillfully contrived that the engine once being set going—all things proceed according to the artificer's first design, and the motions . . . do not require the peculiar interposing of the artificer, or any intelligent agent employed by him.[3]

Robert Lenoble, in the selections from *Mersenne, ou la Naissance de Mécanisme* discusses the impact of the mechanistic science on the religious belief of Marin Mersenne, a Minim friar who was one of the key intellectual figures of the seventeenth century. He argues that the scientific model for his thinking led him little by little, almost against his own will, to "replace the

---

[2]Ch. Frish, ed., *Johannes Kepler Opera, Vol. 2* (Frankfort: Heyder and Zimmer, 1859), pp. 83-84.

[3]From Edwin Burtt, *The Metaphysical Foundations of Modern Physical Science* (Garden City, N.Y.: Doubleday & Co., 1954), p. 202.

imitation of Jesus Christ by the imitation of the Divine Engineer as a basis for Christian piety." And he discusses how this emphasis on an artisan God both encouraged a democratization of religion and a decline in its other-worldly and spiritual emphasis.

Whereas the mechanistic astronomy of Kepler and Galileo had its greatest influence on religious belief in the seventeenth century, the mechanistic biology and psychology of René Descartes played a far more pervasive role in seventeenth-century thought. Descartes, as we saw earlier, believed that the bodily aspects of man were purely mechanical and that even his sensations and passions could be mechanistically explained. Few of his contemporaries followed Descartes so far as to deny the possibility of nonmechanistic aspects of the natural world and of man—although in the eighteenth century Julian O. La Mettrie published his *L'Homme, Machine (Man, A Machine)* which carried the Cartesian mechanism to its logical conclusion in total materialism—but the Cartesian view did have immediate impact in philosophy and in social and political theory. Even Benedictus Spinoza, one of the least materialistic thinkers of the seventeenth century, could not avoid the influence or the contamination.

Spinoza, who had been one of Descartes' proteges early in his career, revolted against the master, but he could not escape the mechanistic influences. In his *Short Treatise on God, Man, and his Well-being* of 1662, for example, Spinoza discusses the analysis of complex ideas:

> A thing composed of different parts must be such that the parts thereof, taken separately, can be conceived and understood separately without the composite whole being necessary thereto.[4]

This image of Spinoza's indicates an extremely important aspect of the mechanistic metaphor, for it is dealing with the very logic to be used in philosophical discussions and stating that this logic must be patterned on the kind of analysis that one can make of mechanical systems. The analytic approach to understanding, as opposed to the earlier mystical or magical approaches and to the later organistic and gestalt approaches, limits itself by the assumption that knowledge must be somehow atomic—isolatable into small unitary and individually meaningful chunks—and not holistic. And it seems clear that the analytic approach was strongly reinforced by the patterns of mechanical analyzable systems. This analytic approach can be seen not only in the continental tradition of philosophy deriving from Descartes, but also in the rival English empiricist tradition stemming from the works of John Locke, who received his initiation into the mechanistic philosophy as an assistant to Robert Boyle.

By far the most influential practitioner of the analytic mechanistic

---

[4]John Wild, ed., *Spinoza Selections* (New York, 1930) p. 56.

model of understanding in the seventeenth century was Thomas Hobbes, whose political ideas have dominated Western social thought for nearly three hundred years. Hobbes' introduction to his most widely read work, *Leviathan; or the Matter, Forme, and Power of a Commonwealth, Ecclesiastical and Civil*, demonstrates both the use of mechanistic imagery and the associated tendency to break down complex social problems into small isolated units which can be considered one at a time.

Science Pervades Religion

Robert Lenoble

If, in the scientific work of [Marin] Mersenne, it is rewarding to discover the love of progress and the kind of sincerity which brightens any personal endeavor whatever the unavoidable errors may be, it is just as disappointing and sometimes painful to read the pages in which he deals with morals. Indeed, he is in the grip of a purely legal and mechanical concept of spiritual life, and this concept forces him finally to subordinate religion and morals to a policy of social defense lacking in generosity.

We are saying now, and we shall say it again when necessary, that his intentions are not suspect to us. He thought very sincerely that he was working only for God, and no doubt his prayer reached the God of Christianity. But he could not avoid the fact that his piety and his morals bear the imprint of his personality. We have noted already quite frequently that he is not interested in theology, and that, much more than Descartes, he deserved the blame ascribed by Pascal: namely that he thought mostly about the God of the philosophers and of the scientists. Moreover, writing as he does after the wars of religion, he distrusts a philosophy which could once more render doubtful the premises of the "moral order." Thus he asks of morals and of religion that they offer stability to a social organization within which the scientist could work freely. Criticism stops at the precise point where it could weaken "those great institutions which, once shaken, are so hard to retain." In this aspect, once more, he represents perfectly a considerable part of the opinion of his era. We can even notice in his thought the beginnings of a crisis which matures only one-half century later: theology is aban-

---

Source: Robert Lenoble, *Mersenne, ou la Naissance du Mecanisme* (Paris: Librairie Philosophique J. Vrin, 1943), pp. 532-541. Translation by Elie Cohen and Richard Olson. Translated and printed by permission of the publisher.

doned, and religious thought becomes too schematic and loses its mystical impulse. Mersenne, like his friend Hobbes, is already turning to science in order to define the content of the essential dogmas of civilization. With some surprise, we shall discover in this zealous defender of orthodoxy traces of "indifferentism" which worried some of his friends and sometimes the heretics themselves.

The Ideal of the Scholar

Though it is false, in spite of Haureau's statement, that Mersenne had entered the cloister to escape from the troubles of secular life, and even though some pages written by him about religious life bear the proof of sincere emotion, still it is true that it was probably the demon of science which suggested to Mersenne that he could serve God efficiently by becoming a physicist. The most eloquent pages of his ethics are, to be sure, those in which he writes about the beauty of study. Studies calm the passions and insure peace among men. They provide the real benefits as opposed to those which tempt superficial minds: ". . . long studies soften the mind of men and do not often allow that they be brave, bold, strong, and courageous, as is required of those who must protect the people from all dangers and all enemies. In any case, the pleasure provided by good reading is so keen that the men who have sampled it despise that which is provided by great honors, which are often accompanied by more thorns than roses." Lastly, studies make man almost similar to pure spirits: "Thus it is easy to infer that scientists are greatly in debt to divine Providence which set them in a level higher than that of the ignorant and which prepared them a situation in this world close to that of the angels, so that by using their enlightened knowledge, they can teach others and attract everyone to God." Consequently, if in this world the scientists "are often mistreated, being forced to survive from bread crumbs whereas others eat the most exquisite dishes," the scholar rightly feels that this is the result of a disorder. However, he still trusts Providence and respects the world just as it is.

Thus Mersenne continues the tradition of the medieval monks who felt that knowledge was the basis of a very special dignity which raised them well above the common man. But in the hectic beginning of the seventeenth century, such a superior status could in no way be translated into political power. On the other hand, although society remained brutal, it did become sufficiently gentler, so that the scientist could lead a comfortable life under the protection of powerful and cultured noblemen. Between the forceful claims of monastic autonomy and the impatience of the enlightenment, Mersenne represents the very model of the seventeenth century domestic scholar.

The Scientists' Invasion: Piety, Ethics and Mathematics

He thus continues the tradition of the scholarly monks: God is the real truth and studies represent the means that lead to it.

But the same problem never appears twice in the same way after several centuries. The kind of science to which Mersenne refers when he states that it leads to God is no longer the scholastic physics which flows naturally into the affirmation and the contemplation of the First Principle—a science which is in reality totally metaphysical and quite contemptuous of material applications. First of all, he no longer believes in Aristotelian physics: his pragmatism created a gap between positivistic science and a realistic metaphysical system. Moreover, he is living at a time when men finally come to grips with the problem of the meaning which befits material civilization: the "art" of the hydraulic scientists, the "nice inventions" of the engineers, adopted by mathematical physics, enter the domain of science on their own merits and, even more, become "the true physics" and push back the speculations on atoms, matter, and form into the realm of the unmeasurable. Thus while he continues to claim that studies lead to God, the one element which he places in the center of moral life is science in the modern meaning of the word: the knowledge of phenomena and the practice of the mechanical arts.

Pascal, who had the same concept of science as Mersenne, vehemently refused to make it a means of sanctification, and he broke with Descartes whose metaphysics gave him the right to continue in his own way the tradition of the scholarly monks. Mersenne did not have the foresight nor the scruples of Pascal, and he set positive science at the center of his piety.

Of course, he thought at first that it was useful to religion that the faithful be "both good Catholics and good mathematicians." But this justification of science (still extrinsic and traditional in any case) does not constitute what is most interesting in his theories. We noted already that, with flagrant naiveté, he states that Heaven is the abode of mathematical and physical truths, in which abode we shall finally get to know the secret of Gyges' ring and the exact value of π (pi).

We have already cited his prosopopoeia to the God of the scientists of which we repeat that it is doubtless unique in all of Christian literature: "How great the God of Israel, who wished to make the semidiameters of the planets visible so that optics might arrive at a knowledge of the magnitude and distance of the stars. . . ." We noted that in the text of the Bible he seeks, above all, material for indefinite scientific developments. His manuscripts have another surprise for us.

While he pursued his endless work on *Genesis*, he had begun to work on a commentary on the Gospels. This new work was barely begun, but what we have left of it still remains a very edifying text. If there ever was a text which seems to discourage in advance the curiosity of the physicist because it has no other end than to focus attention on the "unique necessity" of interior life, that text has to be the Gospel. However, Mersenne went beyond his usual powers and succeeded by an incredible feat in dealing with the Gospel as if it were an anthology of physics problems! The most beautiful excerpts of the Sermon on the Mount inspire no commentary from him. At most the passage, "Reach an agreement with your brother rather than drag him in front of a judge who will force you to part with your last penny" provides him the opportunity to write two pages to explain currency and measures. "Behold the lilies of the fields; Solomon in his splendor was never dressed as any one of them"—on this text, whose subtle poetry moved the most intellectual scholars, Mersenne asks only two questions. Why was it not possible for Solomon to make garments as beautiful as the lilies? This, Mersenne answered, was a problem for the embroiderers. Why did Christ speak of lilies and not of roses? No doubt this is a problem for botanists! When he transcribes the advice about the fast, "When you fast, do not be like the hypocrites who look so sad; be happy and wear perfume," he directs the reader, for commentary worth considering, to the specialists on perfumes, on their nature and usages. The healing of the possessed is, above all, a problem for the physicians, just like all the miracles as well. Jesus chose twelve apostles; let us not believe, as do the cabalists, in the power of numbers; and yet, we are allowed to think that the Master wanted to honor the peculiar arithmetical properties of the number twelve, about which Mersenne without fail, gives the reader a brief summary. "The hairs on your head are numbered. . . ." In theory, it is possible to count them; but in practice one encounters great difficulties, especially because their roots are not very apparent, and because their thickness varies; we can say, however, that on the average there are 187,024 hairs on a normal head. Let us not have contempt for arithmetic since we can see by this example that Christ did practice it!

The reader sees what I mean. And above all, let us note that these are not some *obiter dicta* [incidental remarks]. We can find such examples only in this draft of the most extraordinary commentary on the Gospel ever conceived in the mind of man.

We can thus understand that not once do we read in the whole work of this religious scholar a single prayer to Christ, to the Trinity, to the Holy Ghost, or to the Virgin Mary. All the prayers which he saw fit to print, because we know that his personal piety is not to be doubted, are addressed to the Inventor of Mathematics and of Physics, or, as we shall see, to the keeper of the social order. Thus, when he writes that the knowledge of parallaxes can lead the deist to desire a better life, or that for the Virgin Mary the number of acts that she posed grew like the logarithm of the graces that she received, his intention is not only to convert by the display of his erudition. Even when he stopped seeing in such exercises any efficient proof, he once more repeated them with love: in his mind, indeed, they represented the purest emanation of the piety of the scholarly Christian.

Still not doubting the purity of his intentions, we must say that he gradually tended to replace *The Imitation of Jesus Christ* by *The Imitation of the Divine Engineer*. We know that, for him, physics is not the image of metaphysics: its function is to construct mechanical models which produce the same appearances as the creative act of God, the divine plan being a mystery as far as we are concerned. But we also know that, by giving us this science completely limited to phenomena, God wanted not merely to assure our dominion over the world; with this science, he also gives us the means of imitating him "in external productions." The art of the engineer thus received a spiritual value which Christianity had never thought of giving it, and which no Christian had ever understood in the sense which Mersenne gave it. Indeed, not only does he write that "statics, hydraulics, and pneumatics produce such prodigious effects that it seems that men can imitate the most admirable works of God,"—this sentence, isolated from its remote meaning, might appear like a pure rhetorical figure—but, in a text which is no less startling in the history of Christianity than his prosopopeia to the God of the astronomers and which serves well as a résumé of his entire philosophy, he writes clearly that, provided that one can purify one's intentions —that is, if one wishes, through physics, to unite with the great Physicist of the universe—the knowledge of phenomena is the means to salvation. If God amused himself in creating the diversity of beings, we unite with him when we delight in this world which is so intriguing: "Thus we can deserve eternal life by the pleasure which we take in observing all the wealth and all the diversity which is in the world, if however, we relate them to God's, and if we make it conform to that of which ours is a reflection." Saint Augustine, in a very beautiful text, had offered a magnificent commentary on these words of Christ: "Learn from me that I am soft and humble at heart." If he calls us to him, it is not in order to teach us how to create the world, nor in order to create the marvels of nature; he tells us nothing but that he is humble and soft. There is quite a difference between the piety of Saint Augustine and that of Mersenne. . . .

Of course, the mechanical arts benefit from the consecration given to science. Doubtless, we still can find in Mersenne's writing a few rare texts in which there appears again the traditional distrust of the moralist toward the works which can take man away from meditation. But these arts are too closely tied to science for him to resign himself to abandon them. On the contrary, his desire to defend them is so evident that even before creation of a coherent doctrine in their favor, he uses any means whatsoever. Can't the Gospel show us that Jesus Christ himself was interested in wine-growers, since he said: "*Vinea mea electa. Ego sum vitis et vita* [This is my select vintage. I am the vine and the life]." Wine production and the related arts, like agriculture, are thus perfectly legitimate. Likewise, Jesus was the first apothecary because he suggested to the sons of Zebedia the excellent remedy of his bitter chalice. We have to study hydrostatics since the Gospel praises the tears of Saint Peter and of Magdalene, and because falling tears can be understood only by physicists. As for the miraculous journey of Mary Magdalene and of her companions from Palestine to Provence, sailing in a canoe without sails or oars, any reasonable man will find this a sufficient reason to lead him to become interested in the art of navigation. In summary, all the arts, like all the sciences, can be "made to fit the Gospel."

When he founded his doctrine of ***Deus artifex scientiam tribuens hominibus*** [God the artisan, who offers science to man], he added to these horrendous sophistications, which he never abandoned, some arguments of more worth. We know his doctrine and we have already quoted from his texts: the engineer imitates God, and moreover, he makes himself useful to his peers. We have no reason to doubt that Mersenne was deeply sincere when in his work of improving the technique of organs or of drawing the plans for a submarine, he thought that he was acquiring holiness by the imitation of the wine-making God, the apothecary of the Gospel!

Among what we now call the fine arts, he was familiar only with music. When he writes about the pleasure which we feel when we hear some beautiful songs, we notice however, an odd scruple in him: the saint who is deep in contemplation needs no music. But quickly he is reassured by the fact that, besides the perfect saints, there are the others, and to them music is quite useful as "a diversion worthy of a scholar," which can prod them to loving God and help them in converting vulgar souls; music is common to angels and to men, and God himself finds pleasure in it.

What can we draw from all this but that Mersenne candidly fashioned his piety according to his personal preferences and his love of positive science?

We cannot deny that this syncretism is appealing in certain aspects. It gives to his ethics a pleasant and casual look. One can be Christian and a scientist and, because most men cannot live all the time in the contemplation of the Eternal, we might as well direct them toward the sciences and the arts, which are at least the creations of God. But, above all, his preference for the mechanical arts (which are at the same time the achievement of science, as defined by his pragmatic artificialism, and also a means by which one can be useful to others) fosters in him a real liking for the common people. He is perhaps one of the first men who did free himself from the intellectual mandarin elite feeling which typifies the thought of the medieval scholars.

To those who despise the fiddlers because supposedly they are good for nothing, Mersenne answers that in any case these good people "harm no one and that everyone can receive a part of the innocent pleasure which comes from their sounds and harmony." He knows that the shepherds who play the bagpipe and the flute are not necessarily lacking in intelligence and that they sometimes understand some principles of music; and Mersenne gives them some advice relating to this art: "in order to give them some relief, for they deserve that one work for their benefit, because they have the honor of being the first to be notified of the Nativity of the Son of God by the music of the

Angels." When he sees "the common people who play the fiddle," he is much too interested in their instrument to show them the least contempt. Finally, as the mechanical arts become integrated with science, the craftsmen acquire the right to belong to the community of fine minds. This is certainly an important date in the history of moral ideas, and according to us, one could not quote many preceding or contemporaneous tests in which a scholar prides himself on working for the good of the shepherds.

The two problems which existed at the beginning of the seventeenth century thus receive a solution: the science of phenomena becomes a legitimate endeavor and even a highly worthy one; material civilization acquires a spiritual meaning in allowing us to master the forces of nature and to use them in order to reduce the burden of the common people. Religion and science, in spite of the radical change undergone by the latter, can go on coexisting and can even help each other. . . .

Introduction to Leviathan

Thomas Hobbes

Nature, the art whereby God hath made and governs the world, is by the art of man, as in many other things, so in this also imitated, that it can make an artificial animal. For seeing life is but a motion of limbs, the beginning whereof is in some principal part within; why may we not say, that all automata (engines that move themselves by springs and wheels as doth a watch) have an artificial life? For what is the heart, but a spring; and the nerves, but so many strings; and the joints, but so many wheels, giving motion to the whole body, such as was intended by the artificer? Art goes yet further, imitating that rational and most excellent work of nature, man. For by art is created that great Leviathan called a Commonwealth, or State, in Latin *Civitas*, which is but an artificial man; though of greater stature and strength than the natural, for whose protection and defence it was intended; and in which the sovereignty is an artificial soul, as giving life and motion to the whole body; the magistrates, and other officers of judicature and execution, artificial joints; reward and punishment, by which fastened to the seat of the sovereignty every joint and member is moved to perform his duty, are the nerves, that do the same in the body natural; the wealth and riches of all the particular members, are the strength; salus populi, the people's safety, its

Source: Thomas Hobbes, *The English Works of Thomas Hobbes of Malmesbury; Now First Collected and Edited by Sir William Molesworth, Bart.*, Vol. III (London: J. Bohn, 1839), pp. ix-xii.

business; counsellors, by whom all things needful for it to know are suggested unto it, are the memory; equity, and laws, an artificial reason and will; concord, health; sedition, sickness; and civil war, death. Lastly, the pacts and covenants, by which the parts of this body politic were at first made, set together, and united, resemble that fiat, or the let us make man, pronounced by God in the creation.

To describe the nature of this artificial man, I will consider

First, the matter thereof, and the artificer; both which is man.

Secondly, how, and by what covenants it is made; what are the rights and just power or authority of a sovereign; and what it is that preserveth or dissolveth it.

Thirdly, what is a Christian commonwealth.

Lastly, what is the kingdom of darkness.

Concerning the first, there is a saying much usurped of late, that wisdom is acquired, not by reading of books, but of men. Consequently whereunto, those persons, that for the most part can give no other proof of being wise, take great delight to show what they think they have read in men, by uncharitable censures of one another behind their backs. But there is another saying not of late understood, by which they might learn truly to read one another, if they would take the pains; that is, *nosce teipsum*, read thyself: which was not meant, as it is now used, to countenance, either the barbarous state of men in power, towards their inferiors; or to encourage men of low degree, to a saucy behaviour towards their betters; but to teach us, that for the similitude of the thoughts and passions of one man, to the thoughts and passions of another, whosoever looketh into himself, and considereth what he doth, when he does think, opine, reason, hope, fear, &c. and upon what grounds; he shall thereby read and know, what are the thoughts and passions of all other men upon the like occasions. I say the similitude of passions, which are the same in all men, desire, fear, hope, &c; not the similitude of the objects of the passions, which are the things desired, feared, hoped, &c: for these the constitution individual, and particular education, do so vary, and they are so easy to be kept from our knowledge, that the characters of man's heart, blotted and confounded as they are with dissembling, lying, counterfeiting, and erroneous doctrines, are legible only to him that searcheth hearts. And though by men's actions we do discover their design sometimes; yet to do it without comparing them with our own, and distinguishing all circumstances, by which the case may come to be altered, is to decypher without a key, and be for the most part deceived, by too much trust, or by too much diffidence; as he that reads, is himself a good or evil man.

But let one man read another by his actions never so perfectly, it serves him only with his acquaintance, which are but few. He that is to govern a whole nation, must read in himself, not this or that particular man; but mankind: which though it be hard to do, harder than to learn any language or science; yet when I shall have set down my own reading orderly, and perspicuously, the pains left another, will be only to consider, if he also find not the same in himself. For this kind of doctrine admitteth no other demonstration.

4 Newtonian Science: The Fleeting Triumph of Natural Theology and the Search for a Social Physics

Sir Isaac Newton's *Mathematical Principles of Natural Philosophy* appeared in 1687; and in spite of the fact that no more than a handful of men and probably one woman—the Marquise du Châtelet—were able to understand all of its complex geometrical arguments, this book, more than any other, moved the Western world-view in the direction of scientism. In a work entitled *Elements of Philosophy*, Jean D'Alembert, scientist and co-editor of the famous *Encyclopedie*, gave a superb description of the way in which Newtonian science informed the fundamental attitudes of the enlightenment. He wrote:

> Natural science from day to day accumulates new riches. Geometry, by extending its limits, has borne its torch into the regions of physical science which lay nearest at hand. The true system of the world has been recognized.... In short, from the earth to Saturn, from the history of the heavens to that of insects, natural philosophy has been revolutionized; and nearly all other fields of knowledge have assumed new forms ... the discovery and application of a new method of philosophizing, the kind of enthusiasm which accompanies discoveries, a certain exaltation of ideas which the spectacle of the universe produces in us; all these causes have brought about a lively fermentation of minds. Spreading throughout nature in all directions, this fermentation has swept everything before it which stood in its way with a sort of violence, like a river which has burst its dams.... Thus, from the principles of the secular sciences to the foundations of religious revelation, from metaphysics to matters of taste, from music to morals, from the scholastic disputes of theologians to matters of commerce, from natural law to the arbitrary laws of nations ... everything has been discussed, analyzed, or at least mentioned. The fruit or sequel of this general effervescence of minds has been to cast new light on some matters and new shadows on others, just as the ebb and flow of the tides leaves some things on the shore and washes others away.[1]

D'Alembert hints in this statement that the major impact of scientific

---

[1]Ernst Cassirrer, *The Philosophy of the Enlightenment* (Boston: Beacon Press, 1955), pp. 46-47.

thought—particularly that of Newton and of John Locke, who extended the scientific method into the domain of psychology—was to open up a new conception of the very function and capabilities of the human mind. Newton's success in proving that the methods which he explicitly laid down in the *Principia*—methods derived from Galileo, Descartes, and Bacon—were capable of solving such a great riddle as that of universal gravitation had a great psychological impact on the age. His success was seen not merely as the reduction of a limited number of physical phenomena to some kind of order; it was seen as establishing once and for all ***The Cosmic Universal Law.*** It was a great victory for human knowledge. And it appeared, during a period of intense philosophical, religious, and political turmoil, to demonstrate that human beings were capable of penetrating the deepest secrets of nature and might hope to solve even their moral and political problems as well.

The scientific-mathematical-empirical method seemed to receive almost pontifical authority by its association with Newton's accomplishments in natural philosophy; and while it may be an oversimplification, I think it is not really a major distortion to say that the enlightenment of the eighteenth century was in large measure an attempt to extend the domain of application of scientific method to include not only inanimate nature, but also the laws of human thought, morality, society, and religion. One may believe that this attempt was misguided, but it was most certainly not misanthropic. The conviction of the men of the early eighteenth century was that *now* was the time to unlock all of the secrets of nature through scientific analysis. And as the models for how this end should be accomplished, many took the *Principia* and the *Opticks* of Isaac Newton.

One of the first fields to which Newtonian methods were applied was religion. Newton himself had argued that his natural philosophy could be used to support religious belief; and he was followed by a number of important English clergymen. Perry Miller's "Bentley and Newton" and the four letters from Newton to Bentley illustrate the initial use of Newton's thought by Anglican churchmen.

During the first decades of its existence, Newtonian natural theology seemed to eliminate the specter of the retired God implied by the earlier mechanical universe analogy. Newton demanded the continuing involvement of God in the world both to provide a cause for gravitational attraction and to adjust certain imperfections in the system, and this ensured God's immanence in the universe. Religious arguments depending on Newtonian science suffered a series of blows during the eighteenth century, however, as theoretical astronomers gradually reduced the range of phenomena which could not be explained without extraordinary interventions.

The second area in which scientific examples provided a model for enlightenment developments was political or social philosophy. Some attempts to graft Newtonianism to politics were very naive and direct, as was J. T. Desagulier's *The Newtonian System of the World: The Best*

Model of Government, An Allegorical Poem of 1728. Many, like the attempts of Baron d'Holbach, showed little true understanding of the science that they claimed to be using as a basis for thought; but others, like the works of Voltaire and of Condorcet, illustrated a deep and penetrating sense of scientific method.

Henry Guerlac, in his "Three Eighteenth-Century Social Philosophers: Scientific Influences on Their Thought" points up some of the problems in assessing the Newtonian influence on eighteenth-century social ideas, warning that while Newton might have symbolized science for enlightenment figures, the science which they borrowed was often cartesianism or a crude mechanism that would have been anathema to Newton. Nonetheless, he does discuss how Voltaire in particular based some of his beliefs on a sophisticated understanding of Newtonian science.

Certainly the most important long-range attempt to introduce scientific method into social and political considerations was made by a group of Parisian social philosophers including the important Socialist theorists Saint-Simon and Charles Fourier and the founder of Positivism, Auguste Comte. The Newtonian stimulus to the work of these men is attested to in Saint-Simon's early attempts to urge a Religion of Newton to replace traditional Christianity and by Fourier's statement that, "My theory is the continuation of Newton's on attraction. In this new mine he exploited only the material vein, I exploit the industrial. I am a continuator and I have never countenanced the name Fourieriste." But far more important than the personal Newtonian cast of some of this thinking was the attempt to construct a science of society patterned on the natural sciences. Auguste Comte's call for a science of sociology is one which still guides the great bulk of social thinkers in the Western world:

> We possess now a celestial physics, a terrestrial physics, either mechanical or chemical, a vegetable physics, and an animal physics; we still count one more and last one, social physics to complete the system of our knowledge of nature. I understand by social physics the science which has for its subject the study of social phenomena considered in the same spirit as astronomical, physical, chemical, or physiological phenomena, that is, subject to natural, invariable laws, the discovery of which is the special object of investigation.[2]

F. A. Hayek's "The Source of the Scientistic Hubris: L'Ecole Polytechnique," provides a penetrating and unsympathetic account of the institution through which science was firmly established as a dominant source of political ideology in the movement that surrounded Saint-Simon, Comte, and Fourier.

---

[2]See *Positive Politics* (Paris: Carilan-Goevry and Vor Dalmont, 1854), vol. IV, Appendix, pp. 148-150.

Newton and Bentley

Perry Miller

Richard Bentley was born in 1662 in a family of substantial Yorkshire yeomen. He achieved fame (and left an impress on British scholarship that still is felt) as a classical scholar of prodigious erudition, and also a certain infamy as the tempestuous Master of Trinity College, Cambridge, which he ruled from 1700 until his death in 1742 with so tyrannical a hand that he excited repeated insurrections of the Fellows. He was a massive philologist, who found the supreme felicity of life in the emendation of a corrupted text or in the exposure of a forgery. He made a sensation among the learned in 1699 by demonstrating that a body of letters long attributed to a Sicilian tyrant of the 6th century B.C., named Phalaris, was a fabrication made some five or ten centuries later. These epistles had been publicly admired by gentlemen such as Sir William Temple who believed that the writers of antiquity were far superior to all moderns, including Shakespeare and Milton. By showing the letters to be spurious, Bentley impeached both the acumen and the taste of these "ancients." The greatest classicist of his time thus appeared a barbarous and ruthless modernist, and so was furiously attacked in a squib called *The Battle of the Books*, written by an erstwhile secretary of Sir William Temple, one Jonathan Swift.

In 1691, Bentley, having taken his degree at St. John's College, was chaplain to Bishop Stillingfleet of Worcester, a leader of liberal theologizing, who early said of Bentley that "had he but the gift of humility, he would be the most extraordinary man in Europe." On December 30 died Robert Boyle, a great physicist and chemist, a gentleman, and one who devoutly believed the new science to be a bulwark against the "atheism" so widely affected during the Restoration by the wits of the taverns and coffee-houses. He left funds sufficient to yield £ 50 a year for endowing an annual lectureship of eight discourses on the evidences of Christianity. There were four trustees, one of whom was John Evelyn; another was Bishop Tenison of Lincoln, who had encountered and appraised the chaplain of his colleague in Worcester. The trustees took what seemed a long chance, and nominated Bentley. He threw himself into the challenge with the same energy he expended upon Greek manuscripts or in opposing dons.

The principal source of the atheism Bentley had to counteract was

---

Source: Perry Miller, "Newton and Bentley," in *Isaac Newton's Papers and Letters on Natural Philosophy*, edited by I. Bernard Cohen, assisted by Robert E. Schofield (Cambridge, Mass.; Harvard University Press, 1958), pp. 271-278.

Thomas Hobbes, who had been under fire from the pious and the orthodox for forty years. Platonists like Ralph Cudworth had belabored him with preexistent ideas, Richard Cumberland with inherent moral law, and ecclesiastical authoritarians, most notably John Dryden, with general abuse. But, so far, it seemed to the guardians of Christianity that the tide of atheism had not been checked; clearly a new method was required. Bentley was exactly the man for the occasion, because he was one of the first to grasp the importance of a book published in 1690 by John Locke, *An Essay Concerning Human Understanding*; Bentley saw at once that thereafter nobody in the age would give credence to the notion of an innate idea. If his *A Confutation of Atheism* was really going to confute, it would need *proofs*. Bentley was the sort of bulldog who, ordered to find proofs, would bring back dozens of them between his jaws.

A mind that operated in this fashion would already have been thinking that if theological propositions were now to rest their defense exclusively on demonstrations satisfactory to reason, the defender would have to know something about a book that Isaac Newton of Trinity College had published in 1687, the *Principia*. So far, it appeared, few if any were able to understand it, and many said it was nonsense, but Bentley had to see for himself. However, he was a linguist and a literary scholar, and needed help; in the summer of 1691, before the lectureship was instituted, he had asked a Scottish mathematician, John Craige, to tell him what books he would need to master in order to qualify himself for following the *Principia*. Craige sent back, as an essential minimum, a bibliography so tremendous that even a Bentley was aghast; but characteristically he began to look about for short cuts, and, taking his courage in his hands, addressed Newton directly. From Trinity College came a much shorter list, encouraging directions, and apparently full sympathy. "At the first perusal of my Book," said Newton, "it's enough if you understand the propositions with some of the demonstrations which are easier than the rest." He thought Bentley should read the first sixty pages, then skip to the third book and get the design of that; then he might at leisure go back to such propositions as he had a desire to know. With the task thus cut down to manageable proportions, Bentley rapidly comprehended (so he thought) the whole design. When the call came, he was ready. He devoted the first six of his Boyle Lectures to proving the existence of God from such data as the faculties of the human soul and the structure of the body, but he triumphantly expounded in the last two on the new, difficult, and mathematically irrefutable physics. His success was immense, and in the opinion of many (including Bentley himself), *A Confutation of Atheism* so routed the atheists that they did not dare any longer show their faces openly, and so took refuge in the pretense of "deism."

The two sermons are important in the history of Western thought not only because they were the first popular attempt to lay open the "sublime discoveries" of Newton, but because they set the precedent for the entire Enlightenment. So far, neither the infidel nor the believer had been able to cope with the new wisdom; Bentley seized the initiative, and gave believers the assurance (or perhaps one should say the illusion) that the Newtonian physics, by conclusively showing that the order of the universe could not have been produced mechanically, was now the chief support of faith. Whether employed by Christians or deists, Bentley's technique for deducing religious propositions out of the equations of the *Principia* became an indispensable ingredient in the whole complex of 18th-century optimism.

But, for our purposes, the sermons are still more important because, whatever their merits as expositions of the system, they called forth from the great man himself four letters which are major declarations in modern history of the method and of the mentality of the scientist. While the manuscript was being printed, Bentley found himself worried for fear he had not sufficiently disposed of the theory of Lucretius (from whom Hobbes derived) that the cosmic system began with chance bumpings together of descending atoms, each endowed with an innate power of gravity. He wrote to Newton for further clarification, so that he could make last-minute changes in his proof. It took Newton four letters, from December 10 to February 25, to set Bentley straight (in fact, we may wonder whether Bentley fully got the point!), and Bentley appreciated their importance. He carefully preserved them, so that his executor could publish them in 1756. Dr. Johnson, observing that the questions had caused Newton to think out further consequences of his principles than he had yet anticipated, said of them about the finest thing that can be said, that they show "how even the mind of Newton gains ground gradually upon darkness."

The sermons show that Bentley had indeed perceived the general thesis, though the letters suggest that in the printed form Bentley made it more precise than he had done in the pulpit. This is the argument that, had gravity been the only force active at the moment of creation, the planets of our system would have fallen quickly into the sun. Hence must be assumed a specific intervention of force (only a divine force would do) which arrested the descents at the appropriate places and sent the planets spinning on their transverse orbits. Likewise, when one considers the spacing of these orbits, no principle of science will determine the relations of the distances except that "The Author of the system thought it convenient." Bentley seemed to Newton on the right track insofar as he argued that the operations of gravity over empty spaces could mean only that an "agent" was constantly guiding the stars and planets according to certain laws. Assuredly, this agent must have a volition, and must be "very well skilled in mechanics and geometry." Bentley was eager to call the agent God; Newton had no objection.

But evidently, either in the first draft of the sermons or in a letter, Bentley said something which implied that gravity was in some sense an inherent property of matter, implicit in the very substance, a sort of "occult quality," or a kind of eternal magnetism. The vehemence with which Newton rejects any such opinion is striking. Between the letters numbered II and III in this printing, Bentley wrote back a worried answer: he was so fully aware that in Newton's system universal gravitation could never be solved "mechanically" that he was surprised to have Newton warn him against the heresy. "If I used that word, it was only for brevity's sake." Well, brevity to a philologist might be one thing, but another to Newton. He wanted language exact, and certainly in the printed version Bentley took care that not even for brevity's sake should there be any suggestion that gravity is synonymous with material existence. Thus corrected, Bentley was able to conclude that mutual gravitation can operate at a distance only because it is simultaneously regulated by the "agent" and not by the system itself; here then was what he and the age most wanted, "a new and invincible argument for the being of God." From this point the sailing was clear, and Bentley goes ahead like a ship in full rig, to the joyous conclusion that everything concerning this system and particularly this globe, including the inclination of its axis and the irregular distribution of land and ocean, has been appointed for the best by a divine intelligence.

The letters show that Newton wanted to be helpful, and he was eager that Bentley should not misrepresent him; yet they are not prolix, they do not volunteer anything beyond replies to particular questions, and the careful reader does not get the impression of an outgoing enthusiasm. Newton was human enough to be eager for fame and almost pathologically jealous for his reputation; but he was shrewd enough to be able to utilize Bentley without being taken in by him. For years after the Boyle Lectures, Bentley made a public parade of his friendship with Newton, and took upon himself the office of urging a second edition of the *Principia*. When Newton at last consented in 1708 to allow Roger Cotes, Fellow of Trinity College (in whom Newton did have confidence), to prepare the text, Bentley officiously acted as middleman—and pocketed the profits! John Conduitt records that he was disgusted, and asked Newton point-blank why he let Bentley "print his *Principia*" when Bentley obviously did not understand it. "Why," replied the lordly Newton, "he was covetous, and I let him do it to get the money."

In the light of this revelation we may wonder what, back in February of 1693, Bentley made, if anything, of the extraordinary clause in the third letter, where Newton says that whether the agent who is the cause of gravity "be material or immaterial, I have left to the consideration of my readers." This hardly seems the tone of one who has joined a crusade against materialistic atheism! But still more startling is the sentence that comes in the next paragraph, where Newton shows Bentley that mathematical language may seem to common sense an impropriety of speech, still "those things which men understand by improper and contradictious phrases may be sometimes really in nature without any contradiction at all." There is no suggestion in Bentley's two sermons that he had even a dim sense of what Newton tried in this passage to convey to him. For Bentley, the Newtonian system was clear, rational, simple; it could be translated at once and throughout into declarative affirmations of natural theology. That was its beauty and its utility. That there was any incongruity between the process of the human mind and those of the universe would henceforth be unthinkable. Newton had, Bentley was assured, linked them indissolubly.

This conviction, as I have said, became the major premise of the Age of Reason. Bentley's tactics were taken over by Voltaire in 1738 when he conquered the mind of the Continent by popularizing Newton. Actually, the assumption remained undisturbed—or indeed strengthened—by that revolution in sensibility which we call the Romantic movement. Even after the character of reason had been radically transformed from Bentley's solid prose to the inward intuitions of the poet, the assumption that there is a perfect "correspondence" between the structures of the psyche and those of physics endured. Emerson summarized the Romantic optimism by declaring that the laws of nature answer to those of mind as image in the mirror. Only recently, and mainly in our own distracted time, has science freed itself from the literary incubus that Bentley fastened upon it. But we should find this worth meditating upon, that Newton explicitly warned him that what men are apt to consider self-contradictory may, nevertheless, be the rule in nature.

There is a mystery in these letters—the enigma that is Newton himself. Nobody in 1692, nor for a century thereafter, noted that, when Bentley confidently brought God into action as the diverter of falling bodies into "this transverse and violent motion," the Creator became, in a sense, only a half-creator of the system. God's action was made once and once for all; it was that "first impulse impressed upon them, not only for five or six thousand years, but many millions of millions." But the gravitating motion, the descent toward the sun, is continuous; despite his effort to make clear his agreement with Newton, Bentley still calls it "a constant energy infused into matter by the Author of all things." Did Newton, in his secret heart, have the wit, which no contemporary possessed, to see that such toying with the notion that gravity was a constant energy infused into matter raised the question of whether the infusion really had been made by the author of all things? Might this not be only a gratuitious addition, made by a mind precommitted to the thesis, by one incapable properly of dealing with the meaning of the infinite? Whether Newton had read Pascal we do not know, but assuredly Bentley never had!

If the letters mean anything, then, they mean that Newton was not quite a Newtonian. He was holding something in reserve, not giving himself entirely to his own discoveries, stupendous as he realized them to be. As for ultimate causes, he knew how to say that he did not know. Our curiosity is aroused, but never shall be satisfied, by the evasive ending of the first letter: Isaac Newton had still another argument to prove the existence of God, potentially very strong, but because the principles on which it was grounded were not yet widely enough received, "I think it more advisable to let it sleep." What were those principles? Perhaps he meant simply the realm of optics into which he was now venturing, already musing upon ideas he was to let see the light only in the form of a question at the end of Query 31 in the 1717 edition of the *Opticks*, when even then he was "not yet satisfied about it for want of experiments." But we cannot help asking if in his subtle consciousness there was a sense of still more complex principles which would need to wait still longer before becoming "better received." And were these withheld principles possibly just those dark and inexplicable discrepancies between the mind of the creature and the methods of the creation which he could dare to contemplate, but of which the Bentleys of this world never attain even a rudimentary awareness?

Newton's Letters to Bentley

Isaac Newton

LETTER I

To the Reverend Dr. Richard Bentley, at the Bishop of Worcester's House, in Park Street, Westminster

Sir,

When I wrote my treatise about our system, I had an eye upon such principles as might work with considering men for the belief of a Deity; and nothing can rejoice me more than to find it useful for that purpose. But if I have done the public any service this way, it is due to nothing but industry and patient thought.

As to your first query, it seems to me that if the matter of our sun and planets and all the matter of the universe were evenly scattered throughout all the heavens, and every particle had an innate gravity toward all the rest, and the whole space throughout which this matter was scattered was but finite, the matter on the outside of this space would, by its gravity, tend toward all the matter on the inside and, by consequence, fall down into the middle of the whole space and there compose one great spherical mass. But if the matter was evenly disposed throughout an infinite space, it could never convene into one mass; but some of it would convene into one mass and some into another, so as to make an infinite number of great masses, scattered at great distances from one to another throughout all that infinite space. And thus might the sun and fixed stars be formed, supposing the matter were of a lucid nature. But how the matter should divide itself into two sorts, and that part of it which is fit to compose a shining body should fall down into one mass and make a sun and the rest which is fit to compose

---

Source: *Four Letters from Sir Isaac Newton to Doctor Bentley, Containing Some Arguments in Proof of a Deity* (London: R. and J. Dodsley, 1756). Spelling and punctuation have been modernized.

an opaque body should coalesce, not into one great body, like the shining matter, but into many little ones; or if the sun at first were an opaque body like the planets or the planets lucid bodies like the sun, how he alone should be changed into a shining body whilst all they continue opaque, or all they be changed into opaque ones whilst he remains unchanged, I do not think explicable by mere natural causes, but am forced to ascribe it to the counsel and contrivance of a voluntary Agent.

The same Power, whether natural or supernatural, which placed the sun in the center of the six primary planets, placed Saturn in the center of the orbs of his five secondary planets and Jupiter in the center of his four secondary planets, and the earth in the center of the moon's orb; and therefore, had this cause been a blind one, without contrivance or design, the sun would have been a body of the same kind with Saturn, Jupiter, and the earth, that is, without light and heat. Why there is one body in our system qualified to give light and heat to all the rest, I know no reason but because the Author of the system thought it convenient; and why there is but one body of this kind, I know no reason but because one was sufficient to warm and enlighten all the rest. For the Cartesian hypothesis of suns losing their light and then turning into comets, and comets into planets, can have no place in my system and is plainly erroneous; because it is certain that, as often as they appear to us, they descend into the system of our planets, lower than the orb of Jupiter and sometimes lower than the orbs of Venus and Mercury, and yet never stay here, but always return from the sun with the same degrees of motion by which they approached him.

To your second query, I answer that the motions which the planets now have could not spring from any natural cause alone, but were impressed by an intelligent Agent. For since comets descend into the region of our planets and here move all manner of ways, going sometimes the same way with the planets, sometimes the contrary way, and sometimes in crossways, in planes inclined to the plane of the ecliptic and at all kinds of angles, it is plain that there is no natural cause which could determine all the planets, both primary and secondary, to move the same way and in the same plane, without any considerable variation; this must have been the effect of counsel. Nor is there any natural cause which could give the planets those just degrees of velocity, in proportion to their distances from the sun and other central bodies, which were requisite to make them move in such concentric orbs about those bodies. Had the planets been as swift as comets, in proportion to their distances from the sun (as they would have been had their motion been caused by their gravity, whereby the matter, at the first formation of the planets, might fall from the remotest regions toward the sun), they would not move in concentric orbs, but in such eccentric ones as the comets move in. Were all the planets as swift as Mercury or as slow as Saturn or his satellites, or were their several velocities otherwise much greater or less than they are, as they might have been had they arose from any other cause than their gravities, or had the distances from the centers about which they move been greater or less than they are, with the same velocities, or had the quantity of matter in the sun or in Saturn, Jupiter, and the earth, and by consequence their gravitating power, been greater or less than it is, the primary planets could not have revolved about the sun nor the secondary ones about Saturn, Jupiter, and the earth, in concentric circles, as they do, but would have moved in hyperbolas or parabolas or in ellipses very eccentric. To make this system, therefore, with all its motions, required a cause which understood and compared together the quantities of matter in the several bodies of the sun and planets and the gravitating powers resulting from thence, the several distances of the primary planets from the sun and of the secondary ones from Saturn, Jupiter, and the earth, and the velocities with which these planets could revolve about those quantities of matter in the central bodies; and to compare and adjust all these things together, in so great a variety of bodies, argues that cause to be, not blind and fortuitous, but very well skilled in mechanics and geometry.

To your third query, I answer that it may be represented that the sun may, by heating those planets most of which are nearest to him, cause them to be better concocted and more condensed by that concoction. But when I consider that our earth is much more heated in its bowels below the upper crust by subterraneous fermentations of mineral bodies than by the sun, I see not why the interior parts of Jupiter and Saturn might not be as much heated, concocted, and coagulated by those fermentations as our earth is; and therefore this various density should have some other cause than the various distances of the planets from the sun. And I am confirmed in this opinion by considering that the planets of Jupiter and Saturn, as they are rarer than the rest, so they are vastly greater and contain a far greater quantity of matter, and have many satellites about them; which qualifications surely arose, not from their being placed at so great a distance from the sun, but were rather the cause why the Creator placed them at a great distance. For, by their gravitating powers, they disturb one another's motions very sensibly, as I find by some late observations of Mr. Flamsteed; and had they been placed much nearer to the sun and to one another, they would, by the same powers, have caused a considerable disturbance in the whole system.

To your fourth query, I answer that, in the hypothesis of vortices, the inclination of the axis of the earth might, in my opinion, be ascribed to the situation of the earth's vortex before it was absorbed by the neighboring vortices and the earth turned from a sun to a comet; but this inclination ought to decrease constantly in compliance with the motion of the earth's vortex, whose axis is much less inclined to the ecliptic, as appears by the motion of the moon carried about therein. If the sun by his rays could carry about the planets, yet I do not see how we could thereby effect their diurnal motions.

Lastly, I see nothing extraordinary in the inclination of the earth's axis for proving a Deity, unless you will urge it as a contrivance for winter and summer, and for making the earth habitable toward the poles; and that the diurnal rotations of the sun and planets, as they could hardly arise from any cause purely mechanical, so by being determined all the same way with the annual and menstrual motions they seem to make up that harmony in the system which, as I explained above, was the effect of choice rather than chance.

There is yet another argument for a Deity, which I take to be a very strong one; but till the principles on which it is grounded are better received, I think it more advisable to let it sleep.

I am your most humble servant to command,

Is. Newton

Cambridge, December 10, 1692

LETTER II

For Mr. Bentley, at the Palace at Worcester

Sir,

I agree with you that if matter evenly diffused through a finite space, not spherical, should fall into a solid mass, this mass would affect the figure of the whole space, provided it were not soft, like the old chaos, but so hard and solid from the beginning that the weight of its protuberant parts could not make it yield to their pressure; yet, by earthquakes loosening the parts of this solid, the protuberances might sometimes sink a little by their weight, and thereby the mass might by degrees approach a spherical figure.

The reason why matter evenly scattered through a finite space would convene in the midst you conceive the same with me, but that there should be a central particle so accurately placed in the middle as to be always equally attracted on all sides, and thereby continue without motion, seems to me a supposition fully as hard as to make the sharpest needle stand upright on its point upon a looking glass. For if the very mathematical center of the central particle be not accurately in the very mathematical center of the attractive power of the whole mass, the particle will not be attracted equally on all sides. And much harder it is to suppose all the particles in an infinite space should be so accurately poised one among another as to stand still in a perfect equilibrium. For I reckon this as hard as to make, not one needle only, but an infinite number of them (so many as there are particles in an infinite space) stand accurately poised upon their points. Yet I grant it possible, at least by a divine power; and if they were once to be placed, I agree with you that they would continue in that posture without motion forever, unless put into new motion by the same power. When, therefore, I said that matter evenly spread through all space would convene by its gravity into one or more great masses, I understand it of matter not resting in an accurate poise.

But you argue, in the next paragraph of your letter, that every particle of matter in an infinite space has an infinite quantity of matter on all sides and, by consequence, an infinite attraction every way, and therefore must rest *in equilibrio*, because all infinites are equal. Yet you suspect a paralogism in this argument, and I conceive the paralogism lies in the position that all infinites are equal. The generality of mankind consider infinites no other ways than indefinitely; and in this sense they say all infinites are equal, though they would speak more truly if they should say they are neither equal nor unequal, nor have any certain difference or proportion one to another. In this sense, therefore, no conclusions can be drawn from them about the equality, proportions, or differences of things; and they that attempt to do it usually fall into paralogisms. So when men argue against the infinite divisibility of magnitude by saying that if an inch may be divided into an infinite number of parts the sum of those parts will be an inch; and if a foot may be divided into an infinite number of parts the sum of those parts must be a foot; and therefore, since all infinites are equal, those sums must be equal, that is, an inch equal to a foot.

The falseness of the conclusion shows an error in the premises, and the error lies in the position that all infinites are equal. There is, therefore, another way of considering infinites used by mathematicians, and that is, under certain definite restrictions and limitations, whereby infinites are determined to have certain differences or proportions to one another. Thus Dr. Wallis considers them in his *Arithmetica Infinitorum*, where, by the various proportions of infinite sums, he gathers the various proportions of infinite magnitudes, which way of arguing is generally allowed by mathematicians and yet would not be good were all infinites equal. According to the same way of considering infinites, a mathematician would tell you that, though there be an infinite number of infinite little parts in an inch, yet there is twelve times that number of such parts in a foot; that is, the infinite number of those parts in a foot is not equal to but twelve times bigger than the infinite number of them in an inch. And so a mathematician will tell you that if a body stood *in equilibrio* between any two equal and contrary attracting infinite forces, and if to either of these forces you add any new finite attracting force, that new force, howsoever little, will destroy their equilibrium and put the body into the same motion into which it would put it were those two contrary equal forces but finite or even none at all; so that in this case the two equal infinites, by the addition of a finite to either of them, become unequal in our ways of reckoning; and after these ways we must reckon, if from the considerations of infinites we would always draw true conclusions.

To the last part of your letter, I answer, first, that if the earth (without the moon) were placed anywhere with its center in the *orbis magnus* and stood still there without any gravitation or projection, and there at once were infused into it both a gravitating energy toward the sun and a transverse impulse of a just quantity moving it directly in a tangent to the *orbis magnus*, the compounds of this attraction and projection would, according to my notion, cause a circular revolution of the earth about the sun. But the transverse impulse must be a just quantity; for if it be too big or too little, it will cause the earth to move in some other line. Secondly, I do not know any power in nature which would cause this transverse motion without the divine arm. Blondel tells us somewhere in his book of Bombs that Plato affirms that the motion of the planets is such as if they had all of them been created by God in some region very remote from our system and let fall from thence toward the sun, and so soon as they arrived at their several orbs their motion of falling turned aside into a transverse one. And this is true, supposing the gravitating power of the sun was double at that moment of time in which they all arrive at their several orbs; but then the divine power is here required in a double respect, namely, to turn the descending motions of the falling planets into a side motion and, at the same time, to double the attractive power of the sun. So, then, gravity may put the planets into motion, but without the divine power it could never put them into such a circulating motion as they have about the sun; and therefore, for this as well as other reasons, I am compelled to ascribe the frame of this system to an intelligent Agent.

You sometimes speak of gravity as essential and inherent to matter. Pray do not ascribe that notion to me, for the cause of gravity is what I do not pretend to know and therefore would take more time to consider of it.

I fear what I have said of infinites will seem obscure to you; but it is enough if you understand that infinites, when considered absolutely without any restriction or limitation, are neither equal nor unequal, nor have any certain proportion one to another, and therefore the principle that all infinites are equal is a precarious one.

Sir, I am your most humble servant,

Is. Newton

Trinity College, January 17, 1692/3

LETTER III

For Mr. Bentley, at the Palace at Worcester

Sir,

Because you desire speed, I will answer your letter with what brevity I can. In the six positions you lay down in the beginning of your letter, I agree with you. Your assuming the *orbis magnus* 7,000 diameters of the earth wide implies the sun's horizontal parallax to be half a minute. Flamsteed and Cassini have of late observed it to be about 10 minutes, and thus the *orbis magnus* must be 21,000, or, in a round number, 20,000 diameters of the earth wide. Either computation, I think, will do well; and I think it not worth while to alter your numbers.

In the next part of your letter you lay down four other positions, founded upon the six first. The first of these four seems very evident, supposing you take attraction so generally as by it to understand any force by which distant bodies endeavor to come together without mechanical impulse. The second seems not so clear, for it may be said that there might be other systems of worlds before the present ones and others before those, and so on to all past eternity, and by consequence that gravity may be coeternal to matter and have the same effect from all eternity as at present, unless you have somewhere proved that old systems cannot gradually pass into new ones or that this system had not its original from the exhaling matter of former decaying systems but from a chaos of matter evenly dispersed throughout all space; for something of this kind, I think you say, was the subject of your Sixth Sermon, and the growth of new systems out of old ones, without the mediation of a divine power, seems to be apparently absurd.

The last clause of the second position I like very well. It is inconceivable that inanimate brute matter should, without the mediation of something else which is not material, operate upon and affect other matter without mutual contact, as it must be if gravitation, in the sense of Epicurus, be essential and inherent in it. And this is one reason why I desired you would not ascribe innate gravity to me. That gravity should be innate, inherent, and essential to matter, so that one body may act upon another at a distance through a *vacuum*, without the mediation of anything else, by and through which their action and force may be conveyed from one to another, is to me so great an absurdity that I believe no man who has in philosophical matters a competent faculty of thinking can ever fall into it. Gravity must be caused by an agent acting constantly according to certain laws, but whether this agent be material or immaterial I have left to the consideration of my readers.

Your fourth assertion, that the world could not be formed by innate gravity alone, you confirm by three arguments. But in your first argument you seem to make a *petitio principii*; for whereas many ancient philosophers and others, as well theists as atheists, have all allowed that there may be worlds and parcels of matter innumerable or infinite, you deny this by representing it as absurd as that there should be positively an infinite arithmetical sum or number, which is a contradiction *in terminis*, but you do not prove it as absurd. Neither do you prove that what men mean by an infinite sum or number is a contradiction in nature, for a contradiction *in terminis* implies no more than an impropriety of speech. Those things which men understand by improper and contradictious phrases may be sometimes really in nature without any contradiction at all: a silver inkhorn, a paper lantern, an iron whetstone, are absurd phrases, yet the things signified thereby are really in nature. If any man should say that a number and a sum, to speak properly, is that which may be numbered and summed, but things infinite are numberless or, as we usually speak, innumerable and sumless or insummable, and therefore ought not to be called a number or sum, he will speak properly enough, and your argument against him will, I fear, lose its force. And yet if any man shall take the words 'number' and 'sum' in a larger sense, so as to understand thereby things which, in the proper way of speaking, are numberless and sumless (as you seem to do when you allow an infinite number of points in a line), I could readily allow him the use of the contradictious phrases of 'innumerable number' or 'sumless sum,' without inferring from thence any absurdity in the thing he means by those phrases. However, if by this or any other argument you have proved the finiteness of the universe, it follows that all matter would fall down from the outsides and convene in the middle. Yet the matter in falling might concrete into many round masses, like the bodies of the planets, and these, by attracting one another, might acquire an obliquity of descent by means of which they might fall, not upon the great central body, but upon the side of it, and fetch a compass about and then ascend again by the same steps and degrees of motion and velocity with which they descended before, much after the manner that comets revolve about the sun; but a circular motion in concentric orbs about the sun they could never acquire by gravity alone.

And though all the matter were divided at first into several systems, and every system by a divine power constituted like ours, yet would the outside systems descend toward the middlemost; so that this frame of things could not always subsist without a divine power to conserve it, which is the second argument; and to your third I fully assent.

As for the passage of Plato, there is no common place from whence all the planets, being let fall and descending with uniform and equal gravities (as Galileo supposes), would, at their arrival to their several orbs, acquire their several velocities with which they now revolve in them. If we suppose the gravity of all the planets toward the sun to be of such a quantity as it really is, and that the motions of the planets are turned upward, every planet will ascend to twice its height from the sun. Saturn will ascend till he be twice as high from the sun as he is at present, and no higher; Jupiter will ascend as high again as at present, that is, a little above the orb of Saturn; Mercury will ascend to twice his present height, that is, to the orb of Venus; and so of the rest; and then, by falling down again from the places to which they ascended, they will arrive again at their several orbs with the same velocities they had at first and with which they now revolve.

But if, so soon as their motions by which they revolve are turned upward, the gravitating power of the sun, by which their ascent is perpetually retarded, be diminished by one half, they will now ascend perpetually, and all of them at equal distances from the sun will be equally swift. Mercury, when he arrives at the orb of Venus, will be as swift as Venus; and he and Venus, when they arrive at the orb of the earth, will be as swift as the earth; and so of the rest. If they begin all of them to ascend at once and ascend in the same line, they will constantly, in ascending, become nearer and nearer together, and their motions will constantly approach to an equality and become at length slower than any motion assignable. Suppose, therefore, that they ascended till they were almost contiguous and their motions inconsiderably little, and that all their motions were at the same moment of time turned back again or, which comes almost to the same thing, that they were only deprived of their motions and let fall at that time; they would all at once arrive at their several orbs, each with the velocity it had at first, and if their motions were then turned sideways and at the same time the gravitating power of the sun doubled, that it might be strong enough to retain them in their orbs, they would revolve in them as before their ascent. But if the gravitating power of the sun was not doubled, they would go away from their orbs into the highest heavens in parabolical lines. These things follow from my *Principia Mathematica*, Book I, Propositions XXXIII, XXXIV, XXXVI, XXXVII.

I thank you very kindly for your designed present, and rest

Your most humble servant to command,

Is. Newton

Cambridge, February 25, 1692/3

LETTER IV

To Mr. Bentley, at the Palace at Worcester

Sir,

The hypothesis of deriving the frame of the world by mechanical principles from matter evenly spread through the heavens being inconsistent with my system, I had considered it very little before your letters put me upon it, and therefore trouble you with a line or two more about it, if this comes not too late for your use.

In my former I represented that the diurnal rotations of the planets could not be derived from gravity, but required a divine arm to impress them. And though gravity might give the planets a motion of descent toward the sun, either directly or with some little obliquity, yet the transverse motions by which they revolve in their several orbs required the divine arm to impress them according to the tangents of their orbs. I would now add that the hypothesis of matters being at first evenly spread through the heavens is, in my opinion, inconsistent with the hypothesis of innate gravity, without a supernatural power to reconcile them; and therefore it infers a Deity. For if there be innate gravity, it is impossible now for the matter of the earth and all the planets and stars to fly up from them, and become evenly spread throughout all the heavens, without a supernatural power; and certainly that which can never be hereafter without a supernatural power could never be heretofore without the same power.

You queried whether matter evenly spread throughout a finite space, of some other figure than spherical, would not, in falling down toward a central body, cause that body to be of the same figure with the whole space, and I answered yes. But in my answer it is to be supposed that the matter descends directly downward to that body and that that body has no diurnal rotation. This, sir, is all I would add to my former letters.

I am your most humble servant,

Is. Newton

Cambridge, February 11, 1693

Three Eighteenth-Century Social Philosophers: Scientific Influences on Their Thought

Henry Guerlac

We have been asked to discuss "the manner in which the rise of science in the seventeenth century affected the culture and world view of the eighteenth century." I should like to raise the question of how great a 'common understanding of science,' and of its meaning for man, the eighteenth century possessed, by taking up briefly three well-known social philosophers of the Enlightenment: Montesquieu, Voltaire, and the Baron d'Holbach. I propose to examine whether an interest in science was in each case a chief influence upon, or a principal source of, their social philosophy; and to ask whether science meant the same thing to each of them, and to what extent Newton, whose name and accomplishments are so frequently invoked by the *philosophes* and their historians, provided a common inspiration.

Of Montesquieu, the first of my triad, I shall speak less fully than he deserves, for he was surely one of the most original, if not perhaps the most influential, minds of his century. Montesquieu's scientific interests were the avocation of his early manhood, when as a young magistrate of Bordeaux in the 1720's he was active in the affairs of the town's Academy. From the scientific juvenilia of this period; from the later entries in his workbook, the *Pensées*, and from his masterpiece, the *Esprit des lois*, a rather clear picture emerges of the scientific warp of his thought and of the motives that engendered this passing enthusiasm.

Like his contemporaries, the young Montesquieu was moved by pride and wonder at the scientific accomplishments of the seventeenth century, which had cleared away so many superstitious beliefs and offered such a precise and simple picture of the world around us. But to Montesquieu, science was less significant as a remedy for the disease of credulity and superstition, less valuable because of the useful applications which might flow from it, than as a dramatic proof of the power of the human mind. For the man of humanistic training like himself, he saw a task of special urgency: to civilize and domesticate the recent discoveries of science, and fit them into place in the great heritage of humane learning, by removing that artificial barrier which technical jargon, poor writing, a dry and forbidding language, had erected. But for the bolder purpose which began to shape itself in his mind, he might well have sought to emulate Fontenelle, whom he so much

Source: Henry Guerlac, "Three Eighteenth-Century Social Philosophers: Scientific Influences on Their Thought," *Daedalus* 87 (1958), 8-24.

admired, and have anticipated Voltaire as a humane expositor of the natural sciences. Instead he took a more fruitful path, that of attempting to demonstrate in a work of great erudition and creative power that science and humane learning could be combined in a search for the *Spirit of the Laws*. For nothing is more characteristic of Montesquieu than his conviction that all spheres of man's knowledge are compatible and harmonious; that "all the sciences are good and support each other."

If the facts and conclusions of science, as Montesquieu understood them, contributed less to the *Spirit of the Laws* than he surely hoped, the general spirit of scientific inquiry was, for perhaps the first time in history, invoked in the study of man in society. Cool objectivity, detached relativism, painstaking accumulation of fact and observation in support of his generalizations—these seem quite clearly to reflect the example of the students of nature. "Observations," Montesquieu once wrote, "are the history of physical science; systems are its fables." If natural science had been remade by adhering to such principles as these—and he might well have drawn them in part from Thomas Sprat's *History of the Royal Society*, a work he admired—could not a science of man emerge from the application to the study of man of these tenets of the New Experimental Philosophy?

It is when we turn to the substance of his scientific borrowings that we are disappointed. Montesquieu remained to his death an unwavering disciple of Descartes, though by the time the *Spirit of the Laws* appeared in 1748 Newtonian views had long since triumphed in France. Montesquieu's physical philosophy—I do not say his physics, for an understanding of the mathematical nature of physics was beyond his capabilities—was Cartesian, strictly mechanistic and deterministic. In natural history, physiology and medical theory—the scientific fields which attracted him most—he accepted without question the teachings of the iatro-physical school which had emerged in the wake of Descartes' speculations. Employed in the pages where Montesquieu discusses his theory of the role of climate, this already obsolescent physiology contributes nothing but confusion.

Much has been made of Montesquieu's sweeping definition of laws, "Laws, in the widest sense, are the necessary relations which derive from the nature of things," and of the fact that Montesquieu is the first modern thinker to apply deterministic principles, not only to physical nature, but to man in his social and political life. For this—as much as for his more empirical side—the founders of sociology, Comte and Durkheim, mark him as a precursor.

Yet it is Cartesian dualism, the dichotomy between mind (or soul) and body, between *l'homme physique* and *l'homme moral*, which is the key to Montesquieu's view of natural law as it regulates the conduct of men. While all beings, living or inert, are subject to the inexorable operation of physical law—the laws of the physical world—man is also under a law peculiar to himself; the natural moral law, the law of God, which is perceived by human reason and regulates man's behavior as a rational being. The diverse positive laws of different peoples are the special applications of the general moral law; or, better still, of the moral law as modified by time and place, by historical tradition and local custom, by the wisdom and moral frailty of law-givers, and by the circumstantial operation of physical laws on the minds and bodies of men.

In the sections which are devoted to his famous theory of the influence of climate and topography, Montesquieu tries to show how physical laws interact with and modify the interpretations and applications of moral law. But this scientific ingredient really plays a very small part in Montesquieu's total plan. It is the *moral* law of nature which really concerns him, and this (as we well know) lay ready at hand, expounded in the writings of Stoic philosophers, Roman jurists and Christian theologians.

Montesquieu found the intellectual and moral forces, illustrated by examples of the traditional sort and familiar to anyone steeped in the classics, most useful in explaining the variations of positive law in different times and places. Psychological and social causes—*les causes morales*—are far more important, he feels he must warn his reader, than physical and environmental causes, for it is these that mainly determine the character of a people and the laws and institutions under which they live. Just as in an earlier work Montesquieu found the central cause for the rise and fall of the Roman Republic in the changing character of the Roman people, so in his description of different forms of government he discovers for each a psychological determinant or principle: for the Monarchy, 'honor'; for the Republic, 'virtue'; and for Despotism, 'fear.' It is these so-called principles which determine the pattern of laws a people shall have and their decay which threatens the stability of a society.

I believe Montesquieu's scientific veneer contributed less to the *Esprit des lois* than some have supposed. It was from the ready resources of the humanistic scholar—from omnivorous reading in the literature of travel, from his own perceptive observation of men and manners, and above all from his devotion to the Greek and Roman classics—that he drew his wealth of illustration and most of his inspiration. It was from the old, prescientific natural-law tradition and from the bookish riches of 'moral philosophy'—from St. Thomas Aquinas, from Plutarch, Seneca and his beloved Cicero, and of course from Aristotle and the historians of Greece and Rome—that he drew his conceptual framework and his basic outlook. Steeped in the classics from his school days (as were all Frenchmen of this supposedly scientific century) he lived at the Château de La Brède the life of a Roman gentleman, close to his fields and his vines, and inseparable from his magnificent library. It is too much to suppose that the scientific interests of his youth—in which he maintained a desultory, if receding, curiosity all his life—could outweigh his deeper allegiance to the humanities. Montesquieu did not need Pope to tell him that the proper study of mankind is man.

It comes as something of a surprise that Voltaire—incredibly versatile man of letters, tireless pamphleteer, and sibylline court jester—should have taken as one of his strongest claims to fame the fact that, as much as any man, he had introduced the French public to the discoveries, the philosophy, and the method of Sir Isaac Newton.

The trip to England in 1726-1729 first brought to Voltaire's attention the two English thinkers who were to influence so profoundly his life and thought: Newton and Locke. But his full conversion to the physical doctrines of Newton and his rejection of the prevailing Cartesianism took place only after his return to France and was due in large measure to the influence of that gifted blue-stocking and competent Newtonian, the Marquise du Chatelet. Under the eye of the *immortelle Emilie*, and indeed with considerable assistance from the Minerva of France, Voltaire published in 1738 his *Elements of Newton's Philosophy*. Sharing Montesquieu's conviction that scientific discoveries should be humanized and spread abroad, yet disdainful of Fontenelle's cloying sentimentality, Voltaire set out to "remove the thorns" from Newton's writings, "but without loading them with flowers which do not suit them."

The *Elements of Newton's Philosophy* is not precisely the sort of book we might expect. It is not just a simple exposition of Newton's discoveries: of his *System of the World*, of the doctrine of attraction, and of the experiments on light and color. There is, at least in the later revisions, considerable space given to the religious implications of Newtonian thought and the support it accorded to Voltaire's deistical convictions. The chapters on light are not confined to Newton's own discoveries, but include some historical information and a good deal of later material. Most striking is the interest Voltaire displays in the physiology of vision. A chapter on the human eye as an optical instrument is followed by chapters treating the psychology of visual perception, a subject which fascinated Voltaire. Here he draws upon Locke and upon Bishop Berkeley's *New Theory of Vision*. He recounts the famous case of Dr. Cheselden's successful operation on the young man born blind, and elaborates at unnecessary length upon Newton's casual comparison of the colors of the spectrum with the tones of the diatonic scale, a subject of timely interest because of Father Castel's invention of his color organ.

Voltaire's devotion to the sensationalist philosophy of John Locke apparently finds an echo in these passages, and Voltaire seems concerned to emphasize, here as elsewhere, the essential harmony of the doctrines of these two great English thinkers.

To Voltaire, Newton was supremely important for having demonstrated the effectiveness of a new method of mathematico-experimental discovery in science, the famous method of analysis and composition. To dissect nature —even, as Hermann Weyl pointed out, the perceptively or intuitively simple, like a beam of white light or the elliptical path of a planet—into constituent elements, and then to confirm this dissection by successfully recombining the elements to restore the original phenomenon: this was the method that Newton not only practiced but here and there elucidated in the *Opticks* and the *Principia*.

Locke (so it seemed to Voltaire) had applied as far as he could the same method. He had laid open to man the anatomy of his mental processes "just as a skilled anatomist explains the workings of the human body." And the conclusions Locke reached—that all knowledge depends upon sense experience; that accordingly most synthetic or a priori thinking is groundless; that much verbal discourse is without meaning—all this supplied Voltaire with a sharp and devastating implement with which to cut down the luxuriant overgrowth of theological assertion and philosophical system.

That Locke had set bounds to the pretensions of the human intellect was for Voltaire his outstanding accomplishment. But it was easy to pursue Locke's clues and end in skepticism, as Hume was to demonstrate. It was possible also to derive from Locke—as did the Abbé Buffier, whom Voltaire read approvingly—the conclusion that all the basic truths to which the human mind habitually gives consent are mere probabilities, excepting only the knowledge of our own existence. But if in philosophy we may perhaps enjoy the indolent luxury of suspended judgment, life requires action, and men—from all time—have acted in accord with probability, which comes nearest to Truth. Voltaire was obviously attracted to this early form of the 'philosophy of common sense'; and I believe he found in Newton's accomplishments a remarkable exemplification of it.

Locke did not draw these consequences except in one significant instance. With respect to the moral sciences, he was confident that man could find a pathway to demonstrative knowledge; but as to the sciences of nature he was wholly pessimistic:

> I am apt to doubt that, how far soever human industry may advance useful and experimental philosophy in physical things, *scientifical* will still be out of our reach. . . . *Certainty* and *demonstration* are things we must not, in these matters, pretend to.

Elsewhere, he is still more emphatic, asserting that natural philosophy can never become a science, that is, never completely certain, like a theorem in geometry:

> Experiments and historical observations we may have, from which we may draw advantages of ease and health, and thereby increase our stock of conveniences for this life; but beyond this I fear our talents reach not, nor are our faculties, as I guess, able to advance.

The situation with Newton was similar, but also in a manner different. Those eighteenth-century devotees of Newtonian thought who believed that because Newton used a mathematical method he held his conclusions to be eternally valid and perfectly true, like the demonstrations of geometry, had not read their Newton. But Voltaire had, and he well knew that the knowledge gained through the method of analysis did not have the cogency of a geometrical proof. And he was familiar with those passages where Newton admits that the propositions arrived at by induction are "very nearly true," and that "although the arguing from Experiments and Observations by Induction be no Demonstration of general conclusions; yet it is the best way of arguing which the Nature of Things admits of. . . ."

To Voltaire it must have seemed that Newton—so serenely confident, so indifferent to this scandalous state of affairs—was a useful corrective to Locke and the implied threat of a sterilizing skepticism. Newton's dramatic demonstration of the power of the new science—its success in ordering our knowledge of the physical world; its striking ability to predict the unforeseen—was accompanied by his frank avowal that the entire edifice rested upon probabilities. Here was a superlative illustration of Buffier's pragmatic, common-sense theory; and it convinced Voltaire that philosophical doubt need not paralyze action but in fact was the road to successful action.

I cannot agree with those who have suggested that Voltaire's brief excursion into the physical sciences had little influence on his subsequent thought and writing and that he returned without regret to literary pursuits. Some have thought *Micromégas*, his little science fiction tale, to be his swan song, recording his disillusionment with the sciences. Instead, it seems to me merely another assertion of his basic outlook: another assault on the makers of systems, whose futile answers he contrasts—citing the achievements of Huygens, Maupertuis, Leeuwenhoek, Swammerdam and Réaumur—with the success of the tiny humans in scientific observation and measurement.

Voltaire's later work shows in a number of places the impress of his early contact with science; nor did he ever abandon his scientific interests altogether. Allusions to the superiority of the analytical method are frequent in his writings, always with appropriate references to Newton and Locke. And numerous are the reminders that absolute truth is a chimera, that our knowledge can only be probable knowledge, which is all that man in his middling state can aspire to. The true method of thought, in philosophy and human affairs alike, is that which Newton introduced, with such success, in natural science. As Cassirer has summarized it so well, analysis was to Voltaire "the staff which a benevolent nature has placed in the blind man's hands. Equipped with this instrument he can feel his way forward among appearances, discovering their sequences and arrangement; and this is all he needs for his intellectual orientation to life and knowledge."

There was a matter of deep import to Voltaire, which commanded his chief attention for nearly a decade late in life, and on which he brought to bear the ideas we have been discussing. This was his campaign on behalf of those men—Calas, Sirven and the others—who were the victims of religious intolerance and judicial oppression. In the 1760's—aroused to fighting pitch by the juridical murder of Calas—Voltaire turned his attention, with the same single-mindedness he had devoted earlier to science, to the study of French criminal law, with particular reference to the strange rules of evidence employed in capital trials. His grasp of the technical side of the problem has earned him the admiration of historians of French law, Esmein for example.

The logical force of the arguments Voltaire directed against some of the bizarre legal practices of his time stems directly from his views about the certitude of our knowledge. Behind the formalized rules of evidence employed by the judges, strictly codified during the previous century, he discerned the same intellectual arrogance, the same confidence in the power of the human mind to attain absolute truth, that he pilloried in theology and philosophy.

"There is no year," he wrote in this period, "when some provincial judges do not condemn to a frightful death some innocent father of a family; and that peacefully, gaily even, as one slits the throat of a turkey in the farm-yard." The judges, he remarked, do not suffer from doubts or misgivings and think that guilt can be proved like a theorem in geometry. But can we attain in human affairs such a certitude as will allow "seven men to enjoy legally the amusement of putting an eighth man to death in public?" Voltaire is convinced that we cannot.

"I am certain," Voltaire says in the *Philosophical Dictionary*, "I have friends; my fortune is secure; . . . my lover will be faithful; these are phrases which any man with some experience in life strikes from his lexicon."

The history of the human mind throws a revealing light on this problem of certitude. Before Copernicus, everyone was *certain* that the earth was at rest and that the sun rose and set. Not long ago witchcraft, divination, possession by devils, were the things deemed most certain in the eyes of all men. Yet today this certitude has, to say the least, somewhat diminished. Indeed there is no certitude so long as it is physically or morally possible that things might be otherwise.

Yet many judicial cases, he continues in the same vein, have been settled as certainties when on further examination they have turned out to be errors. "When the judges condemned Langlade, Lebrun, Calas, Sirven . . . and so many others, all later shown to be innocent, they were certain, or should have been, that all these unfortunate man were guilty; and yet they were wrong. . . . If such is the misfortune of humanity that one must be content with extreme probabilities, one should at least take into consideration the age, social position, the bearing of the accused, the motive he might have had to commit the crime."

Against the quasi-mechanical system of proof used in the criminal courts—where two half-proofs, as they were called, or four quarter-proofs, added together were deemed to constitute full proof—Voltaire had nothing but scorn. The whole system appeared archaic, fallacious and cruel, especially so in the case of capital offenses. When life is at stake, he insisted, even the greatest probability ought not to be thought sufficient for conviction.

It is perhaps Voltaire's least-known contribution that he sought to introduce this principle of uncertainty into the rules of criminal evidence; and that in pamphlet after pamphlet during this legal debate he reiterated his conviction that in the realm of human affairs, probabilities and not certainties are the most we can expect. His mature opinion is to be found temperately and gravely stated in a little-read but important piece, his *Essay on Probabilities Applied to the Law* (1772). Here, after insisting upon his main point, he writes:

> One must take a stand, but one should not take it at random. It is therefore necessary to our weak and blind human nature, always subject to error, to study probability with as much care as we learn arithmetic and geometry. To judges, this study is peculiarly important. . . . A judge passes his life in weighing probabilities against each other, in calculating them, in evaluating their force.

I have not introduced this topic to suggest that Voltaire is a forgotten pioneer in probability theory, for I am not sure how much he knew of this very important current in eighteenth-century science. I doubt whether he bothered his head with the mortality tables of Antoine Deparcieux, the first in France to continue the pioneer work of Graunt, Petty and Edmund Halley, though he cites him approvingly on one occasion. And he certainly would have made little headway with the work of James Bernoulli or Abraham Demoivre on the mathematics of probability. Yet it is perhaps no accident that a devoted disciple of Voltaire, the mathematician Condorcet, attempted to apply probability theory to the sort of problems which interested Voltaire—the decisions of courts and legislatures—and urged that statistics and probability might some day form the basis of a reliable science of human affairs.

I have tried to illustrate here, by some concrete examples, what Voltaire may have got from his study of Newton. Paradoxical as it must seem, Voltaire appears—for all his technical limitations and his mathematical ineptitude—to have grasped the implications of Newtonian thought and method better than any other French man of letters of the eighteenth century. From Newtonians and Lockeians of the strict observance—chiefly from Voltaire and Condillac—the much-abused *ideologists* of the Revolutionary period took their start. Yet this tradition, culminating in the *Idéologues*, was buried for a time under an avalanche of materialist systems of moral and social philosophy; this line, which extends from Voltaire and Condillac through Turgot and Condorcet, seems to me to embody the deepest and most persistent elements of the Enlightenment's faith in science. At least, so the eighteenth-century scientists appeared to believe; for it was to this current of the Age of Reason that men like Lavoisier and Laplace, among other scientific figures who wished to play their role as renovators of society, appear to have lent their allegiance.

It is only fair and proper to introduce an out-and-out materialist, if only in contrast to those more moderate men I have already discussed. My candidate is the Baron d'Holbach, whose powerful *Système de la nature* is frequently referred to and, quite evidently, not often read with attention.

The *Système de la nature* is not a work of science or natural philosophy, or even primarily a work of social theory, but a thundering attack on all supernatural beliefs. Having devoted a decade of his life to assaulting Christianity in a series of books and pamphlets, most of them anonymous, d'Holbach was now in 1770 prepared to take the next step and give to the world his refutation of all deists and theists, Voltaire and Newton included. Somewhat incidental to his main purpose d'Holbach pretended to advance a rigorously secular view of man and society, derived—as he sincerely hoped—from the axioms of materialism and of science. From a set of assertions about the natural world—confident, combative and uncompromising—d'Holbach draws out his view of man and his formula for man's secular redemption. Man is unhappy because he is out of tune with Nature; Nature—*le grand tout* as d'Holbach calls it—consists only of material bodies in motion; and man is not only a part of nature, as Montesquieu would have agreed, but is a purely physical being. Just as belief in a realm of spirits, or even in a single spiritual creator, is illusory, so also is the habitual distinction between *l'homme physique* and *l'homme moral. L'homme physique* is man acting according to those laws of cause and effect which our senses can observe; *l'homme moral* is man acting in response to physical causes which our ignorance conceals from us.

In the early chapters d'Holbach sets forth his view of physical nature from which all this is supposedly derived:

> It is to physics and to experience that man must have recourse in all his investigations: he must consult them in matters of religion, ethics, legislation, political government, the sciences and the arts, even in his pleasures and sufferings. Nature acts by simple, uniform and invariable laws which experience enables us to know.

In view of these words, we should expect d'Holbach to share with Voltaire some knowledge of contemporary physics, or at least of the popularized Newtonian physics of the eighteenth century. Yet the "immortal Newton" is mentioned seldom and then with scarcely veiled impatience. And the physics we are offered would scarcely pass as Newtonian. Newton, to be sure, is praised for having done away with the chimerical causes which had been invoked to explain the motion of the planets, as for example the vortices of Descartes; but d'Holbach is unhappy with Newton and his followers for regarding the cause of gravitation as unexplained or inexplicable. D'Holbach suggests that gravitation may be merely a special case of that propensity toward motion which is a property of all matter, and which depends upon "the inner and outer configuration of the bodies," a view of motion which is as far from Newton as it is from Descartes. Elsewhere, with bland inconsistency, he cites the followers of Newton as identifying attraction and repulsion with sympathies and antipathies, with affinities and *rapports*. But, once again, this is not the language of eighteenth-century physics, any more than is d'Holbach's definition of inertia as "gravitation on self" or his use of Newton's Third Law as an example of *nisus*, the urge of bodies at rest to move.

It can be shown that d'Holbach's physical theory owes a good deal to his earlier experience as a chemist, more indeed than to his knowledge, or lack of knowledge, of physics. Chemistry had in truth been his first love; he was the author of many of the chemical and mineralogical articles that appeared in Diderot's *Encyclopédie* and like Diderot he was a disciple and close friend of Rouelle, the outstanding chemist of the day. D'Holbach's ideas concerning matter—his pan-energism, if I may call it that—owe a good deal to the long line of chemical philosophers, latter-day followers of Paracelsus, of whom one of the most important is G. E. Stahl, d'Holbach's acknowledged hero and a man whose writings on the phlogiston theory he had been responsible in part for introducing into France.

But there is another current still more evident in d'Holbach's thought: his debt to classical antiquity. This was clearly perceived by the prosecuting attorney who read the indictment that led to the condemnation and public burning of the *Système de la nature*. The author of this nefarious and sacrilegious work is not accused of having misused the results of contemporary science, but of having revived and extended the materialistic system of Epicurus and Lucretius.

That d'Holbach was thoroughly steeped in the Greek and Latin classics is evident from the most cursory inspection of the *Système de la nature*. It is, in fact, upon classical authorities rather than scientific ones that his paganism and materialism are explicitly based; and it is just in the pretentiously scientific preliminaries of his book that the classical references are most evident. Plutarch, Lucretius, Cicero—especially the *De divinatione* and the *De natura deorum*—and above all Seneca, provide the heaviest field pieces. Democritus and Aristoxenes are cited to show that in antiquity the soul was thought of as material. Not merely Lucretius but also Manilius and Pythagoras (the latter courtesy of Ovid) are authorities for the eternity of matter. Lucretius is quoted, as one would expect, on the evils that flow from the fear of death; while Seneca provides him with arguments in defense of suicide and with support for the principle that man is inherently neither good nor evil, as well as with useful quotations on fatalism.

The ancients are even made to share in the glory of modern scientific discoveries and doctrines, much in the spirit of Dutens' recent history. Citing Diogenes Laërtius as his source, d'Holbach asserts that the Newtonian "system of attraction" is very old, having been anticipated by Empedocles in his doctrine of Love and Strife. Again he rather labors the point that Aristotle, long before the "profound Locke," insisted that nothing enters the mind except through the senses. Indeed, he seems to take a special pleasure in claiming that the two most popular philosophic doctrines of his day were of ancient origin.

As much as by any other influence, including that of the Greek materialists, d'Holbach's thought is shaped by the teachings of the Roman followers of the Stoa: Cicero, Seneca and Marcus Aurelius. It is from these men, rather than primarily from Newton or Descartes, that he draws his deep conviction of an inexorable order of physical nature, an order to which man, as a part of nature, must submit:

> Let man cease to look outside of the world in which he lives for beings who will give him a happiness which nature refuses him: let him study nature, learn its laws, contemplate its energy and the immutable way in which it acts. Let him apply his discoveries to his own happiness . . . and submit in silence to those laws from which nothing can shield him. Let him agree to remain in ignorance of those causes which for him are enveloped in an impenetrable veil; let him suffer without complaint the decrees of a universal force which cannot turn back or ever depart from the rules which its essence has imposed on it.

If d'Holbach makes his bow to the science of his day, it is both awkward and perfunctory. He genuflects appropriately before the busts of Locke and Newton, mumbles the suitable incantations, and passes on to more important matters, and to the search for more convincing authority. It is as if he contented himself with showing that contemporary findings of science appeared to confirm the insight of the ancients that there is an inexorable order of material nature.

If the method and spirit of Newtonian science meant anything to d'Holbach, it is not apparent, not even in his manner of exposition. While Montesquieu's method is comparative and historical, and whereas other writers, like the physiocrats, affected an abstract, pseudomathematical method of presentation, such as Tocqueville and Taine deplored, d'Holbach's method can best be described as rhetorical, dialectical and hortatory. The style moves in whirlpools and eddies: it is a series of loosely-linked affirmations, challenges, questions and indictments. It is the work of a prosecuting attorney familiar with the forms and devices of Roman forensic. It begins, not really with axioms, but with an accusation; it ends, not with a *quod erat demonstrandum*, but with a summation of the case and a final, impassioned charge to the jury.

Regrettably, I shall end with neither, but only with some tentative suggestions. If my account of these three social thinkers of the eighteenth century has any validity, it may perhaps suggest that historians of the Age of Reason have been rather too sanguine in treating the thinkers of this complex age as though they did indeed possess a 'common understanding of science.' I am not too concerned to stress the point that of these three men only one, Voltaire, seems really to have got the drift of Newton. Newton's name was after all only a symbol or catchword of the age, though one of great evocative power. It can do little harm, when treating the eighteenth century in general terms, to use the Newtonian symbol, as did the men of that century, to stand for the complex framework of science itself.

But if we have been wrong in making science so exclusively the intellectual force which shaped the eighteenth-century mind (and so in consequence much of our modern thought) our interpretations will have to be radically altered. Perhaps we can no longer lay solely at the door of science all the social nostrums, all the flights of fancy, all the naive confidence in the power of reason so characteristic of that century. Science may have to share the credit and the responsibility with certain less spectacular forces, and one of these I have hinted at. I referred several times to the deep-rooted classical formation of the French mind of the eighteenth century; and I would have liked to develop more fully, especially in my discussion of d'Holbach, the implications of the 'new paganism': that peculiar movement which has its roots in the open or clandestine writings of the earlier *libertins* and which is so marked a characteristic of the last years of the century. The vision of an age before the advent of Christianity—a golden age of free thought and free inquiry when men were men and not fallen angels—certainly served to express the secular aspirations of the Enlightenment fully as well as the vision of a world of the future, guided and improved by the light of science.

The Rise of Scientism

F. A. von Hayek

Never will man penetrate deeper into error than when he is continuing on a road which has led him to great success. And never can pride in the achievements of the natural sciences and confidence in the omnipotence of their methods have been more justified than at the turn of the eighteenth and nineteenth centuries, and nowhere more so than at Paris where almost all the great Scientists of the age congregated. If it is true, therefore, that the new attitude of man towards social affairs in the nineteenth century is due to the new mental habits acquired in the intellectual and material conquest of nature, we should expect it to appear where modern science celebrated its greatest triumphs. In this we shall not be disappointed. Both the two great

---

Source: F. A. von Hayek, "The Counter-Revolution of Science—I. The Source of the Scientistic Hubris: L'Ecole Polytechnique," *Economica* 8 (1941), 9-23. Reprinted by permission of the author and the London School of Economics and Political Science.

intellectual forces which in the course of the nineteenth century transformed social thought—modern socialism and that species of modern positivism, which we prefer to call scientism, spring directly from this body of professional scientists and engineers which grew up in Paris, and more particularly from the new institution which embodied the new spirit as no other, the *Ecole polytechnique.*

It is well known that French Enlightenment was characterised by a general enthusiasm for the natural sciences as never yet known before. Voltaire is the father of that cult of Newton which later was to be carried to ridiculous heights by Saint-Simon. And the new passion soon began to bear great fruits. At first the interest concentrated on the subjects connected with Newton's great name. In Clairault and d'Alembert, with Euler the greatest mathematicians of the century, Newton soon found worthy successors who in turn were followed by Lagrange and Laplace, likewise giants.

And with Lavoisier, not only the founder of modern chemistry but also a great physiologist, and, to a lesser degree, with Buffon in biological science, France began to take the lead in all important fields of natural knowledge.

The great *Encyclopaedia* was a gigantic attempt to unify and popularise the achievements of the new science, and d'Alembert's "Discours préliminaire" (1754) to the great work, in which he attempted to trace the rise, progress and affinities of the various sciences, may be regarded as the Introduction not only to the work but to a whole period. This great mathematician and physicist did much to prepare the way for the revolution in mechanics by which towards the end of the century his pupil Lagrange finally freed it from all metaphysical concepts and restated the whole subject without any reference to ultimate causes or hidden forces, merely describing the laws by which the effects were connected. No other single step in any science expresses more clearly the tendency of the scientific movement of the age or had greater influence or symbolic significance.

Yet while this step was still gradually preparing in the field where it was to take its most conspicuous form, the general tendency which it expressed was already recognised and described by d'Alembert's contemporary Turgot. In the amazing and masterly discourses which the latter as a young man of 23 delivered at the opening and the closing of the session of the Sorbonne in 1750, and in the sketch of a *Discourse on Universal History* of the same period, he outlined how the advance of our knowledge of nature was accompanied throughout by a gradual emancipation from those anthropomorphic concepts which first led man to interpret natural phenomena after his own image as animated by a mind like his own. This idea, which was later to become the leading theme of positivism and was ultimately misapplied to the science of man himself, was soon afterwards widely popularised by President C. de Brosses under the name of fetishism, the name under which it remained known till it was much later replaced by the expressions anthropomorphism and animism. But Turgot went even further and, completely anticipating Comte in this point, described how this process of emancipation passed through three stages where, after supposing that natural phenomena were produced by intelligent beings, invisible and resembling ourselves, they began to be explained by abstract expressions such as essences and faculties, till at last "by observing reciprocal mechanical action of bodies hypotheses were formed which could be developed by mathematics and verified by experience."

It has often been pointed out that most of the leading ideas of French Positivism had already been formulated by d'Alembert and Turgot and their friends and pupils Lagrange and Condorcet. For most of what is valid and valuable in that doctrine this is unquestionably true, although their positivism differed from that of Hume by a strong tinge of French rationalism. And, as there will be no opportunity to go into this aspect more fully, it should perhaps be specially stressed at this stage that throughout the development of French positivism this rationalist element, probably due to the influence of Descartes, continued to play an important role.

It must be pointed out however that these great French thinkers of the eighteenth century showed scarcely any trace yet of that illegitimate extension to the phenomena of society of scientistic methods of thought which later became so characteristic of that School—excepting perhaps certain ideas of Turgot about the philosophy of history and still more so some of Condorcet's last suggestions. But none of them had any doubt about the legitimacy of the abstract and theoretical method in the study of social phenomena and they were all staunch individualists. It is particularly interesting to observe that Turgot, and the same is true of David Hume, was at the same time one of the founders of positivism and of abstract economic theory, against which positivism was later to be employed. But in some respects most of hese men unwittingly started trains of thought which produced views on social matters very different from their own.

This is particularly true of Condorcet. Mathematician like d'Alembert and Lagrange, he definitely turned to the theory as well as to the practice of politics. And although to the last he understood that "meditation alone may lead us to general truths in the science of man", he was not merely anxious to supplement this by extensive observation but occasionally expressed himself as if the method of the natural sciences were the only legitimate one in the treatment of the problems of society. It was particularly his desire to apply his beloved mathematics, especially the newly developed calculus of probability, to his second sphere of interest, which led him to stress more and more the study of those social phenomena which would be objectively observed and measured. As early as 1783, in the oration at his reception into the *Académie*, he gave expression to what was to become a favourite idea of positivist sociology, that of an observer to whom physical and social phenomena would appear in the same light, because, "a stranger to our race, he would study human society as we study those of the beavers and bees". And although he admits that this is an unattainable ideal because "the observer is himself a part of human society", he repeatedly exhorts the scholars "to introduce into the moral sciences the philosophy and the method of the natural sciences".

The most seminal of his suggestions however occurs in his *Sketch of a Historical Picture of the Progress of the Human Mind*, the famous Testament of the Eighteenth Century as it has been called, in which the unbounded optimism of the age found its last and greatest expression. Tracing human progress in a great outline through all history, he conceives of a science which might foresee the future progress of the human race, accelerate and direct it. But to establish laws which will enable us to predict the future, history must cease to be a history of individuals and must become a history of the masses, must at the same time cease to be a record of individual facts but must become based on systematic observation. Why should the attempt to base on the results of the history of the human race a picture of its future destiny be regarded as chimerical? "The only foundation for the knowledge of the natural sciences is the idea that the general laws, known or unknown, which regulate the phenomena of the Universe, are necessary and constant; and why should that principle be less true for the intellectual and moral faculties of man than for the other actions of nature?" The idea of natural laws of historical development and the collectivist view of history were born, merely as bold suggestions, it is true, but to remain with us in a continuous tradition to the present day.

Condorcet himself became a victim of the Revolution. But his work guided to a large extent that same Revolution, particularly its educational reforms, and it was only as a result of these that towards the beginning of the new century the great institutionalised organisation of science arose which saw not only one of the most glorious periods of scientific advance but also the birth of that scientism which is more particularly our concern. As is so often the case with similar movements, it was only in the second or third generation that the mischief was done by the pupils of the great men exaggerating and misapplying the ideas of their masters beyond their proper limits.

In three respects the direct consequences of the Revolution are of special interest to us. In the first place the very collapse of the existing institutions called for immediate application of all the knowledge which appeared as the concrete manifestation of that Reason which was the Goddess of the Revolution. As one of the new scientific journals which spring up at the end of the Terror expressed it "The Revolution has razed everything to the ground. Government, morals, habits, everything has to be rebuilt. What a magnificent site for the architects! What a grand opportunity of making use of all the fine and excellent ideas that had remained speculative, of employing so many materials that could not be used before, of rejecting so many others that had been obstructions for centuries and which one had been forced to use."

The second consequence of the Revolution which we must briefly consider is the complete destruction of the old and the creation of an entirely new scholastic system which had profound effects on the outlook and general views of the whole next generation. The third is more particularly the foundation of the *Ecole Polytechnique*.

The Revolution had swept away the old educational system of *collèges* and *universités* based largely on classical education, and after some short-lived experiments replaced them in 1795 by the new *écoles centrales* which became the sole centres of secondary education. In conformity with the ruling spirit and by an over-violent reaction against the older schools, the teaching in the new institutions was for some years almost entirely confined to the scientific subjects. Not only the ancient languages were reduced to minimum and in practice almost entirely neglected, even the instruction in literature, grammar, and history was very inferior and moral and religious instruction, of course, completely absent. Although after some years a new reform endeavoured to make good some of the gravest deficiencies, the interruption of the instruction in those subjects for a series of years was sufficient to change the whole intellectual atmosphere. Saint-Simon described this change in 1812 or 1813: "Such is the difference in this respect between the state of . . . even thirty years ago and that of to-day that while in those not distant days, if one wanted to know whether a person had received a distinguished education, one asked: 'Does he know his Greek and Latin authors well?', to-day one asks: 'Is he good at mathematics? Is he familiar with the achievements of physics, of chemistry, of natural history, in short, of the positive sciences and those of observation?'"

Thus a whole generation grew up to whom that great storehouse of social wisdom, the only form indeed in which an understanding of the social processes achieved by the greatest minds is transmitted, the great literature of all ages, was a closed book. For the first time in history that new type appeared which as the product of the German *Realschule* and of similar institutions was to become so important and influential a figure in the later nineteenth and the twentieth century: the technical specialist who was regarded as educated because he had passed through difficult schools but who had little or no knowledge about society, its life, growth and problems and its values, which only the study of history, literature and languages can give.

Not only in secondary education but still more so in higher education the Revolutionary Convention had created a new type of institution which was to become permanently established and a model imitated by the whole world, the *Ecole Polytechnique*. The wars of the Revolution and the help which some of the scientists had been able to render in the production of essential supplies had led to a new appreciation of the need of trained engineers, in the first instance for military purposes. But industrial advance also created a new interest in machines. Scientific and technological progress created a widespread enthusiasm for technological studies, which expressed itself in the foundation of such societies as the *Société philotechnique* and the *Société polytechnique*. Higher technical education had till then been confined to specialised schools such as the *Ecole des Ponts et Chaussées* or the various military schools. It was at one of the latter that G. Monge, the founder of descriptive geometry, Minister of Marine during the Revolution and later friend of Napoleon, taught. He sponsored the idea of a single great school in which all classes of engineers should receive their training in the subjects they had in common. He communicated that idea to Lazare Carnot, the "organiser of victory", an old pupil of Monge and himself no mean physicist and engineer. These two men impressed their stamp on the new institution which was created in 1794. The new *Ecole Polytechnique* was (against the advice of Laplace) to be devoted mainly to the applied sciences—in contrast to the *Ecole Normale*, created at the same time and devoted to theory—and remained so during the first ten or twenty years of its existence. The whole teaching centred, to a much higher degree than is still true of similar institutions, around Monge's subject, descriptive geometry, or the art of blue-print making, as we may call it without injustice to show its special significance for engineers. First organised on essentially civilian lines, the school was later given a purely military organisation by Napoleon who also, however much he favoured it otherwise, persistently resisted any attempt to liberalise its curriculum and had even the permission to provide a course in so harmless a subject as literature wrung from him only with some difficulty.

Yet in spite of the limitations as to the subjects taught, and the even more serious limitation on the previous education of the students in its early years, the *Ecole* commanded from the very beginning a teaching staff probably more illustrious than any other institution in Europe has had before or since. Lagrange was among its first professors, and although Laplace was not a regular teacher there, he was connected with the school in many ways, including the office of chairman of its council. Monge, Fourier, Prony, and Poinsot were among the first generation of teachers of mathematical and physical subjects; Berthellot, who continued the work of Lavoisier, and several others hardly less distinguished, taught chemistry. The second generation which began to take over early in the new century included such names as Poisson, Ampère, Gay-Lussac, Thénard, Arago, Cauchy, Fresnel, Malus, to mention only the best known, incidentally nearly all ex-students of the *Ecole*. The institution had only existed for a few years when it had become famous all over Europe, and the first interval of peace in 1801-2 brought Volta, Count Rumford and Alexander von Humbold on pilgrimage to the new temple of science.

This is not the place to speak at length of the conquest of nature associated with these names. We are only concerned with the general spirit of exuberance which they engendered, with the feeling which they created that there were no limits to the powers of the human mind and to the extent to which man could hope to harness and control all the forces which so far had threatened and intimidated him. Nothing perhaps expresses more clearly this spirit than Laplace's bold idea of a world formula which he expressed in a famous passage of his *Essai philosophique sur les Probabilités:* "A mind that in a given instance knew all the forces by which nature is animated and the position of all the bodies of which it is composed, if it were vast enough to include all these data within his analysis, could embrace in one single formula the movements of the largest bodies of the Universe and of the smallest atoms; nothing would be uncertain for him; the future and the past would be equally before his eyes." This idea, which exercised so profound a fascination on generations of scientistically-minded people is, as is now becoming apparent, not only a conception which describes an unattainable ideal, but in fact a quite illegitimate deduction from the principles by which we establish laws for particular physical events. It has been shown by modern logical analysis to be itself a piece of "metaphysical speculation".

It has been well described how the whole of the teaching at the *Ecole Polytechnique* was penetrated with the positivist spirit of Lagrange and all the courses and the textbooks used were modelled on his example. Perhaps even more important, however, for the general outlook of the polytechnicians was the definite practical bent inherent in all the teaching, the fact that all the sciences were taught mainly in their practical applications and that all the pupils looked forward to using their knowledge as military or civil engineers. The very type of the engineer with his characteristic outlook, ambitions, and limitations was here created. That synthetic spirit which would not recognise sense in anything that had not been deliberately constructed, that love of organisation that springs from the twin sources of military and engineering practices, the aesthetic predilection for everything that had been consciously constructed over anything that had "jest growed", was a strong new element which was added to—and in the course of time even began to replace—the revolutionary ardour of the young polytechnicians. The peculiar characteristics of this new type who, as it has been said, "prided themselves on having more precise and more satisfactory solutions than anyone else for all political, religious and social questions" and who "ventured to create a religion as one learns at the *Ecole* to build a bridge or a road" was early noticed, and their propensity to become socialists has often been pointed out. Here we must confine ourselves to mentioning that it was in this atmosphere that Saint-Simon conceived some of the earliest and most fantastic plans for the reorganization of society, and that it was at the *Ecole Polytechnique* where, during the first twenty years of its existence,

Auguste Comte, Prosper Enfantin, Victor Considerant and some hundreds of later Saint-Simonians and Fourierists received their training, followed by a succession of social reformers throughout the century down to Georges Sorel.

But, whatever the tendencies among the pupils of the institution, it must again be pointed out that the great scientists who built the fame of the *Ecole Polytechnique* were not guilty of illegitimate extensions of their technique and habits of thought to fields which were not their own. They little concerned themselves with problems of man and society. This was the province of another group of men, in their time no less influential and admired, but whose efforts to continue the eighteenth century traditions in the social sciences were in the end to be swamped by the tide of scientism and silenced by political persecution. It was the misfortune of the *idéologues*, as they called themselves, that their very name should be perverted into a catchword describing the very opposite from what they stood for, and that their ideas should fall into the hands of the young engineers who distorted and changed them beyond recognition.

It is a curious fact that the French scholars of the time of which we are speaking should have been divided into two "distinct societies which had only one single trait in common, the celebrity of their names." The first were the professors and examiners at the *Ecole Polytechnique* and the *Collège de France* which we already know. The second was the group of physiologists, biologists and psychologists, mostly connected with the *Ecole de Médecine* and known as the Ideologues.

Not all of the great biologists of which France could boast at the time belonged to this second group. At the *Collège de France*, Cuvier, the founder of comparative anatomy and probably the most famous of them, stood close to the pure scientists. The advances of the biological sciences as expounded by him contributed perhaps as much as anything else to create the belief in the omnipotence of the methods of pure science. More and more problems that had seemed to evade the powers of exact treatment were shown to be conquerable by the same methods. Other biologists, who at least since have become equally famous, Lamarck and St. Hilaire, remained at the periphery of the ideologist group and did not concern themselves much with the study of man as a thinking being. But Cabanis and Main de Biran, with their friends Destutt de Tracy and Degérando, made the latter the central problem of their labours.

Ideology, in the sense in which the term was used by that group, meant simply the analysis of human ideas and of human action, including the relation between man's physical and mental constitution. The expression was indeed used in very much the same wide sense as their German contemporaries used the term anthropology. The inspiration of the French group came mainly from Condillac and the field of their studies was outlined by Cabanis, one of the founders of physiological psychology, in his *Rapports du physique et du moral de l'homme* (1802). And although there was much talk among them about applying the methods of natural science to man, this meant no more than that they proposed to study man without prejudices and without nebulous speculations about his end and destiny. But this prevented neither Cabanis nor his friends from devoting a large part of their life work to that analysis of human ideas which gave ideology its name. Nor did it occur to them to doubt the legitimacy of introspection. If the second head of the group, Destutt de Tracy, proposed to regard the whole of ideology as part of zoology, this did not prevent him from confining himself entirely to that part of it which he called *idéologie rationelle*, in contrast to the *idéologie physiologique*, and which consisted of logic, grammar and economics.

It cannot be denied that in all this, out of their enthusiasm for the pure sciences, they used many misleading expressions which were grossly misunderstood by Saint-Simon and Comte. Cabanis in particular stressed repeatedly that physics must be the basis of the moral sciences; but with him too this meant no more than that account must be taken of the physiological bases of mental activities, and he always recognised the three separate parts of the "science de l'homme", physiology, analysis of ideas, and morals. But, in so far as the problems of society are concerned, while Cabanis' work remained mainly programmatic in character, Destutt de Tracy made very important contributions. We need here concern ourselves only with one: his analysis of value and its relation to utility, where, proceeding from the foundations laid by Condillac, he went very far in providing what classical English political economy lacked and what might have saved it from the impasse into which it got—a correct theory of value. Destutt de Tracy (and Louis Say, who later continued his work) may indeed be said to have anticipated by more than half a century what was to become one of the most important advances of social theory, the subjective (or marginal utility) theory of value.

It is true that others outside their circle went much further in the application of the technique of the natural sciences to social phenomena, particularly the *Société des Observateurs de l'Homme*, which, largely under Cuvier's influence, went some way in confining social study to a mere recording of observations reminiscent of similar organisations of our own day. But on the whole there can be no doubt that the ideologues preserved the best tradition of the eighteenth century *philosophes*. And while their colleagues at the *Ecole Polytechnique* became the admirers and friends of Napoleon and received from him all possible support, the ideologues remained staunch defenders of individual freedom and incurred the wrath of the despot.

It was Napoleon who gave currency to the word ideologue in its new sense by using it as a favourite term of contempt for all those who ventured to defend freedom against him. And he did not content himself with abuse. The man who understood better than any of his imitators that "in the long run the sword is always beaten by the spirit" did not hesitate to carry his "repugnance for all discussion and the teaching of political matters" into practice. The economist J. B. Say, a member of the ideologist group and for some years editor of its journal, the *Décade philosophique*, was one of the first to feel the strong hand. When he refused to change a chapter in his *Traité d'économie politique* to suit the wishes of the dictator, the second edition was prohibited and the author removed from the *tribunat*. In 1806 Destutt de Tracy had to appeal to President Jefferson to secure the publication of at least an English translation of his *Commentaire sur l'esprit des lois* which he was not allowed to publish in his own country. A little earlier (1803) the whole of the second class of the *Académie*, that of the moral and political science, had been suppressed. In consequence, these subjects remained excluded from the great *Tableau de l'état et des progrès des sciences et des arts depuis 1789* which the three classes of the *Académie* had been ordered to furnish in 1802. This was symbolic of the whole position of these subjects under the Empire. The teaching of them was prevented and the whole younger generation grew up in ignorance of the achievements of the past. The door was thus opened to a new start unencumbered by the accumulated results of earlier study. Social problems were to be approached from a new angle. The methods, which since d'Alembert had so successfully been used in physics, whose character had now become explicit, and which more recently had led to similar results in chemistry and biology, were now to be applied to the science of man. With what results we shall gradually see.

5 Darwinism: The Decline of Religion and the Deification of Struggle

To write of the cultural impact of Darwinian biology is an immensely more difficult task than to write of the influence of Copernicus or Newton or of Freud or Einstein. This is not because Darwin's impact was any less direct. In fact, the transfer of Darwinian notions—natural selection, survival of the fittest, man as a natural animal closely related to the apes—to political, social, and religious realms has been a far more open, direct, and widespread phenomenon than the use of any other scientific theory. The problem arises because the uses of evolutionary theory were so much influenced by the prior ideologies of the men who read or heard of Darwin's work and adopted, or better yet, adapted his notions for their purposes. As Bert Loewenberg wrote:

> Charles Darwin was all things to all men. Evolution was transmuted into a sanction for progress while it was employed as a footnote to the doctrine of pessimism. Phrases such as "the struggle for existence" and "survival of the fittest" were gayly ripped from the mosaic of evolution and made to serve the wishful thinking of aristocratic theorists. While evolution gave comfort to thinkers of this stripe, it also inspired collectivists' dreamers with scientific hope. Advocates of laissez-faire, mindful of vested interests under attack by proletarian philosophics, pointed to the immutable sway of natural law, but reformers preached social sermons based on identical texts. Pleaders for universal peace and the brotherhood of man were stirred by instructions found in evolutionary monographs, but militarists studied the same manuals with opposite results. Certain philosophers claimed Darwin as their authority for materialism and hedonism; others found metaphysical and ethical idealism confirmed by Darwinian biology.[1]

In spite of the ambiguous interpretations of Darwin—the contrary implications drawn from the analogies between biological evolution and

[1]Bert James Loewenberg, "Darwinism Comes to America, 1859-1900," *Mississippi Valley Historical Review* 28 (1941), 363.

innumerable other social, political, and humanistic concerns—it is not true that such arguments merely cancelled one another out so that the world would have been much the same if evolutionary theory had never been developed. Darwinian arguments seemed to legitimize and lend the authority of scientific truth to doctrines that demanded such support. They inflamed passions and provided a popular and emotional appeal which helped to push men into actions they might not otherwise have countenanced.

Furthermore, it seems to me that there was a distinct drift of human beliefs and attitudes under the influence of evolutionary ideas. In spite of the earlier criticisms of the argument-from-design and the blows received by orthodox Christianity at the hands of post-Newtonian mechanists, there is a strong sense in which the ***Origin of Species*** "shattered the simple faith of thousands." Similarly, although one of the immediate responses to Darwin was optimistic, seeing in evolution a guarantee of continuous progress, the longer-term effect was to accelerate the growing nineteenth-century pessimism so powerfully portrayed by George Bernard Shaw in his introduction to ***Back to Methuselah***:

> . . . the Darwinian process may be described as a chapter of accidents. As such, it seems simple, because you do not at first realize all that it involves. But when its significance dawns on you, your heart sinks into a heap of sand within you. There is a hideous fatalism about it, a ghastly and damnable reduction of beauty and intelligence, of strength and purpose of honor and aspiration, to such casually picturesque changes as an avalanche may make in a mountain landscape, or a railway accident in a human figure. To call this Natural Selection is blasphemy, possible to many for whom nature is nothing but a casual aggregation of inert and dead matter, but eternally impossible to the spirits and souls of the righteous. If it be no blasphemy, but a truth of science, then the stars of heaven, the showers and dew, the winter and summer, the fire and heat, the mountains and hills, may no longer be called to exalt the Lord with us by praise: their work is to modify all things by blindly starving and murdering everything that is not lucky enough to survive in the universal struggle for hogwash.[2]

More specifically, in the political realm, although Darwinian ideas led a handful of writers to see hope in the cooperation among individuals that seemed to be a result of evolutionary pressures, political theorists generally saw in Darwinism a justification for the tyranny of the strong over the weak. Many politicians even saw in Darwin a justification for Imperialist and racist policies. Though Darwin himself often opposed extending the concept of race or species to different groups of men, his theories could easily be adapted to justify such phenomena as slavery in the Americas and the Germanic master-race theory.

---

[2]***Back to Methuselah: A Metabiological Pentateuch*** (New York: Brentano's, 1921), pp. xlv-xlvi.

Because of the overwhelming mass of literature on the cultural and social impact of Darwinism, I have been forced to be extremely restrictive in my selections. In particular, I have not included any material dealing with Social Darwinism in America even though the penultimate social impact of Darwin may have been on the American scene where John D. Rockefeller contended that, "The growth of a large business is merely the survival of the fittest. . . . The American Beauty Rose can be produced in the splendor and fragrance which brings cheer to its beholder only by sacrificing the early buds which grow up around it. This is not an evil tendency in business. It is merely the working out of a law of nature and a law of God." Here Darwinian arguments provided the rationale for supporting the most glaring injustices in society. My rationale is that American Social Darwinism is widely and well treated in other books that are easily available; for example, in Richard Hofstadter's *Social Darwinism in American Thought.*

Alfred Russell Wallace's essay of 1858, "On the Tendency of Varieties To Depart Indefinitely from the Original Type," which introduces the selections on Darwinism serves two purposes. It provides a most succinct statement of the biological theory of evolution through natural selection, so that one has a clear idea of the biological doctrine from which so many social and moral implications were drawn. In addition, Wallace's essay, which appeared in the same issue of the *Journal of the Proceedings of the Linnaean Society* as Darwin's first publication on evolution, exemplifies the fact that Darwin was not the sole source of the evolutionary principles to which his name inevitably became attached. We speak and write of Darwinian ideas because Darwin provided the painstaking and detailed evidence for a set of notions which had existed as speculations for at least a half century and because he was able to create a coherent and convincing system which somehow gave those notions a new respectability.

In his chapter, "Science and Religion: A Mid-Victorian Conflict" from *Darwin and the General Reader*, Alver Ellegard discusses the English religious context within which the implications of evolutionary theory were analyzed, showing how conservative and liberal clergy alike were seriously troubled by its challenges to scripture and to natural theology.

George Nasmyth's discussion of the causes of the success of the philosophy of force is a detailed account and critique of the use of evolutionary theory to justify militaristic, imperialistic, and racist political policies by a social Darwinist who emphasized the element of cooperation rather than conflict in biological development. Nasmyth raises the very critical question of how the analogy from natural phenomena was misunderstood and misused by virtually all social Darwinists, and he reminds us again that much of what we call Darwinism existed before Darwin.

The final selection of this section raises a new issue; the intimate connection between Darwinist arguments and socialism, the political ideology which has motivated dramatic changes in the world. Karl Marx was a critical but appreciative reader of Darwin who saw evolutionary theory as "a basis in natural science for the class struggle in history." And Friedrich Engels acknowledged the connection between Marxism and Darwinism by arguing in his eulogy on Marx that "just as Darwin discovered the law of evolution in organic nature, so Marx discovered the law of evolution in human history." The complex relations between Darwinian and early socialist thought have been analyzed over and over again, but seldom with more clarity than by the German biologist and political conservative, Oscar Schmidt. Schmidt's essay, "Science and Socialism" from the *Deutsche Rundshau* (translated and published separately in 1879 in *Popular Science Monthly*) denies the propriety of the socialists' appeal to Darwin but demonstrates very clearly how evolutionary theory was used to support socialist contentions.

On The Tendency of Varieties To Depart Indefinitely from the Original Type

Alfred R. Wallace

One of the strongest arguments which have been adduced to prove the original and permanent distinctness of species is, that *varieties* produced in a state of domesticity are more or less unstable, and often have a tendency, if left to themselves, to return to the normal form of the parent species; and this instability is considered to be a distinctive peculiarity of all varieties, even of those occurring among wild animals in a state of nature, and to constitute a provision for preserving unchanged the originally created distinct species.

In the absence or scarcity of facts and observations as to *varieties* occurring among wild animals, this argument has had great weight with naturalists, and has led to a very general somewhat prejudiced belief in the stability of species. Equally general, however, is the belief in what are called "permanent or true varieties,"—races of animals which continually propagate their like, but which differ so slightly (although constantly) from some other race, that the one is considered to be a *variety* of the other. Which is the *variety* and which the original *species*, there is generally no means of determining, except in those rare cases in which the one race has been known to produce an offspring unlike itself and resembling the other.

Source: Alfred R. Wallace, "On the Tendency of Varieties To Depart Indefinitely from the Original Type," *Journal of the Proceedings of the Linnaean Society of London* (Zoology) 3 (1858), 52-62.

This, however, would seem quite incompatible with the "permanent invariability of species," but the difficulty is overcome by assuming that such varieties have strict limits, and can never again vary further from the original type, although they may return to it, which, from the analogy of the domesticated animals, is considered to be highly probable, if not certainly proved.

It will be observed that this argument rests entirely on the assumption, that *varieties* occurring in a state of nature are in all respects analogous to or even identical with those of domestic animals, and are governed by the same laws as regards their permanence or further variation. But it is the object of the present paper to show that this assumption is altogether false, that there is a general principle in nature which will cause many *varieties* to survive the parent species, and to give rise to successive variations departing further and further from the original type, and which also produces, in domesticated animals, the tendency of varieties to return to the parent form.

The life of wild animals is a struggle for existence. The full exertion of all their faculties and all their energies is required to preserve their own existence and provide for that of their infant offspring. The possibility of procuring food during the least favourable seasons, and of escaping the attacks of their most dangerous enemies, are the primary conditions which determine the existence both of individuals and of entire species. These conditions will also determine the population of a species; and by a careful consideration of all the circumstances we may be enabled to comprehend, and in some degree to explain, what at first sight appears so inexplicable—the excessive abundance of some species, while others closely allied to them are very rare.

The general proportion that must obtain between certain groups of animals is readily seen. Large animals cannot be so abundant as small ones; the carnivora must be less numerous than the herbivora; eagles and lions can never be so plentiful as pigeons and antelopes; the wild asses of the Tartarian deserts cannot equal in numbers the horses of the more luxuriant prairies and pampas of America. The greater or less fecundity of an animal is often considered to be one of the chief causes of its abundance or scarcity; but a consideration of the facts will show us that it really has little or nothing to do with the matter. Even the least prolific of animals would increase rapidly if unchecked, whereas it is evident that the animal population of the globe must be stationary, or perhaps, through the influence of man, decreasing. Fluctuations there may be; but permanent increase, except in restricted localities, is almost impossible. For example, our own observation must convince us that birds do not go on increasing every year in a geometrical ratio, as they would do, were there not some powerful check to their natural increase. Very few birds produce less than two young ones each year, while many have six, eight, or ten; four will certainly be below the average; and if we suppose that each pair produce young only four times in their life, that will also be below the average, supposing them not to die either by violence or want of food. Yet at this rate how tremendous would be the increase in a few years from a single pair! A simple calculation will show that in fifteen years each pair of birds would have increased to nearly ten millions whereas we have no reason to believe that the number of the birds of any country increases at all in fifteen or in one hundred and fifty years. With such powers of increase the population must have reached its limits, and have become stationary, in a very few years after the origin of each species. It is evident, therefore, that each year an immense number of birds must perish—as many in fact as are born; and as on the lowest calculation the progeny are each year twice as numerous as their parents, it follows that whatever be the average number of individuals existing in any given country, *twice that number must perish annually*,—a striking result, but one which seems at least highly probable, and is perhaps under rather than over the truth.It would therefore appear that, as far as the continuance of the species and the keeping up the average number of individuals are concerned, large broods are superfluous. On the average all above *one* become food for hawks and kites, wild cats and weasels, or perish of cold and hunger as winter comes on. This is strikingly proved by the case of particular species; for we find that their abundance in individuals bears no relation whatever to their fertility in producing offspring. Perhaps the most remarkable instance of an immense bird population is that of the passenger pigeon of the United States, which lays only one, or at most two eggs, and is said to rear generally but one young one. Why is this bird so extraordinarily abundant, while others producing two or three times as many young are much less plentiful? The explanation is not difficult. The food most congenial to this species, and on which it thrives best, is abundantly distributed over a very extensive region, offering such differences of soil and climate, that in one part or another of the area the supply never fails. The bird is capable of a very rapid and long-continued flight, so that it can pass without fatigue over the whole of the district it inhabits, and as soon as the supply of food begins to fail in one place is able to discover a fresh feeding-ground. This example strikingly shows us that the procuring of a constant supply of wholesome food is almost the sole condition requisite for ensuring the rapid increase of a given species, since neither the limited fecundity, nor the unrestrained attacks of birds of prey and of man are here sufficient to check it. In no other birds are these peculiar circumstances so strikingly combined. Either their food is more liable to failure, or they have not sufficient power of wing to search for it over an extensive area, or during some season of the year it becomes very scarce, and less wholesome substitutes have to be found; and thus, though more fertile in offspring, they can never increase beyond the supply of food in the least favourable seasons. Many birds can only exist by migrating, when their food becomes scarce, to regions possessing a milder, or at least a different climate, though, as these migrating birds are seldom excessively abundant, it is evident that the countries they visit are still deficient in a constant and abundant supply of wholesome food. Those whose organization does not permit them to migrate when their food becomes periodically scarce, can never attain a large population. This is probably the reason why woodpeckers are scarce with us, while in the tropics they are among the most abundant of solitary birds. Thus the house sparrow is more abundant than the redbreast, because its food is more constant and plentiful —seeds of grasses being preserved during the winter, and our farm-yards and stubble-fields furnishing an almost inexhaustible supply. Why, as a general rule, are aquatic, and especially sea birds, very numerous in individuals? Not because they are more prolific than others, generally the contrary; but because their food never fails, the sea-shores and river-banks daily swarming with a fresh supply of small mollusca and crustacea. Exactly the same laws will apply to mammals. Wild cats are prolific and have few enemies; why then are they never as abundant as rabbits? The only intelligible answer is, that their supply of food is more precarious. It appears evident, therefore, that so long as a country remains physically unchanged, the numbers of its animal population cannot materially increase. If one species does so, some others requiring the same kind of food must diminish in proportion. The numbers that die annually must be immense; and as the individual existence of each animal depends upon itself, those that die must be the weakest—the very young, the aged, and the diseased—while those that prolong their existence can only be the most perfect in health and vigour —those who are best able to obtain food regularly, and avoid their numerous enemies. It is, as we commenced by remarking, "a struggle for existence," in which the weakest and least perfectly organized must always succumb.

Now it is clear that what takes place among the individuals of a species must also occur among the several allied species of a group—viz. that those which are best adapted to obtain a regular supply of food, and to defend themselves against the attacks of their enemies and the vicissitudes of the seasons, must necessarily obtain and preserve a superiority in population; while those species which from some defect of power or organization are the least capable of counteracting the vicissitudes of food, supply, &c., must diminish in numbers, and, in extreme cases, become altogether extinct. Between these extremes the species will present various degrees of capacity for ensuring the means of preserving life; and it is thus we account for the abundance or rarity of species. Our ignorance will generally prevent us from accurately tracing the effects to their causes; but could we become perfectly acquainted with the organization and habits of the various species of animals, and could we measure the capacity of each for performing the different acts necessary to its safety and existence under all the varying circumstances by which it is surrounded, we might be able even to calculate the proportionate abundance of individuals which is the necessary result.

If now we have succeeded in establishing these two points—1st, *that the animal population of a country is generally stationary, being kept down by a periodical deficiency of food, and other checks;* and 2nd, *that the comparative abundance or scarcity of the individuals of the several species is entirely due to their organization and resulting habits, which, rendering it more difficult to procure a regular supply of food and to provide for their personal safety in some cases than in others, can only be balanced by a difference in the population which have to exist in a given area*—we shall be in a condition to proceed to the consideration of *varieties*, to which the preceding remarks have a direct and very important application.

Most or perhaps all the variations from the typical form of a species must have some definite effect, however slight, on the habits or capacities of the individuals. Even a change of colour might, by rendering them more or less distinguishable, affect their safety; a greater or less development of hair might modify their habits. More important changes, such as an increase in the power or dimensions of the limbs or any of the external organs, would more or less affect their mode of procuring food or the range of country which they inhabit. It is also evident that most changes would affect, either favourably or adversely, the powers of prolonging existence. An antelope with shorter or weaker legs must necessarily suffer more from the attacks of the feline carnivora; the passenger pigeon with less powerful wings would sooner or later be affected in its powers of procuring a regular supply of food; and in both cases the result must necessarily be a diminution of the population of the modified species. If, on the other hand, any species should produce a variety having slightly increased powers of preserving existence, that variety must inevitably in time acquire a superiority in numbers. These results must follow as surely as old age, intemperance, or scarcity of food produce an increased mortality. In both cases there may be many individual exceptions; but on the average the rule will invariably be found to hold good. All varieties will therefore fall into two classes—those which under the same conditions would never reach the population of the parent species, and those which would in time obtain and keep a numerical superiority. Now, let some alteration of physical conditions occur in the district—a long period of drought, a destruction of vegetation by locusts, the irruption of some new carnivorous animal seeking "pastures new"—any change in fact tending to render existence more difficult to the species in question, and tasking its utmost powers to avoid complete extermination; it is evident that, of all the individuals composing the species, those forming the least numerous and most feebly organized variety would suffer first, and, were the pressure severe, must soon become extinct. The same causes continuing in action, the parent species would next suffer, would gradually diminish in numbers, and with a recurrence of similar unfavourable conditions might also become extinct. The superior variety would then alone remain, and on a return to favourable circumstances would rapidly increase in numbers and occupy the place of the extinct species and variety.

The *variety* would now have replaced the *species*, of which it would be a more perfectly developed and more highly organized form. It would be in all respects better adapted to secure its safety, and to prolong its individual existence and that of the race. Such a variety *could not* return to the original form; for that form is an inferior one, and could never compete with it for existence. Granted, therefore, a "tendency" to reproduce the original type of the species, still the variety must ever remain preponderant in numbers, and under adverse physical conditions *again alone survive*. But this new, improved, and populous race might itself, in course of time, give rise to new varieties, exhibiting several diverging modifications of form, any of which, tending to increase the facilities for preserving existence, must, by the same general law, in their turn become predominant. Here, then, we have *progression and continued divergence* deduced from the general laws which regulate the existence of animals in a state of nature, and from the undisputed fact that varieties do frequently occur. It is not, however, contended that this result would be invariable; a change of physical conditions in the district might at times materially modify it, rendering the race which had been the most capable of supporting existence under the former conditions now the least so, and even causing the extinction of the newer and, for a time, superior race, while the old or parent species and its first inferior varieties continued to flourish. Variations in unimportant parts might also occur, having no perceptible effect on the life-preserving powers; and the varieties so furnished might run a course parallel with the parent species, either giving rise to further variations or returning to the former type. All we argue for is, that certain varieties have a tendency to maintain their existence longer than the original species, and this tendency must make itself felt; for though the doctrine of chances or averages can never be trusted to on a limited scale, yet, if applied to high numbers, the results come nearer to what theory demands, and, as we approach to an infinity of examples, become strictly accurate. Now the scale on which nature works is so vast—the numbers of individuals and periods of time with which she deals approach so near to infinity, that any cause, however slight, and however liable to be veiled and counteracted by accidental circumstances, must in the end produce its full legitimate results.

Let us now turn to domesticated animals, and inquire how varieties produced among them are affected by the principles here enunciated. The essential difference in the condition of wild and domestic animals is this,—that among the former, their well-being and very existence depend upon the full exercise and healthy condition of all their senses and physical powers, whereas, among the latter, these are only partially exercised, and in some cases are absolutely unused. A wild animal has to search, and often to labour, for every mouthful of food—to exercise sight, hearing, and smell in seeking it, and in avoiding dangers, in procuring shelter from the inclemency of the seasons, and in providing for the subsistence and safety of its offspring. There is no muscle of its body that is not called into daily and hourly activity; there is no sense or faculty that is not strengthened by continual exercise. The domestic animal, on the other hand, has food provided for it, is sheltered, and often confined, to guard it against the vicissitudes of the seasons, is carefully secured from the attacks of its natural enemies, and seldom even rears its young without human assistance. Half of its senses and faculties are quite useless; and the other half are but occasionally called into feeble exercise, while even its muscular system is only irregularly called into action.

Now when a variety of such an animal occurs, having increased power or capacity in any organ or sense, such increase is totally useless, is never called into action, and may even exist without the animal ever becoming aware of it. In the wild animal, on the contrary, all its faculties and powers being brought into full action for the necessities of existence, any increase becomes immediately available, is strengthened by exercise, and must even slightly modify the food, the habits, and the whole economy of the race. It creates as it were a new animal, one of superior powers, and which will necessarily increase in numbers and outlive those inferior to it.

Again, in the domesticated animal all variations have an equal chance of continuance; and those which would decidedly render a wild animal unable to compete with its fellows and continue its existence are no disadvantage whatever in a state of domesticity. Our quickly fattening pigs, short-legged sheep, pouter pigeons, and poodle dogs could never have come into existence in a state of nature, because the very first step towards such inferior forms would have led to the rapid extinction of the race; still less could they now exist in competition with their wild allies. The great speed but slight endurance of the race horse, the unwieldly strength of the ploughman's team, would both be useless in a state of nature. If turned wild on the pampas, such animals would probably soon become extinct, or under favourable circumstances might each lose those extreme qualities which would never be called into action, and in a few generations would revert to a common type, which must be that in which the various powers and faculties are so proportioned to each other as to be best adapted to procure food and secure safety,—that in which by the full exercise of every part of his organization the animal can alone continue to live. Domestic varieties, when turned wild, *must* return to something near the type of the original wild stock *or become altogether extinct*.

We see, then, that no inferences as to varieties in a state of nature can be educed from the observation of those occurring among domestic animals. The two are so much opposed to each other in every circumstance of their existence, that what applies to the one is almost sure not to apply to the other. Domestic animals are abnormal, irregular, artificial; they are subject to varieties which never occur and never can occur in a state of nature: their very existence depends altogether on human care; so far are many of them removed from that just proportion of faculties, that true balance of organization, by means of which alone an animal left to its own resources can preserve its existence and continue its race.

The hypothesis of Lamarck—that progressive changes in species have been produced by the attempts of animals to increase the development of their own organs, and thus modify their structure and habits—has been repeatedly and easily refuted by all writers on the subject of varieties and species, and it seems to have been considered that when this was done the whole question has been finally settled; but the view here developed renders such an hypothesis quite unnecessary, by showing that similar results must be produced by the action of principles constantly at work in nature. The powerful retractile talons of the falcon- and the cat-tribes have not been produced or increased by the volition of those animals; but among the different varieties which occurred in the earlier and less highly organized forms of these groups, *those always survived longest which had the greatest facilities for seizing their prey.* Neither did the giraffe acquire its long neck by desiring to reach the foliage of the more lofty shrubs, and constantly stretching its neck for the purpose, but because any varieties which occurred among its antitypes with a longer neck than usual *at once secured a fresh range of pasture over the same ground as their shorter-necked companions, and on the first scarcity of food were thereby enabled to outlive them.* Even the peculiar colours of many animals, especially insects, so closely resembling the soil or the leaves or the trunks on which they habitually reside, are explained on the same principle; for though in the course of ages varieties of many tints may have occurred, *yet those races having colours best adapted to concealment from their enemies would inevitably survive the longest.* We have also here an acting cause to account for that balance so often observed in nature,—a deficiency in one set of organs always being compensated by an increased development of some others—powerful wings accompanying weak feet, or great velocity making up for the absence of defensive weapons; for it has been shown that all varieties in which an unbalanced deficiency occurred could not long continue their existence. The action of this principle is exactly like that of the centrifugal governor of the steam engine, which checks and corrects any irregularities almost before they become evident; and in like manner no unbalanced deficiency in the animal kingdom can ever reach any conspicuous magnitude, because it would make itself felt at the very first step, by rendering existence dfficult and extinction almost sure soon to follow. An origin such as is here advocated will also agree with the peculiar character of the modifications of form and structure which obtain in organized beings—the many lines of divergence from a central type, the increasing efficiency and power of a particular organ through a succession of allied species, and the remarkable persistence of unimportant parts such as colour, texture of plumage and hair, form of horns or crests, through a series of species differing considerably in more essential characters. It also furnishes us with a reason for that "more specialized structure" which Professor Owen states to be a characteristic of recent compared with extinct forms, and which would evidently be the result of the progressive modification of any organ applied to a special purpose in the animal economy.

We believe we have now shown that there is a tendency in nature to the continued progression of certain classes of *varieties* further and further from the original type—a progression to which there appears no reason to assign any definite limits—and that the same principle which produces this result in a state of nature will also explain why domestic varieties have a tendency to revert to the original type. This progression, by minute steps, in various directions, but always checked and balanced by the necessary conditions, subject to which alone existence can be preserved, may, it is believed, be followed out so as to agree with all the phenomena presented by organized beings, their extinction and succession in past ages, and all the extraordinary modifications of form, instinct, and habits which they exhibit.

Darwinism and Religion

Alvar Ellegård

When reading contemporary discussions of the Darwinian theory, one is struck by the fact that both opponents and supporters often vigorously insisted that the doctrine must be judged on its scientific merits alone. Still, there were few critics indeed who refrained from offering, in addition to the scientific arguments, comments on the religious bearing of the new views—even if only to deny that they possessed any religious significance. This very denial may be taken as proof that they believed that their readers needed such an assurance, which was certainly a correct interpretation of the situation.

We have observed in a previous chapter that there was a clear correlation between attitudes towards the Darwinian theory on one hand, and religious opinions on the other. It would indeed have been surprising if it had been otherwise. The Biblical cosmogony as set forth in Genesis was still the prevalent belief, not only among the masses, but also among the educated. It is true that the discoveries of geology, chiefly in the first half of the nineteenth century, had shaken these beliefs, and necessitated quite an extensive reinterpretation of the Mosaic account of Creation—not to speak of earlier reinterpretations called forth by the Copernican revolution in astronomy.

Source: Alvar Ellegård, "Science and Religion: A Mid-Victorian Conflict," *Darwin and the General Reader: The Reception of Darwin's Theory of Evolution in the British Periodical Press: 1859-1872* (Göteborg, Sweden: Elanders Boktrycheri Aktiebolag, 1958), pp. 93-113.

But at each of these advances of natural science the Church had yielded only on the most exposed points: the general accuracy of the Biblical story of the early history of the world, and especially of our race, was not allowed to be called in question.

It is true that many thoughtful Churchmen warned their less liberal brethren that "the Bible was not intended to teach scientific truth." But the Biblical version, even of these historical and scientific matters, had the sanction of universal tradition within the Christian world. Moreover, there was not much that history or science could put in its stead. When Darwin published, Prehistoric Archaeology was in its infancy: such terms as Stone Age, Bronze Age, Iron Age date from the middle of the nineteenth century. Anthropology and ethnology were in an equally rudimentary condition. Geology was barely half a century old as a science. When Voltaire had wished to ridicule the contention that the sea shells deposited in the various rock strata in the Alps were proof of the Deluge, he suggested that they had been dropped by pilgrims or crusaders on their way to the Holy Land, or that they were formed by chemical action.

There was in fact hardly any reason why people should not think that, even if the world as such might date back untold aeons of time, as geologists now claimed, then at any rate the history of mankind stretched only some 6,000 years back, which was the date Archbishop Ussher had arrived at from Biblical data. The Bible and the annals of antiquity took one back, literally, to the dawn of time, to the pristine freshness of mankind's youth.

This view was so deeply rooted and so universal as to seem to most people as nothing but plain common sense. It was not at all necessary to invoke the supernatural sanction of the Divine inspiration of the Bible in order to make the story credible. On the contrary, the fact that the Biblical account seemed so fully in accordance with common sense may even have appeared as an additional reason to believe in the supernatural origin of the Bible. As a writer in the Evangelical *Record* expressed it, "The initial statement of the Bible comes to us with a force, a clearness and a proof such as no conscientious man can repel, for it is backed up not only by all those evidences which prove the Scriptures to be the Word of God, but also by the testimony of every man's conscience and by the universal traditions of the human race." It was easy to say, after the new geological and historical doctrines had been assimilated, that the religious beliefs which had to be changed on account of the scientific discoveries had no religious significance. But how was the common man to tell which of his beliefs were truly religious, and which were not? To most people the body of knowledge that they obtained from the Bible must have appeared at least to some extent religious. Not for nothing was Biblical history sacred history.

The Copernican conflict between science and religion had concerned the constitution of the universe around us. The recent conflict between Geology and Genesis had touched the history of the earth, and the animals and plants living on it, but not man himself, of whom no trace was found below the most recent strata. Darwin's theory struck much nearer the heart of religious beliefs, for now the history of man himself was brought into the center of the discussion. The blow fell the more heavily as concessions made by theologians to geologists had often been coupled with assertions that what was important in Genesis was not the account of the earth's creation, but the creation of man and his early history. In the face of the inroads of geology, theologians had built up a sort of inner defence line around the Bible's version of man's history. In this sphere even the chronology of the Bible was adhered to with extraordinary tenacity. But in the conflicts over Darwinism much more than chronology was of course involved: much more even than a mere reinterpretation of single passages in Genesis. For man's early history, as told in the Bible, was closely bound up with the important religious concepts of the Fall, Original Sin, Atonement, and Redemption. These ideas were explicated in terms of events in sacred history. If those events were to prove fictitious, the concepts themselves would appear to hang in the air.

No sincere Christian could envisage the disintegration of those fundamental concepts of his religion without the gravest misgivings. A man's religion may be said to be an expression of the scale of values that determines his attitudes and actions in various situations in life. The particular beliefs connected with each religion appear to achieve a quasi-symbolical representation of that scale of values or those attitudes: by their verbal and conceptual form they can be easily communicated, and thus serve to create a certain amount of conformity as regards fundamental attitudes within a community. Though the actual concepts, or quasi-symbols, or dogmas, may be of little importance in themselves—what is important is the scale of values they stand for—the connection between the symbol and the thing symbolized becomes to most people so close that one tends to be equated with the other.

Now it is obvious that to many Mid-Victorians, the Biblical account of man's creation and history was part of a system of religious concepts. Because of the confusion between symbol and thing symbolised, acceptance of the Darwinian theory seemed to them to necessitate a complete spiritual revolution, a total change of outlook and attitude towards life. Such a change can hardly be expected to take place without a potent incentive: and to most people, obviously, the possible truth or falsehood of an abstruse scientific theory, which few were in a position to judge on its merits, was not at all a strong enough incentive. They therefore retained their old-established attitude towards life, and with it the concepts and symbols they felt to be connected with this attitude. They by-passed the question of true or false as regards the Darwinian theory, treating it instead as a question of good and bad. The theory was regarded as a religious question, not as a scientific one.

That this was the case, to a great extent, is apparent to any reader of the periodical press of the time. In the popular papers especially, the Darwinian theory was hardly referred to at all except in its relation to religion. And the amount of space and attention given to the theory in the religious publications was immeasurably greater than was normally afforded to scientific questions in those organs.

In some instances, opposition to the theory was openly and squarely based on its possible effect on religion, and ultimately on the well-being of society. To argue in this way was certainly effective from a propagandistic point of view, especially for a public unable to follow the niceties of a scientific argumentation. The underlying attitude was a natural one to take: namely, that the spiritual and moral welfare of the community was a more important consideration than the freedom of scientists to divulge their theoretical conclusions—a freedom which, for instance, the Catholic Church had never granted. One Roman Catholic organ clearly showed that it viewed these matters in quite the same light as the Church had done in Galileo's case. "The salvation of man," it said, "is a far higher object than the progress of science: and we have no hesitation in maintaining that if in the judgment of the Church the promulgation of any scientific truth was more likely to hinder man's salvation than to promote it, she would not only be justified in her efforts to suppress it, but it would be her bounden duty to do her utmost to suppress it . . . .The truth ultimately can do no harm, although, temporarily, injury may follow from an unreasonable application of it."

Protestant writers were not as a rule quite so explicit. They would not lightly admit that the views they combated might be true. And as the question of the truth or falsity of Darwin's theory was a complex one, it was always possible to cite authorities who vouched for its scientific illegitimacy. In regard to Darwinism, religious people therefore were never compelled squarely to face the dilemma outlined by the Catholic writer just quoted. The theory could be declared scientifically deficient: its theological obnoxiousness was an additional reason for repudiating it.

The two aspects were combined in the argument that religious considerations could legitimately be taken into account whenever there was any doubt at all about the truth of a scientific theory. This was eminently the case with Darwin's theory. In practice therefore, those who employed such an argument were taking the same attitude as the Catholic writer just quoted. One Protestant reviewer expressed it thus: "A theory which is incompatible with views long entertained, and of slow growth, which tends to subvert existing notions, and, indirectly at least, to raise harassing doubts on sacred subjects, should be clearly supported by facts far outweighing those which can be brought forward against it." Thus theology was not indeed overtly invoked against the theory, but it was allowed to weight the balance. The Bishop of Oxford endorsed this way of looking at the matter in his *Quarterly* review of the *Origin*: "We cannot, therefore, consent to test the truth of natural science by the Word of Revelation. But this does not make it the less important to point out on scientific grounds scientific errors, when those errors tend to limit God's glory in creation or to gainsay the revealed relations of that creation to Himself." The Darwinians, on their side, recognized that such a theological weighting was bound to occur. Darwin went out of his way in his book to reconcile the feelings of the religious, and his supporters were often at pains to explain that his views harmonized with Christianity.

When it was asserted that the establishment of Darwinism would lead to the destruction of Christianity, it was obvious that no Christian would hesitate in his choice. Those critics who presented the theory in such a light must have been aware that they thereby in effect condemned it on theological grounds. One such writer, vaguely feeling that his argument might recall the Catholic Church's conflict with Galileo, tried to forestall the expected criticism by asserting—with perhaps more heat than justice—that Darwin's case was different. "There are many cases, indeed, in the history of science, where speculations, like those of Kepler, have led to great discoveries ..." he wrote. "It is otherwise, however, with speculations which trench upon sacred ground, and which run counter to the universal convictions of mankind, poisoning the fountains of science, and disturbing the serenity of the Christian world." A Methodist organ painted Darwinism in even direr colours. "We regard this theory, which seeks to eliminate from the universe the immediate, ever-present, all-pervasive action of a living and personal God, which excludes the possibility of the supernatural and the miraculous ... as practically destructive of the authority of divine revelation, and subversive of the foundation of both religion and morality." Nor were such views by any means confined to the religious papers. The *Edinburgh Review* placed a warning as to the religious consequences of Darwin's theories at the very beginning of its review of the *Descent of Man:* "It is impossible to overestimate the magnitude of the issue. If our humanity be merely the natural product of the modified faculties of the brutes, most earnest-minded men will be compelled to give up those motives by which they have attempted to live noble and virtuous lives, as founded on a mistake ... our moral sense will turn out to be a mere developed instinct ... and the revelation of God to us, and the hope of a future life, pleasurable daydreams invented for the good of society. If these views be true, a revolution in thought is imminent, which will shake society to its very foundations by destroying the sanctity of the conscience and the religious sense." The reviewer saw not only religious evil, but also social evil arising out of the theory. The connection was a natural one. If religious belief was affected, the social fabric itself would disintegrate. A writer in the low-brow and somewhat goody *Family Herald* made this point quite bluntly: "Only let our scientific friends show the people, who are quick to learn, that there was no Adam ... that nothing certain is known, and then that chaos which set in during the lower Empire of Rome will set in here; we shall have no laws, no worship, and no property, since our human laws are based upon the Divine." That was written in 1861: ten years later the journal was still of the same opinion: "Society must fall to pieces if Darwinism be true." That the *Times*, in its review of *Descent*, gave prominence to this sort of argument only confirms how widespread was the attitude which gave rise to it. "A man incurs grave responsibility who, with the authority of a well-earned reputation, advances at such a time the disintegrating speculations of this book. He ought to be capable of supporting them by the most conclusive evidence of facts. To put them forward on such incomplete evidence, such cursory investigation, such hypothetical arguments as we have exposed, is more than unscientific—it is reckless."

It must be stressed, however, that unreasoned outbursts of theological zeal were seldom allowed to play the leading part in the attacks against the Darwinian doctrines. More liberal-minded Christians often expressed regret that their co-religionists should reject Darwin on Scriptural grounds. The theory, one said, should be opposed on scientific grounds alone. As a matter of fact, however, this often meant no more than that the religious considerations entered on a more abstract level, in that the scientific arguments were evaluated in terms of a semi-theological philosophy of science. These matters will be discussed more fully in Chapter 9.

Though the incompatibility of Darwin's theory with the traditional beliefs about man's history was the feature that above all drew the general public's attention to the new doctrines, the most serious conflict with religion occurred on another and much more fundamental point than that of the accuracy of the Bible, or the interpretation of certain Christian dogmas,—namely, the idea of Divine Providence. To the Mid-Victorians the conviction that the world was placed under the watchful guidance of a higher power appeared as a fundamental religious belief. Without Divine supervision, one held, everything would disintegrate into chaos, for only chaos could result if the universe were left to the action of chance and blind, inexorable laws. Design and purpose, the attributes of a Mind, were needed to create and sustain a Kosmos. Evidence of such design was found especially in the organic world. Living beings could not have become so perfectly adjusted to their environment as they undoubtedly were, if design and purpose had not been present at their creation.

Now if Darwin was right, the development of the organic world could be explained without recourse to Divine design and purpose. Variations that might be called accidental, and the operation of natural causes, could lead to evolution and progressive change without assuming design. This was achieved by the theory of Natural Selection, which was the scientifically significant part of Darwin's doctrine. But it was also the feature that made his doctrine so difficult to accept for the English theologians, and also for all naturalists of a religious turn of mind, who had been accustomed to look at their study as a parallel to that of theology. Many took quite literally Bacon's words about God's works and God's word. It is true that the more liberal religious schools of thought found little difficulty in assimilating the development theory as such with their religious beliefs. Evolution might be regarded as having occurred through the providential care of God. But not even liberal theologians could accept a history of life, including man, where every reference to Divine Providence appeared superfluous.

The impact of Darwinism on religion would not have been so strong if British theologians had not been so strongly attached to the tenets of Natural Theology. The very success of Natural Theology in Britain—explicable in view of the powerful empiricist tradition of British thought—had led to that close interdependence of science and religion which was going to give rise to serious conflict when science advanced into fields where formerly theology had held exclusive sway.

There is abundant directive evidence that both theologians and scientists in Mid-Victorian Britain did look upon science and religion as closely connected with each other, and dependent on each other. The dependence was logically necessary as long as both science and religion claimed to offer information on matters of fact. Their domains were, so to speak, dovetailed into each other: the limits of each were determined by the other. It is clear that the Darwinian theory acted as a powerful catalyst in exhibiting the dangers of this dependence, and in arousing the latent conflict.

A very marked change in the attitude of theologians towards science, and of scientists towards religion, was taking place in the 'sixties, concurrently with the spread of Darwinian doctrines. The time when Natural History was almost looked upon as a branch of Natural Theology, when every other Church of England clergyman was an amateur naturalist, was passing away. One religious periodical greeted Darwin's *Origin* with the words: "Its publication is a mistake . . . at this time of day, when science has walked in calm majesty out from mists of prejudice and been accepted as a sister by sound theology," and expressed the hope and belief that it would soon be forgotten. Instead, it came to mark a decisive stage in the emancipation of science from theology. A few years later one religious writer had to admit that, "taken on the whole, scientific studies have not a religious but a sceptical tendency," and another, in a searching analysis, concluded that physical science must be held "the present great enemy of religion."

It may be that the undeniable advance of science during these years did not chiefly dismay the theologians themselves. They were not defenceless: after all, the Church had weathered many storms in the past, and possessed a respectable ideological arsenal. But ordinary religious folk—including many scientists—were in a more exposed position. An instance of the concern felt in these quarters was a declaration published in the *Athenaeum*, and later in the *Times*, signed by thirty Fellows of the Royal Society, and forty MDs, expressing the opinion that a scientist "should not presumptuously affirm that his own conclusions must be right, and the statements of Scripture wrong." Many leaders of scientific thought, however, publicly dissociated themselves from this action, which expressed the opinion of a minority of the scientific world only. Unable to rally the support of their fellow-scientists, the religious phalanx later closed their ranks, and in 1865 was formed the Victoria Institute, "to investigate fully and impartially the most important questions of Philosophy and Science, but more especially those that bear upon the great truths revealed in Holy Scripture, with the view of defending these truths against the opposition of Science, falsely so called." Violent anti-Darwinianism was the prevailing characteristic of many prominent members of the Institute.

Another fruit of the acute crisis in the relations between science and religion was the Metaphysical Society, founded in 1869 by a group of earnest liberal-minded Churchmen. Unlike the Victoria Institute, they had, however, been at pains to associate with them several of the leading scientists of the time, including T. H. Huxley. The Society met for discussions regularly for many years, and though the tone was gentlemanly, no real *rapprochement* was achieved. This was not indeed to be expected: the parties differed, and in the end agreed to differ, on fundamental points of the philosophy of science.

Among the general public, the problem of science and religion attracted increasing attention. The 1868 Church Congress at Dublin, and the 1870 Congress at Southampton, made it the central point of their debates. In the press the conflict was a constantly recurring theme, and the hardening climate was noticed. The Nonconformist *British Quarterly* asserted that "the age yearns for religious faith, and is disquieted only because its religious faith is disturbed by the readjustments which the advance of science necessitates." One religious writer complained of the "Scepticism of Science, which has increased so rapidly of late years and still daily progresses," while the freethinkers in the *Westminster Review* found it quite natural that "the theological opinions of the past should be slowly dying out before the scientific opinions of the present."

One did not fail to notice that the advance of science did not concern details only: it was the scientific attitude as such which was inimical to much that had formerly been looked upon as essentially religious. "Physical sciences, when directed against religion," wrote the Roman Catholic *Rambler* in 1862, "only attack it accidentally—in its points of asserted contact with the world of phenomena. But here they wage a war of extermination; they deny the reality of the contact; they account for the phenomena which religion claims as her own upon merely physical laws, and they thus introduce and encourage the suspicion that the claims of religion are due only to the imagination of the pious, or to the imposture of the cheat." It fell to Darwin's theory to bring out the "deep-seated antagonism" between science and religion, and it did so, not chiefly because the evolution theory upset the Biblical creation story, but because the theory of Natural Selection implied the extension of purely scientific methods and procedures to the domain of organic life, and therefore, ultimately, also to that of mind and soul. In these spheres the theological concept of design, and the metaphysical ones of final cause and vital force, had hitherto been accepted as indispensable: but they were irreconcilable with the scientific naturalism of which the Natural Selection theory was a fruit, and to which it gave such a strong impetus. The scientific naturalist claimed for science the whole world of sensory experience, and could tolerate no theological enclaves within that domain. This was one of the fundamental differences between the Darwinian and the earlier conflicts between science and religion. Previous advances of science had indeed diminished the area where theological explanations of natural events were acceptable: the conflict was about the relative size of that area. Now the very existence of any such area at all was disputed.

It was not Darwinism alone that caused the violent crisis in the relations between science and religion in the 1860s. There was at least one other potent factor, namely, the introduction into Britain of more modern methods of Biblical criticism. This "Higher Criticism" had long been known and applied in Germany, but British theology seems to have insulated itself from its Continental counterpart. Strauss' *Life of Jesus*, written in 1835 and translated into English in 1846 (by George Eliot), was regarded by theologians and religious laymen as a book to be abhorred and denounced rather than discussed. Therefore, when in 1860 six prominent liberal clergymen published the *Essays and Reviews*, where some of the ideas of the new school of Biblical study were applied in a mild form, laymen and clerics within the English Church were caught unprepared. In the press the book caused a sensation which, quantitatively at least, was considerably greater than that raised by the *Origin of Species*. Scores of articles and pamphlets written in refutation of the *Essays and Reviews* appeared in 1860 and 1861. A similar reaction was caused by Bishop Colenso's *Pentateuch* where the author showed, among other things, that the chronological statements in the Bible were often incompatible with each other, and therefore could not all be accepted as true.

Both the *Essays and Reviews* and the *Pentateuch* were fruits of an application of the same critical methods to the Bible as to all other historical documents. It was this attitude which was new in British theology, and when one of the Reviewers wrote that he assumed that the Bible should be read like any other book, his statement was treated as a designed impiety. This was the hub of the controversy: whether the evidence of religion should be subjected to scientific tests.

Thus the advance of science in England in the 1860's touched religion in two ways. First, Darwin's theory threatened to oust theological explanations from its last foothold within the world of natural science. Second, the application of scientific principles to the study of the Bible carried over the conflict to the domain of theology itself. Religious people could not but look at the development with grave misgivings. "The danger which to many observers appears to threaten the Christian cause more seriously than any other arises from the application to it of the methods and results of modern science," wrote one Evangelical organ, and in the *Contemporary Review* W. B. Carpenter, that year's President of the British Association, declared in 1872 that "the claims of Science have of late been advanced, not only more strongly, but more aggressively, and some of the positions that have been taken up have been such as apparently to threaten, not the outworks only, but the very citadel, of Religious Faith."

The inroads that the Darwinian theory was making into domains which had previously been regarded as the exclusive preserve of religion led to attempts to effect a separation between what was called the "spheres" of science and religion. But the attempts were hardly successful. It was impossible to agree where the dividing line between the spheres should be drawn. According to the ideology of Natural Theology, in fact, the whole world of natural science was included in the even wider world of religion. One obviously could not say that the natural world proved the truths of religion, and at the same time insist that changes in our knowledge of that world had no religious significance. At the other extreme, some scientific men with positivistic and empiricist leanings claimed for science the whole domain of factual experience, whether physical or mental, leaving to religion nothing but the world of ethics. To accept any of these extreme views was to make the conflict inevitable as long as both science and theology were actively pursued. Scientific men tended to view such a result with equanimity, since it was clear that the victories, in the nineteenth century, were almost constantly on the scientific side. Scientists could delcare that they would judge their results according to the standards of science alone, and leave it to theologians to look after their own house. Those scientists who had no special theological bent therefore solved the problem for themselves by simply ignoring the conflict and its results.

Theologians and religious people could not afford to be so complacent. For while scientific knowledge could be built up and organized completely independently of theology, theology, at least in its 19th century British form, could not be pursued without reference to science, or more precisely, the facts which science possessed the most powerful tools for ascertaining. Therefore, while those scientists who wished to extend the domain of science over the whole area of human knowledge tended to exclude theology, and could afford to exclude it, those religious people who claimed this same domain for theology had perforce to find some accommodation for science within it. Opposition against attempts to separate the spheres of each therefore came mainly from religious quarters. A writer in the Unitarian *National Review*, while recognizing that separation of the spheres was the line "prevailingly assumed both by liberal divines and by reverential and cautious men of science," could not himself recommend that solution. "The *savant* cannot help advancing his lines of thought into human and moral relations, and esteeming them amenable to him. The theologian cannot help applying his faith to the universe, for the supernatural is conceivable only in relation to the natural, and the transcendency of God involves the subordination of the world." That was a logical position to take for a supporter of Natural Theology. And as Natural Theology was so very highly esteemed by British theologians, the violence and extent of the conflict raised by the advance of science is not surprising. It is significant that one religious organ, in its review of the *Origin*, explained that "it would not be dealing fairly by our readers, and especially it would be unmindful of the apologetic value of natural theology, were we to look at this theory from any other point of view than the twofold one of science and theology."

Now if Natural Theology was popular among theologians, it was even more so among a certain class of scientific men. A writer in the *Popular Science Review* took such naturalists to task for their habit of "dabbling in Divine matters," as he said. "It appears to them that, unless they drag the Creator into every second paragraph, their essay will not possess the necessary religious veneering for the public taste." But even this writer blurred the issue by continuing, "Now, when allusion is discriminately and respectfully made to the works of the great First Cause, no fault can be found." As we shall show below, any reference to the First Cause was likely to create misunderstandings when the Darwinian theory was concerned.

An extremely common device used by those religious writers who refused to admit the need for separating the spheres of science and religion was to insist that there was no real conflict between the two. Now it was of course impossible to deny that there was some sort of conflict. There were two ways of accounting for this, either to declare that the conflict was apparent only, or to insist that the conflicting sides did not rightly represent science or religion. Either way was question-begging: to assert that the conflict was apparent only was to state an (unverifiable) conviction that all problems would be satisfactorily solved some time in future; and to say that true science did not clash with religion was of no avail, as long as true science could not be objectively defined.

The first of these alternatives—asserting that the conflict was apparent, and that only a true interpretation was needed to resolve it—was common among Catholics and Anglo-Catholics, which is hardly surprising, since these denominations undoubtedly possessed the most developed theology. The *Dublin Review*, writing in 1858, put the Catholic view clearly enough. "We much fear that a sort of general impression prevails in some portions of society, that the progress of science is inimical to the interests of religion. We cannot wonder indeed that such a fear should disturb the minds of the more religious class of Protestants . . . . [But the Catholic] knows that whatever else may be true or false, his religion is infallibly true . . . . And if in any point they should seem to clash and contradict each other, the Catholic cleaves to that which is certain—his religion, and leaves it to time and inquiry to clear up the difficulty." Less explicitly, Dr. Pusey said in 1865 that "the right interpretation of God's word would never be found to contradict the right interpretation of the facts of physical science." These writers recognized that the problem of interpretation did not concern science only, but theology as well, and could view the advance of science without overmuch fear, since the task of harmonizing it with religion was left to expert theologians, not to the natural lights of the common man.

Protestants, with their more direct dependence on the Bible, were generally less willing to admit any latitude of interpretation on the theological side: they therefore tended to choose the second alternative: to deny that the obnoxious theories were truly scientific. This obviously placed them at a tactical disadvantage, since they had to do battle on the scientists' field, not on their own. Generally, in fact, the argumentation did not pass far beyond bare assertions. "Pseudo-science has assailed the foundations of our faith; we have endeavored to show that true science is a modest but a firm friend to that faith," wrote the ***British Quarterly Review***. Such statements were extremely common, and it seems the phrase "true science" was interpreted in two different ways. One was to equate it with the statements of orthodox scientific authorities: "Some narrow scientific men, with a little knowledge, are full of bigotry and intolerance; not so with the masters of science, who have generally been ardent supporters of revealed religion," was a typical assertion. The other way was to attempt a more theoretical definition of true science, and especially to assert that it was only the theories, or hypotheses, or speculations, of science that were hostile to religious beliefs, but never ascertained facts. The nature of this argument will be discussed in Chapter 9.

Those who were convinced that science and religion were in perfect harmony, had no incentive to attempt a spearation between the spheres of each. In principle, the domain of either was left indefinite and illimited: in practice, scientists were declared to overstep their bounds, or rather, to leave the path of true science, as soon as their conclusions conflicted with traditional religious beliefs. This was the most common attitude among the general public in these matters. The other and more realistic way of solving the conflict was to assign wholly distinct domains to science and religion. This view had adherents chiefly among liberal-minded religious folk.

The most common demarcation line was probably between the physical world, which was given to science, and the spiritual world, which was reserved for theology. Unfortunately the distinction was difficult to uphold, since it could not be made quite clear what was spiritual, and what was physical. To most religious people, there was a causal relation between the spiritual world and the physical one, and if that was the case, it was obviously impossible to assert that the discoveries of religion had no bearing on those of science, and vice versa. This attitude came out clearly in the discussion of one of Professor Tyndall's speeches at the British Association. Tyndall asserted that "the physical philosopher, as such, must be a pure materialist."

So much, said some religious commentators, might be granted, but it only proved that the scientific sphere was a limited one, and theology, which did not limit its purview in this manner, was by implication a higher and nobler study. "If . . . they deny the existence of the spiritual world, they cut off from their view half the field of evidence," wrote the *Guardian*. Theologians were inclined to demand that scientists should admit "spiritual evidence" as being on a par with the "physical evidence" which they were primarily concerned with. Religion might be excluded from the physical world, but it still had a foothold in the mental one. The facts of psychology were not yet given over to science, nor yet the facts of biology: the spheres of science and religion "intersect in the human mind," as one theologian put it. Therefore, it was maintained, the scientist's failure to take into account "spiritual facts" was a deficiency in his science. "The facts of consciousness are as much facts of nature as any others, and the natural philosopher has no business to reject them from his premises," said Canon Mozley, and in the *Contemporary* we read: "The truths of revelation form one connected body of belief based on the wide range of facts and experience which bear their witness to the spiritual world. The assault on them too often rests, not on the assured facts of science, but on the groundless visions of speculation; not on the affirmative proof which is certified by observation, but on the negative suspicion that nothing can exist which the sense-philosophy refuses to recognise." In this way the separation of the spheres was given up, since both domains had to contribute whenever theories were to be constructed. The nature of the argument is analysed more fully in Chapter 9.

A slightly different way of separating the "spheres" was to adopt the style of the Natural theologians, and to maintain that science dealt with secondary causes only, while religion was concerned with the First Cause. The solution was not very satisfactory, since those who propounded it had a tendency to allow explanations in terms of a First Cause as acceptable alternatives to explanations in terms of secondary causes: and this possibility clearly made nonsense of the separation.

The most effective solution of the whole problem was to reserve for religion nothing but the world of morality and ethics. This line had been advocated, among others, by Coleridge, whose authority was now and then invoked in its favour. It may be surmised that this distinction was sometimes intended by writers who employed the looser term of spiritual, but by and large religious people seem to have been reluctant to admit the necessity for such a drastic curtailment of the domain of theology as it implied. The severe treatment meted out to Baden Powell for advocating such a separation is significant: The *North British Review* would have nothing of his main assumption: "His whole theory rests ultimately on an attempt, not only to draw a *distinction*, but to effect a *divorce*—to establish an actual *separation* between the *physical* and *moral* departments of nature." It is also significant that the solution was often commended by empiricist and positivist philosophers and scientists. Religious people and the majority of the general public were not yet prepared to accept that complete divorce of the world of facts from the world of values, which the advance of science and of scientific method made more and more imperative. That ultimate solution, however, was in the air. To scientists, and to Darwinian scientists especially, it became increasingly evident that science had to claim as its sphere not only the world of physical facts, but all facts.

The attempts to separate the spheres of science and religion may be said to be but one aspect of the conflict between the two. The scientists, as the advancing side, were to find no logical halting place until the whole domain of the world of experience was recognized as theirs. The theologians, on the other hand, could not in general admit that religious truths were completely unconnected with the world of experience, which would make nonsense of the idea of Divine Providence, if not of the idea of the Supernatural altogether. Therefore the theologians, as the retreating side, tried to draw the line of demarcation between the two spheres wherever they thought a successful line of defence could be established. The result was continual conflict, as science continued its advance, and the line was pushed further and further back.

A Critique of Social Darwinism

George Nasmyth

What are the causes of the success of the philosophy of force, and how did the distorted form of "social Darwinism," which it claims for its scientific foundation, gain its almost universal acceptance?

In seeking the answer to this question we shall come upon a surprising series of facts.

We shall discover that the doctrine known as "social Darwinism," which finds the cause of social progress in war, universal competition, and the role of struggle and force in human relations, was not created by Darwin; but that he based his whole theory of social progress on the moral law and the social instincts.

We shall find that this doctrine was repudiated in its application as a law of human society by the co-discoverer of the theory of evolution, Wallace; and by Darwin's intimate friend and chief disciple, Huxley.

We shall find that the misapplication of Darwin's biological theory to human society, which is current in the modern world, did not emerge as the

Source: George Nasmyth, *Social Progress and the Darwinian Theory: A Study of Force as a Factor in Human Relations* (New York: G. P. Putnam's Sons, 1916), Chapters II and III. Reprinted by permission of G. P. Putnam's Sons.

result of the thorough discussion of a subject which recent events have shown to be one of the most important questions in applied social science—that is, the place of struggle and force as a factor in human associations. On the contrary, it grew up almost unnoticed, and as an unconscious by-product of a debate between some of the greatest minds of the age, over an entirely irrelevant, and, as the modern world has largely come to regard it, a socially unimportant subject—the theological implications of the Darwinian theory as they shaped themselves in the warfare between science and traditional theology around the issue of evolution *vs.* special creation. Instead of subjecting it to the searching analysis demanded by its practical social importance, the intellectual world and public opinion has accepted "social Darwinism" uncritically and by almost unanimous consent as an integral part of the theory of evolution.

The causes of this almost miraculous success are to be found largely in three factors: (1) the universality of the appeal which "social Darwinism" makes to the human spirit, enlisting both the highest aspirations toward perfection and justice and the lowest instinct of selfish greed and brute force; (2) the intellectual environment in which the social applications of the theory of evolution were developed, and (3) the influence of historical events, especially the Franco-Prussian war, and the rapid growth of Imperialism among all the Western civilized Powers since 1870. . . .

The first day of July, 1858, marks the division between two epochs of human thought; for on this day two papers, one presented by Charles Darwin, and the other by Alfred Russel Wallace, were read before the Linnaean Society at London, and with the reading of these papers, the doctrine of evolution by natural selection was born. On November twenty-fourth of the following year, Darwin published the first instalment of his thought in its fuller development,—his book on *The Origin of Species by Means of Natural Selection.* This was the fruit of thirty years of work and thought by a worker and thinker of genius, and it at once commanded the world's attention by the transparent honesty and judicial fairness with which it presented its wealth of facts, gathered from a world-wide observation; compared with almost infinite patience; and woven into a theory which revealed one of the great unifying principles of the cosmic order.

Darwin had found one of the great secrets at the heart of the evolutionary process for which a long line of investigators from the days of Aristotle had sought in vain,—the thin red line which was to guide him, and after him all workers in the natural sciences, through the labyrinth of the infinite variety of the facts of Nature. The work undoubtedly marks one of the most important events in the history of the human race. Not only was it epoch-making because with its publication Nature re-entered upon a grand and magnificent unity. It was important, too, because it marked the enfranchisement of the human spirit from a mediaeval theology, from outworn traditions, from ancient routines, and the ignorance and superstitions of a barbarous past. Man raised his head; he felt himself master of the world; he saw infinite horizons opening before his eyes, with no authority which henceforth could arrest him in his conquest. We can understand with what enthusiasm this definite liberation of the human mind would be received by the thinkers of a purely scientific spirit.

It would be difficult to exaggerate the changes which have come about in all departments of human thought, as the result of the theory of natural selection. During the past half century, all the sciences, from astronomy to sociology, have been profoundly influenced by Darwin's discovery of evolution. In historical and in political thinking especially, the philosophy of force was greatly strengthened by the discovery of such apparently scientific foundations.

In *The Origin of Species*, Darwin did not apply his theory to human relationships, but confined himself to the field of biology. The only reference which he makes to man is at the end of the book, where he says that in the future "much light will be thrown on the origin of man and his history." Darwin's theory of social progress is contained in *The Descent of Man*, which was not published until twelve years later. During these twelve years Darwin was patiently at work on the application of his theory to human society, and as early as 1864 he wrote to A. R. Wallace:

> The great leading idea is quite new to me viz., that during late ages, the mind will have been modified more than the body; yet I had got as far as to see with you that the struggle between the races of man depended entirely upon intellectual and moral qualities.

But the followers, and especially the popularizers of Darwin's theory could not wait for his own application of the theory of natural selection to social progress. The publication of *The Origin of Species* had acted as a great liberalizing influence upon the minds of men, and the flood of new thought pouring over the world stimulated and nourished research and reasoning in every land. Edition after edition of the book was called for and it was translated even into Japanese and Hindustani. A vast army of young men took up every line of investigation, and epoch-making books appeared in all the great nations. Spencer, Wallace, Huxley, Galton, Tyndall, Tylor, Lubbock, Bagehot, Lewes, in England, and groups of strong men in Germany, Italy, France, and America, published important works in every department of biology. Under these conditions it was inevitable that Darwin's theory should be applied to man.

In order to trace the distortion which Darwin's theories suffered in this application to human society, it is necessary to understand the intellectual world into which they were born, and the philosophical doctrines current in the aristocratic intellectual circles in which they were discussed and developed.

On the one hand can be traced the influence of teachers like Carlyle, Kingsley, and Ruskin, who have done so much to foster the belief in a "divine right" of force. Ruskin's view of the value of war for civilization and art we have already noticed. Charles Kingsley had defended the Crimean War as "a just war against tyrants and oppressors," and had eloquently advocated such a war as in accord with the highest teachings of Christianity and the Bible. The direction of Carlyle's political teachings, which were in accord with his hero worship and "will to power" philosophy, may be judged from the following summary:

> Carlyle condemned democracy, which he identified with laissez-faire, as "a self-cancelling business," a government which only achieved the negation of any government. Representative institutions, a free and broad electorate, in a word all the paraphernalia of democracy, were in his eyes a matter of mere palaver and ballot boxes—"nothing except emptiness" and zero. To get governance, men must turn to those who are able to govern, the silent few, standing aloof and alone in their wisdom, who are nature's appointed Hero-Kings. . . . Wise, and in their wisdom also virtuous, they must guide and even drill their lesser fellows, who shall find in obedience their chief end and highest pleasure. . . .Guidance, regulation, drill became his ideals: military metaphors recur in his writings. He even advanced to the military doctrine that might is the measure of right. If a man be able, wise of heart, strong of will, firm in his resolution to do his duty among his fellows, he must govern according to the measure of his strength, and his right over his fellows is according to his might. "The strong thing is the just thing": rights are "correctly articulated mights."

To men holding this philosophy, "social Darwinism" made an especially strong appeal; it proclaimed the idea of the survival of the fittest; it strengthened their faith in the triumph of the best; it affirmed that Nature practices an incorruptible justice,—that the idea of justice is found even in the biological realm. Thus the philosophy of force enlisted in its service the highest aspirations of the human soul, man's passionate desire for justice and perfection. "Man has an inextinguishable thirst for justice," says Novikov; "it could not be otherwise, because justice is life; injustice death."

On the other hand, Darwin's biological theories were applied to human society in an intellectual world dominated by individualistic scientists like Spencer and by conservative lawyers like Sir Henry Maine.

One of the chief influences in the rise of the philosophy of force was the contribution of Spencer's social philosophy. As early as 1851 we find him recognizing, in *Social Statics*, the

> stern discipline of nature which eliminates the unfit and results in the maintenance of a constitution completely adapted to the surrounding conditions.

And we find a prophecy of the modern "social Darwinism" in the fact that Spencer attacked the system of poor relief in the name of this discipline.

Spencer never became a Darwinian. The first draft of his *Synthetic Philosophy* was made in the beginning of 1858, a few months before Darwin published his first paper, and no essential change was made on account of the publication of the Darwinian theory. Whenever biological principles were needed for his sociology, Spencer adapted to his system the principles which had been suggested by Lamarck as early as 1800. Lamarck had held, (1) that external environment acts on living beings (in adapting this principle, Spencer was undoubtedly much fortified by Buckle's *History of Civilization*, which was published in 1856); (2) that living beings adapt their structures and functions to the external environment, and (3) that such acquired characteristics are inherited (a belief on the basis of which Maine and others defended the hereditary principle of the House of Lords). Darwin, on the other hand, did not believe in the doctrine of purposive adaptation to environment, but he did believe in accidental variations, and that those accidental variations which suited the environment were perpetuated by inheritance. Nevertheless, Spencer's sociological theory, based on struggle, became incorporated as an integral part of the popular understanding of the theory of evolution. There is even some justification for the view that Spencer was more responsible than Darwin himself for the "social Darwinism" which has come to represent the Darwinian theory in public opinion ever since. It is largely through Spencer's contributions that the extreme individualism of an age chiefly under the influence of Adam Smith and Bentham, and in revolt against governmental interference in economic affairs, fell into "social Darwinism." This strong tendency toward the laissez-faire doctrine which was dominant in the aristocratic intellectual atmosphere in which Spencer wrote, was reinforced by Spencer's strong abhorrence to actual government and its ways, a feeling which Spencer says he brought from his "dissenting family, antagonistic to arbitrary control." Thus Spencer's philosophy (and with it the philosophy of force), instead of being established on a scientific basis, had a strong bias of *a priori* conceptions of individual rights and laissez-faire doctrines from the beginning. As Barker says:

> He did not really approach politics through science, without preconceptions drawn from other sources, and with the sole idea of eliciting the political lessons which science might teach; on the contrary, he was already charged with political preconceptions when he approached science, and he sought to find in science examples or analogies to point a moral already drawn and adorn a tale whose plot was already sketched. . . .

The dominance of the teachings of Machiavelli, Bodin, and Hobbes in the aristocratic intellectual circles and among the ruling classes of all countries in Darwin's time contributed greatly to the success of the distorted application of his theory to human society. The new "social Darwinism" was seized upon with enthusiasm by all the men of violence because it permitted them to raise the basest instincts of greed and vandalism to the height of a universal law of nature. Since the feeblest must perish necessarily in the battle for existence, since this is the immutable principle of the living world, then the *vae victis* was of all that one could imagine the most rational and most legitimate course.

We can imagine the effect which this distorted social Darwinism would have upon a man of power like Bismarck. H. Lichtenberger, analysing his character, says:

> Bismarck had in a rare degree the love of force, the joy of exercising and expanding his power and that of his people. He constantly put into practice this "agonistic" conception of existence without remorse and without scruple, without pity for the feeble, and without generosity for the vanquished.

Men of Bismarck's type found in the new doctrines complete justification for their tendencies, a kind of superior sanction for a policy of blood and iron. Political theory in all Europe was based on the new "social Darwinism," and it was proclaimed that might always makes right. Bismarck was the leader of the school in Germany; in England, Chamberlain; in the United States and elsewhere the Imperialists proclaimed with the Iron Chancellor that force alone is noble, beautiful, and respectable. Banditism was raised upon a superb pedestal by the sovereigns, the ministers, and the statesmen with the instinct of conquest.

The historical events of the second half of the nineteenth century contributed greatly to the spread of the philosophy of force as a theory of international relations. Novikov has traced with fine insight the way in which the idea of the struggle for existence and the survival of the fittest was applied to nations under the influence of these historical events.

The development of the Darwinian ideas had been especially marked in Germany, where the first edition of *The Origin of Species* was published in 1860. As early as 1861, Darwin wrote, "my book seems to be exciting much attention in Germany, judging from the number of discussions sent me,"and his son, Francis Darwin, writes, "in a few years the voice of German science was to become one of the strongest of the advocates of evolution."

In the midst of all this discussion came the war of 1870, exerting a profound influence in popularizing the theory of "social Darwinism" as the arbiter of national destiny. It is easy to understand the effect of the war upon the victorious Prussians, and through them upon a large part of the German people, for Prussia had now gained the leadership in the newly-formed German Empire.

Intoxicated by their brilliant victories, they were easily converted to the adoration of brute force. They proclaimed on high that it took precedence of law. They found it entirely natural that it menaced the world. They claimed that the vanquished had no right to protest, that they ought simply to submit to their fate. All the benefits which came from the unity of the German states were ascribed to the victorious war. The great expansion of economic life, following the transition from an agricultural to a predominately industrial state which had set in in the previous decade, was also credited to the war and it was felt that the principle of natural selection could be directly observed at work in the German nation.

It is not surprising that victory should have contributed greatly to the success of the philosophy of force in Germany, but it seems paradoxical that the war of 1870 should have increased the popularity of "social Darwinism" also in *France!* It would seem that this country, having been defeated and subjected to a flagrant violation of its rights, ought to have found force hateful and justice admirable, but nevertheless it did not so happen. How can this apparent contradiction be explained? Novikov has unravelled the complicated causes as follows:

After the defeat of 1870 French public opinion might have followed either of two different directions. The French could have said: "We have suffered a hateful injustice; it is necessary therefore to do everything in our power to insure that such international deeds as this may never be repeated. We must attempt to suppress injustice; in other words we must work for an international union. Might is wrong; Right alone is beautiful. Down with Force; long live the Law!"

But another conclusion was also possible: "The military power of Prussia has inflicted upon us the deepest humiliation and the most cruel torment. If force had been upon our side it is we who would have tasted the sweets of triumph, and the Prussians who would have drained the dregs of defeat. Nothing is more useful than power. Down with Law! Long live Force!"

For many centuries France had been a formidable nation, belligerent, proud, and intoxicated with success. Twice she had exercised an incontestable dominance in Europe, under Louis XIV, and under Napoleon. France had used and abused force. She could not resign herself to defeat. From this we can trace the rise of a *revanche* party and the success of "social Darwinism" in France and we can understand also the growing disfavour which befell "Idealism"—that is to say the political philosophy of justice—in the years immediately following 1870. . . .

Among all the Western nations the unprecedented growth of modern Imperialism, which finds its scientific defence in the application of the Darwinian theory to the struggle between races, has given an immense impulse to the philosophy of force. The leading characteristic of international relations since 1870 has been the competition of rival empires. From 1870 to 1900 Great Britain added to its domains an area of 4,754,000 square miles, with an estimated population of 88,000,000—about forty times the area and double the population of the mother country. The close of the Franco-Prussian war marked the beginning of a new colonial policy for France, and a little later, for Germany, and this policy began to assume practical form after 1880. Since 1880 France has acquired an area of more than 3,500,000 square miles, almost all of it tropical or subtropical, with a native population of about 37,000,000. During the fifteen years following 1884, when Germany entered upon her Imperialist career with a policy of African protectorates and the annexation of Oceanic islands, she acquired an area of about 1,000,000 square miles, with an estimated population of 14,000,000. Almost the whole of her added territory was tropical, and the white population numbered only a few thousands,

Italy, Portugal, and Belgium entered directly into the competition of the new Imperialism between 1880 and 1884. Russia's expansionist policy, though more in the nature of a regular colonial policy of settlement for the purposes of agriculture and industry than the new Imperialism, comes definitely into competition with this in Asia, as in Persia and Manchuria, and has been assuming increasingly an Imperialist nature.

The annexation of Formosa by Japan following the victory over China, and of Korea following the Russo-Japanese war, showed that this rising and progressive Oriental Empire adopted Imperialism with the other characteristics of Occidental civilization.

The entrance of the United States of America upon an Imperialistic career by the annexation of Hawaii, and later, of the Philippines, marks the extension of the competition to the Western Hemisphere.

This unprecedented growth of Imperialism among all the great powers contributed powerfully to the spread of the philosophy of force. On the one hand, "social Darwinism" was enlisted to justify the methods of force which were used so extensively in this process of conquest and subjugation; on the other hand, the results of Imperialism were pointed to as the proofs of the process of the survival of the fittest and the inevitable dominance of the higher civilization, thus contributing to the spread of the pseudo-scientific doctrine.

Even sociologists have shown themselves eager in some cases to accept the philosophy of force as the sufficient justification of Imperialism, and to apply it to defend the necessity, the utility, and even the righteousness of continuing the physical struggle between races and types of civilization to the point of complete subjugation or extermination. Thus Professor Karl Pearson maintains that a constant struggle with other groups or races is demanded for the maintenance and progress of a race or nation. If you abate the necessity of struggle, the vigour of the race flags and perishes. It is to the real interest of a vigorous race, he says, to be

> kept up to a high pitch of external efficiency by contest, chiefly by way of war with inferior races, and with equal races by the struggle for trade routes and for the sources of raw material and of food supply. This is the natural history view of mankind, and I do not think you can in its main features subvert it.

By others, who take a wider, cosmic view, the argument has been put on the ground of "social efficiency." "Human progress," so runs the argument, "requires the maintenance of the race struggle, in which the weakest races shall go under, while the 'socially efficient' races survive and flourish; we are the socially efficient race; therefore our nation must take up the 'white man's burden' and enter upon an Imperialistic career." The principle of social efficiency is described as being "as indisputable as the law of gravitation" by Edmund Demolins, who enunciates it as follows:

> When one race shows itself superior to another in the various externals of domestic life, it *inevitably* in the long run gets the upper hand in public life and establishes its predominance. Whether this predominance is asserted by peaceable means or feats of arms, it is none the less, when the proper time comes, officially established, and afterwards unreservedly acknowledged. I have said that this law is the only thing which accounts for the history of the human race and the revolutions of empires, and that, moreover, it explains and justifies the appropriation by Europeans of territories in Asia, Africa, and Oceania, and the whole of our colonial development.

The gospel of Imperialism, as embodied in the career of Hubert Hervey of the British South African Chartered Company, has been summed up by his fellow-adventurer, Earl Grey, as follows:

> Probably every one would agree that an Englishman would be right in considering his way of looking at the world and at life better than that of the Maori or Hottentot, and no one will object in the abstract to England doing her best to impose her better and higher view on those savages. In so far as an Englishman differs in essentials from a Swede or Belgian, he believes that he represents a more perfectly developed standard of general excellence. Yes, and even those nations nearest to us in mind and sentiment—German and Scandinavian—we regard on the whole as not so excellent as ourselves, comparing their typical characteristics with ours. Were this not so, our energies would be directed to becoming what they are. Without doing this, however, we may well endeavour to pick out their best qualities and add them to ours, believing that our compound will be superior to the foreign stock.
>
> It is the mark of an independent nation that it should feel thus. How far such a feeling is, in any particular case, justified, history alone decides. But it is essential that each claimant for the first place should put forward his whole energy to prove his right. This is the moral justification for international strife and for war, and a great change must come over the world and over men's minds before there can be any question of everlasting universal peace, or the settlement of all international differences by arbitration. More especially must the difficulty caused by the absence of a generally recognized standard of justice be felt in the case of contact between civilized and uncivilized races. Is there any likelihood of the gulf between the white and the black man being bridged within any period of time that we can foresee? Can there be any doubt that the white man must, and will, impose his superior civilization on the coloured races? The rivalry of the principal European countries in extending their influence over other continents should lead naturally to the evolution of the highest attainable type of government of subject races by the superior qualities of their rulers.

This is an excellent statement of the scientific basis of Imperialism, including in its survey the physical struggle between white races, the subjugation of lower races by the white race, the necessity and the utility of this struggle and this subjugation, and finally the right of domination based upon this necessity. The white man believes he is a more excellent type than any other man; he believes he is better able to assimilate any special virtues others may have; he believes that this character gives him a right to rule which no other can possess. Thus, starting from natural history, the doctrine soon takes on the outer garments of ethical and even religious sanctions, and we soon reach the elevated atmosphere of "Imperial Christianity," the "mission of civilization," in which our nation is called upon to teach the "arts of good government," the "dignity of labour." And not only our nation; Mr. Hervey admits that the patriotic Frenchman, the German, the Russian, feels in the same way his own sense of superiority, and the rights it confers on him. So much the better, he says, agreeing with Professor Pearson, for this cross-conviction and these cross-interests insure the survival and the gradual perfection of the fittest through international strife and war. . . .

The primary error of those who have distorted Darwin's theory beyond all recognition is one of stupendous magnitude. It consists in ignoring completely the existence of the physical universe! The cause of progress is assumed to be, not the struggle of man with his environment, from which he gets food, clothing, shelter, and all other necessities, but the struggle of man with man, a struggle which is by its nature unproductive and fruitless.

The infinite error involved in forgetting entirely the existence of the physical universe is due, first, to a common defect of the human mind, which tends to overlook the most commonplace and obvious facts of existence in favour of the unusual and abnormal; and second, to a misunderstanding of the terms "struggle for existence" and "survival of the fittest" as used by Darwin.

Those who believe in the distorted form of "social Darwinism," obsessed by the idea of struggle, forget entirely the greatest struggle of all, the struggle of man against his physical environment, because it is so common and so omnipresent. . . .

It is difficult to imagine a more colossal error than is committed by the philosophy of force when it disregards all this infinite struggle of man against his physical environment and concentrates all its attention upon the struggle of man against man. The biological error involved is like that which was committed by the old political economy, which considered solely the secondary phenomena of exchange between men, and left out of account all considerations of the primary phenomena of production and of the adaptation of the environment to man's needs. . . .

The second error of the philosophy of force is that in which struggle is confused with the extermination of fellow-creatures. A typical example of this confusion may be quoted from Spencer who maintains that without the collective homicides of the past ten thousand years the world would still be inhabited only by cavemen of a feeble type:

> . . . to the unceasing warfare between species is mainly due both growth and organization. Without universal conflict there would have been no development of the active powers. . . . Among predatory animals death by starvation, and among animals preyed upon death by destruction, has carried off the least-favourably modified individuals and varieties. Every advance in strength, speed, agility, or sagacity, in creatures of the one class, has necessitated a corresponding advance in creatures of the other class; and without the never-ending efforts to catch and to escape, with loss of life as the penalty of failure, the progress of neither could have been achieved. . . .
>
> Similarly with social organisms. We must recognize the truth that the struggle for existence between societies has been instrumental to their evolution.

The first objection which presents itself is a biological one. It is strange that Spencer did not realize that his argument was inconsistent. The unceasing warfare between species, he says, is the cause of both growth and organization, that is, of the appearance of more perfect types. Since the Paleozoic Age all the species, without exception, have been subjected to the pressure of the struggle for existence. Why is it then that certain species have evolved to a being as high as man, while others have remained at a more rudimentary stage of life? The struggle for existence cannot be the sole cause of the evolution of species. There must be other causes which we do not know.

From the point of view of sociology, however, it is not this biological objection which is of the greatest interest. It is the immense leap which Spencer makes in applying the definite analogy to human society. He speaks of the struggle between animals and then, without any transition or explanation, says: "Similarly with social organisms." It is astonishing to find an eminent philosopher making such an elementary error. This is an example of the kind of errors of which Darwin complains in one of his letters, in which he said: "How curious it is that several of my reviewers should advance such wild arguments . . . and should bring up the old exploded doctrine of definite analogies. . . ."

In this case Spencer has not exactly forgotten the existence of the universe but he has disregarded one of the most widespread facts which can be observed in it,—that living beings exist in it in relationships of a most extraordinary complexity, ranging from the most irreducible antagonism to the most absolute solidarity. To jump from the conclusion that since certain relations are established between animals of different species, the same relations ought to be found without any modification whatsoever between societies of the human race, is to make an assumption which is not supported by science or reason. Spencer makes two chief conclusions which render his comparison entirely false.

1. He compares the struggle between individuals of *different* species with the struggle between individuals of the same species;

2. He compares the struggle between individuals to the struggle between collectivities. . . .

If Spencer had wished to compare the battles of animals with the battles of human beings, he should have compared the combats of men with the combats between animals of the same species. The combat between a tiger and a bull is not comparable with a duel between two men, because the tiger and the bull are individuals of two species which are not associable and are indeed naturally antagonistic, while the duellists are individuals of the same species. The comparison would have been more exact if Spencer had compared the relations between men with the relations between tigers. In this case he would have been dealing with animals, not indeed associated in any permanent fashion, but at least of the same species. But as soon as we approach the concrete reality we see that tigers do not eat each other and that the relations between individuals of the same species are not the same as those between individuals of different species. Since animals of the same species do not massacre each other, Spencer has no justification from the biological analogy for affirming that human progress would have been impossible if men had not exterminated each other.

A closer analogy would be to compare the relations between men with the relations between individuals of a species capable of association such as bees, apes, beavers, monkeys, etc. But when we come to the real analogy the scene changes entirely. Not only do we see that the individuals of species capable of association do not devour each other, but on the contrary they unite for common work; they exchange services, and as a result they create the group of a higher degree of evolution which is called a society. . . .

Darwinism and Continental Socialism

Oscar Schmidt

I must assume it to be generally known that in last year's Congress of German Naturalists and Physicians, held at Munich, a prominent member incidentally referred to the points of contact between Socialist Democracy and Darwinism, as also to the momentous and redoubtable consequences which might thence ensue. These words of certainly well-meant admonition

---

Source: Oscar Schmidt, "Science and Socialism," *Popular Science Monthly* 14 (1879), 577-591. Translated from the *Deutsche Rundschau* by J. Fitzgerald.

were received with delight by all those who in any event can not tolerate the doctrine of descent, and who accordingly heartily approved of making Darwinism responsible for the most exciting social phenomenon of the time.

It is, of course, all right enough if certain representatives of Socialist Democracy think they can with the aid of Darwinism add force to their opinions; but they jumble together doctrines which either are irrelevant, or which mutually exclude one another.

This fact, indeed, is recognized by another portion of the Socialist-Democratic party, who hold that the socialistic idea must have supplanted the Darwinian principle as applied to the human race, before the new form of society can be realized and made to stand.

The political economists have now for more than a century been studying the "Struggle for Life" in its bearings on the weal or woe of mankind, yet not until the advent of Darwin did they consider the problem understandingly. Under what forms individuals and classes compete with one another; in what way this struggle is to be ennobled for the benefit of the race—these and other like questions are agitated on all sides, as witness one work among many, namely, A. Lange's thoughtful book, "On the Labor Question" ("Ueber die Arbeiterfrage"). It is not, therefore, with this well-known point of contact with Darwinism that we have to do, but with the special application of ostensibly Darwinian results to the justification and the execution of the Socialist-Democratic programme.

Although it has been raining "Quintessences of Socialism" for the instruction of the public, nevertheless we must briefly explain how far Socialist Democracy, as realized in the future, purports to be the ultimate term of a natural development.

Having passed through the period of absolute Inculture, a period which might be roughly characterized by community life in troops and minor family groups under the leadership of strong male individuals, traces of which are found in the mammoth and reindeer caves, man next entered the less rude condition of the hunter and the nomad wherein those minor groups, now developed into clans, rose above the primordial state by their complex organization, their division of labor, and their larger enterprise. But now, as we read in the writings of the Socialist Democrats, with distinction of classes and the institution of personal slavery opens the first epoch of human civilization. In this slavery period the whole man is an instrument of labor, and with all his faculties he is the property of the owner who commands his entire service.

In the second civilization period, characterized by handicraft and the possession of land on the small scale, the still widely diffused custom of socage or compulsory service is a reminiscence of the suppressed institution of slavery. The working people are now burgesses enjoying personal freedom, yet hindered in their spontaneous development by the guild. These craftsmen and workers of the soil are the owners and masters of the implements of their calling. Still they do not attain true enjoyment of life, inasmuch as their whole time is engrossed with mechanical toil, in order to procure a livelihood.

The third estate becomes emancipated from the control of the aristocracy of landlordism, the nobility, and the Church. But already the factory system, the system of production by division of labor, was in operation—a system which strips the workman of his little property, and places him under the control of capital.

The factory system is transformed into machine industry. The simpler tools and machinery of early times, worked by hand, have been in the present century developed into a vast complicity of machines producing motion, transferring motion, and doing the work of mechanical tools. These machines tie the workman fast to themselves, reduce his personal service to a minimum of manipulations, and complete the physical and moral wretchedness of the laboring population by increased employment of women and children.

Machinery is an instrument for the accumulation of capital, and of capital as "unpaid labor" at that.

The development of the various forms of production down to the present time when private capital rules everything, and when small capitalists have no chance as against the great, has been vividly portrayed by Marx. He regards the development of the economic social forms as a natural-history process. ("Capital," page 7.) The proposition on which the whole matter hinges, namely, that regarding "unpaid labor," has the appearance of being so simple and so true, that even to many among the laboring class the grand conclusion appears evident, namely, that "the doom of capitalistic private property is about to be sealed. The expropriators will suffer expropriation." (*Ibid.*, page 793.)

Here we see set forth as the natural consequence of development the violent introduction of the Socialist-Democratic state. With the rise of that form of polity the domination of private capital comes to an end; machinery is no longer private property, but it, with all the other instruments of production, with all that which is understood by the term capital, is transformed into collective or state capital. Private production gives place to collective production. With the abolition, not indeed of all private property, but of all private capital in so far as the same is employed as a means of production, universal participation in the fruits of the collective production, in the means of enjoyment, and in the higher good things of life, is accorded to the whole race.

In this stage of future development the Socialist idea is fully realized. We will not dwell any further on the happy state of things which they say is sure to come about.

If any one would learn more concerning this "picture of fancy," as Socialism is innocently named in one of its chief organs, he may consult

Leopold Jacoby's book, "The Idea of Development" ("Die Idee der Entwickelung," 1874, page 6), where he will find it painted in glowing colors. Or he can get a notion of what it is from Engels's most recent utterances (Eug. Duhring's "Science revolutionized"—"Die Umwalzung der Wissenschaft," 1878).

It would be a great mistake to suppose that in the camp of the Socialist Democrats we must find the state of the future represented as a refined copy of a type belonging to the brute creation: on the contrary, Socialistic writers stoutly maintain that the principle of development implies reconstruction. Nevertheless, for the sake of completeness, we will glance at the animal world and consider there the relations between private and collective capital and private and collective production.

Most animals labor for themselves alone. Their implements of labor (private capital) are represented by their members and their weapons of offense; their gains they employ to sustain life. They gather not into barns. Only among the higher classes of animals do we find association of labor and care on the part of the parents for the welfare of their posterity instincts which can be regarded as confirmed and inherited habits, for all these instinctive actions resolve themselves into labor with gradually accumulated, inherited private capital.

The consociation of higher animals may have the look of work in common, and not instinctive, as for instance when wolves hunt in packs to obtain food, or when animals graze in herds—a habit originating in the need of mutual protection. The colony structures of beavers, the massed nests of the republican bird, are socialistic improvements, in the case of the beaver made under unfavorable outward conditions, and hence *apparently* the result of the animal's voluntary action.

This deceptive appearance is wanting in the structures of the inferior animals, where societies are formed through propagation by budding, the progeny remaining connected with the mother organism. The single individual of a colony of polyps enjoys not only protection against mechanical injuries in the polypidom secreted by all for all, but further, in case its local position in the polypidom is unfavorable for taking in food, it is fed by the collective alimentary canal, into which flows the surplus of the individual production. A still more complicated social condition, with strict division of labor, is seen in the jellyfish known as the "Portuguese man-of-war."

I call attention to these familiar facts in order to show that in the animal kingdom communism and socialism are all the more pronounced the lower the organization of the groups among which they appear; and that, on the other hand, wherever among the higher animals conditions occur which savor of the socialistic principle, in the division of the results of the collective production, the egoism of the individual appears all the stronger. I do not at all mean hence to conclude that the case can not be otherwise in human society.

From the selflessness of the polyp to the egoism of the wolf is a development. How this development has been brought about, and how man must come under its action, Darwin teaches; and Leopold Jacoby tells us that the already quoted gospel of the Social Democracy, namely, Marx's work on "Capital," is "a continuation and complement of Darwin's 'Origin of Species' and his 'Descent of Man.'"

This same opinion was expressed one year earlier in "The People's State" ("Der Volksstaat," 1873, No. 31), and it is now our task to examine into its ground.

The only passage in which Marx himself speaks of any complement of Darwinism, though not in connection with his own researches, but *à propos* of the need of a special history of technology, is where he says: "Darwin has drawn attention to the history of natural technology, that is, the formation of plant and animal organs as production-instruments for the life of animals and plants. Should not equal attention be given to the history of the formation of the production-organs of man in society, seeing that these organs are the material basis of every special society organization?" ("Capital," page 385.)

But the scientific method, that of proving by facts the relations between phenomena, is employed by Marx; and in my opinion he is in the right when he protests against the supposition that his dialectic method is at bottom Hegelian. But neither does he collect all the facts—for example, he knows only on the one hand the oppressor and the extortioner, and on the other their victims reduced to misery—nor does he refrain from gratuitous assumptions, as for instance that of the "Unpaid Labor," the most momentous of them all. Again, it happens that, from his not understanding the results of the doctrine of development, the true and actual relations of sociology to Darwinism are hidden from him. For example, he says that "in reality each special historic mode of production has its own special and historically valid laws of population. An abstract law of population holds for plants and animals only in so far as man does not interfere."

Engels, in the work already quoted (page 491), repeats a similar proposition; but it is not correct, if we use the term "law" in the one sense permitted by exact natural science. The conditions of propagation among plants and animals, the results of their multiplication, vary according to circumstances; and man with his experiments in breeding does not correct Nature, he only copies her. It is plainly out of the question here to speak of laws of population; it were better instead to say that the conditions of population in each period are the effects of special, variable causes peculiar to each stage of development. Cases which are the result of varying circumstances and events are not laws, nor do they justify us in inferring fixed laws.

The attempt has been made, not indeed by Marx himself, but by one of his followers, Leopold Jacoby, to connect logically, in one continuous process, social evolution and its ultimate term, the Socialist-Democratic ideal, with nature's evolution.

He does the impossible with a sophistical argumentation that reminds us of Hegel's dialectic: he is an enthusiast, but it is not for me to pass judgment on his services to Socialist Democracy. Plainly he is an *enfant terrible* for his party. His scientific ideas are of the narrowest kind. Nevertheless, we must reckon with him, since he is the only Socialist-Democrat writer who makes any pretense to observe scientific method in this matter, i.e., the connection between the theory of development and Socialist Democracy. We will later consider Engels's relation to that subject.

Social evolution is nowadays represented by the leaders of Socialist Democracy as being a process of perfectionment necessarily progressing toward a definite end; and as, rightly enough, they do not divorce man from nature, it is plainly their purpose to discover oneness and continuity in social and in natural evolution.

Revolution, say the Social Democrats, is correction of perverted conditions or re-formation for the sake of improvement. Copernicus happily expressed this idea when he gave to his work which upset the astronomical notions of his time the title "De Revolutionibus." It is of no consequence whatever that this is not the true title of the book, but "De Orbium Coelestium Revolutionibus," or that "revolutio" means a turning round and not an overturning.

In short, in these revolutions, as the Socialist-Democratic philosophy further teaches, "we recognize an ever self-perfecting origination and formation of things in the universe": so much we learn from Kant and Laplace. Then came Lamarck with the "doctrine of the continuous and successive development of organic beings on the earth," but for half a century he failed to obtain a hearing till Darwin procured for the doctrine full acceptance. Thus we have to thank Lamarck and Darwin for the fact that we understand the nature of the two great "revolutions," whereof the one produced the existence of organisms in the transition of the inorganic into the organic, while the idea of the other had for its object the appearance of man.

Thus the philosophy of Social Democracy resembles that of Plato, in teaching that the *idea* hovers over the bodily form. It is the *idea* that dominates everywhere, determining the forms of all things. Hence, down to the advent of man there was in nature a steady, thorough, unconscious striving.

And here comes in another great revolution, the third—viz., the universal apparition of the consciousness of the human race, the philosophical establishment of which is the task which Marx has set himself. Thus, then, as we read in the "Volksstaat" (*ubi supra*), "Darwin and Marx have, by their profound and ingenious researches, carried on in totally different scientific fields, attained results of the utmost importance to mankind, and which, being intimately correlated, mutually support and complement one another."

As Social Democracy holds the accomplishment of its ends to be inevitable and necessary, so, as we have seen, it is one of its cardinal principles that all the phenomena which take place in matter, and all developments of matter, are prefigured and predetermined in the idea. The honest workingman thus learns that a statue is less successful, the less conscious the sculptor was of the idea of the work of beauty inherent in the marble block, and the more he suffered himself during his labor to be influenced by considerations of profit and the like. "In this example of the sculptor and his work, we have," says Jacoby, "a direct proof (direct proof!) of the truth of the proposition that ideas are contained in unconscious nature."

I shall be asked whether the utterances just recounted represent the sympathy between Socialist Democracy and Darwinism—whether these abstract and rather curious and confused theses and propositions represent the dangerous elements, that is, the politically dangerous elements imported into Socialist Democracy from the development theory.

With a few additions, which we will make further on, they do!

How, then, does Darwinism stand with respect to these cardinal ideas of the socialistic development doctrine, as laid down by a philosopher of that school?

We find here two ideas wherein Socialist Democracy purports to be at one with the scientific doctrine of Evolution, viz., Development or Revolution is re-formation—i.e., correction of perverted conditions; and All development has for its basis an idea which designates the future goal, and which governs the movement toward the same. In themselves these propositions are plainly innocent enough, and if they were a result of Darwinian research, they need not be disowned. But Darwinism disclaims the honor of having established any such principles.

What we call *origin*, or *development of species*, is in the first place not a reversal of perverted conditions. In such play upon words we have never indulged. The Darwinian principle of development is Natural Selection, and people are not wont to select from perverted types. It is true that the struggle amid which selection goes on includes also the struggle against wrong, when it is waged consciously, but generally it is a *struggle against environment.*

We have first to take into consideration the downfall of the one that struggles. But nature, or if you choose the law of nature, recognizes here no distinction of right and wrong: the question is purely one of might. That one is defeated who possesses the least means, the least amount of fight-capital, however sufficient and abundant the same might be under circumstances different from those here and now prevailing. Certainly no scientific man has ever dreamed of subsuming this case under the Hegelian phrases evil—good, negation—position, perversion—correction, etc.

The opposite of this first instance of the outcome of the struggle for existence is seen where one party, by a process of gradual perfectionment, prevails over its opponents and the environment. To the philosopher who is searching for analogies, this appears to be the practical fulfillment of the idea of perfectionment. Still, these two extremes do by no means exhaust every possible termination of the struggle; for there is another possible issue —one which, though it be overlooked by the Socialist-Democrat philosophers, is nevertheless of enormously frequent occurrence: *the organism that makes the struggle adapts itself to the environment.* In doing this, it must oftentimes pass through such straits that it parts with some of its perfections and falls to a lower grade, like many a European baron who has in America found use for himself as a cook's assistant. Or it so remolds itself and its habits in adapting itself to the environment that, while it in no wise becomes more perfect, it nevertheless, as far as possible, insures for the future its present rank.

Thus, to illustrate by an example, it has been observed that, as a rule, birds of brilliant plumage, which on that very account are more conspicuous objects to their enemies, are far more careful to conceal their nests than birds which are not so conspicuous. This we explain on the theory that the ancestors of the bright-colored species by degrees became wise by experience, and that this experience, reenforced by habit, was transmitted to their progeny by heredity. Natural selection keeps pace with experience and habit. In the case of these birds, the change in nesting is a step of progress, but they have not thereby gained any perfection.

If in the historic evolution of organic nature we saw progress only, we should be strongly tempted to regard progress, pure and simple, as a universal natural law for social development as well. But the lesson taught us by birds of brilliant plumage (to say nothing of the loss of acquired perfection) is repeated throughout the whole world of lowly and lowest organisms. These have stood stock-still and must so remain, despite the perfection attained in many directions. The persistence of the low and the imperfect finds its very simple explanation in the persistence of its universally prevailing life-conditions. Millions and millions of lowly species have perished one after another, giving way to better, i.e., to stronger; and many millions whose slightly variant ancestors escaped from their enemies and rivals have survived. A sea peopled only by fishes, or the land only by mammals, is something unnatural. Thus, the imperfect endures; the perfect, the homogeneous, left to itself alone, becomes self-destructive, as the New-Zealanders began to prey on each other so soon as they had exterminated their only edible wild animal, the dinornis.

In short, Darwinism shows that the evolution of organic life is not to be summed up in a few abstract formulae. It calls attention to the fact that with the gradual succession of species goes on, *with other movements*, a slow perfectionment in different directions; and this necessary but yet only *partial* progress it seeks to explain by "selection of the fittest" in the decay *or* the backwardness of less gifted individuals and species.

In the whole system of Darwinism we find, unfortunately, no hint of a law which shall in advance determine this perfectionment and it is laughable to observe how, on the one side, people are complaining that we set up

Chance on a throne as a universal principle, while, on the other side, people are making the discovery that the guiding principles of the Socialist Democrats, those principles which are shaping the future, are corollaries of the theory of development.

The strength of the Socialist-Democratic teaching lies in this, that the candidates and members of the party, men unpracticed in logical thinking, are sternly schooled in a few principles, and taught to regard the actualization of *the idea* as something necessary *ex necessitate rei.* The fundamental mistake of supposing that Social Democracy has any point of contact with Darwinism arises, as we have explained, out of the supposition that Darwinism too has brought to light ideas which govern organic transformation—such ideas as are supposed to be necessary for calling forth social revolutions.

The Socialist Democrats do away, after their own fashion, with the personal Godhead, and for this, of course, "atheistic science" is held to answer. In L. Jacoby's work we read:

> We call the idea the foreknown existence of the embodied total result of a progressive re-formation. But this foreknowledge can exist in no other way save this, that the thing to which we grant the idea has itself been carried into the progressive re-formation; and from this knowledge follows the other side of the essence of the idea, namely, the being dominated by the idea, the being forcibly moved by the idea in a given direction. If to these first organisms we allow the idea of Man, we *ipso facto* recognize their domination by this idea; in other words, we see how *they have been constrained so to transform themselves as finally to produce from themselves man.*

This thought, stripped of the studied verbiage in which it is here invested, has very generally become rooted in the minds of the Socialist Democrats. In this way they have set up, in the place of a personal God, a sort of infallible bugbear under the guise of an Omnipotent Idea. The whole thing is misty, mystic, supernatural, in no sense scientific, least of all is it a Darwinian explication of facts.

Darwinism holds the exact opposite of all this, maintaining that development does *not* proceed according to ideas. Darwinism sees in nature only forces, laws, causes, and effects. Ideas it must for the present leave to the philosophers; and, moreover, it has absolutely no points of contact with the doctrine of ideas contained in the Socialist-Democrat catechism. Hence, when the Socialist Democracy bases the realization of its ideal on the fact that men who are conscious of the impelling ideas must irresistibly push on the work, and so carries the masses on into a belief in these ideas, it must itself be held responsible. *As for Darwinism, it gives no encouragement to such imaginings*, and hence must, in this respect, be simply indifferent for all, whether they hate Social Democracy or whether they love it.

But there is a category of scientific men who look on the origin of species as a development of the higher out of the lower; who find the Darwinian principle insufficient; who will have nothing unaccounted for, and who therefore conceal their ignorance under such phrases as "tendency to perfection" or "aiming at an end." I might also refer to the "Philosophy of the Unconscious"—now, as I believe, in process of decay—a philosophy which, whenever it knows not how to explain anything, solemnly invokes the aid of its "Unconscious." Between these muddled auxiliaries and the Socialist Democrat's "ideas which govern revolutions and determine the reformation of perverted condition," there exists an unmistakable though perhaps "unconscious" relationship.

There is, then, a point of contact between Socialist Democracy and Darwinism; but, as far as we have examined it, it is seen to rest on erroneous suppositions and ignorance of the essence of the development doctrine. So far we have found it concerned only with a few theoretical propositions; and we have had nothing to say about the practical realization of the Socialistic idea, or of the doctrines which might perhaps be borrowed from Darwinism to add to it strength.

The Socialist Democrats are unanimous in expressing discontent with the social conditions at present existing. But with respect to the specific organization of society in the future their leaders are very reticent. So much is certain, that the great mass of the workingmen, who now have to sell their entire strength for wages that merely suffice to support life, will in the future perform no so-called "unpaid labor." They will have a share in those higher enjoyments the prerequisite condition of which is a higher mental development. Opportunity for attaining this is afforded every one in the Socialist-Democratic state, by a considerable shortening of the hours of purely mechanical toil, and by opening perfectly free schools of every grade. When the whole population has been in this way refined, there is no longer question of the "rude physical struggle for existence." For since in the time to come each individual will develop his reason, and reason can not endure perversion or wrong, each will labor for all and all for each. Whether, instead of the present division of labor, there will be an arrangement having the same effect, but based on personal inclination and personal fitness ("Zukunft," 1878, page 704); or whether the system of alternation of work will be introduced, so that, as Engels would have it (*ubi supra*, page 173), the man who in the forenoon wheels a barrow is an architect in the afternoon; what is to be done when no workmen offer themselves for certain kinds of labor, while for other kinds there are too many; whether and how the workman is to be entitled to a certain measure of the means of enjoyment according to the nature of his work, or the exertions he puts forth, or his individual needs: these are all open questions. The amicable settlement of these questions presupposes, as we have already observed, a general physical and moral elevation of the individual, of which the present mean intellectual status is but a faint foretaste.

Thus Socialist Democracy demands a more equitable general division of the good things of life, that every one shall have a share in life's pleasures, and it aims to demonstrate the justice and the naturalness of the claim. We are interested in it, inasmuch as appeal is made to Darwinism for its establishment.

The revolution which goes on quietly in Christendom has appealed to the equality which exists beyond the grave. The Revolution of 1789 was more practical, and championed the natural right of the Third Estate—the right of equality and fraternity here on earth. What it was that gave this direction to the revolution, and how it expounded the return to nature and to the truth of which so much was then said, is sufficiently well known. Now, again, as the Socialist Democracy is not alone in affirming, and truly enough too perhaps, we stand on the threshold of a social overturning. Again there is question of an appeal to be made to the inborn rights of man. Next after the basic idea controlling revolution, of which we have already spoken, the Socialistic programme, like the programme of all revolutions which proceed from a general state of distress, insists on the restoration of that equality which is the strictly natural right of every man, but which has been lost, owing to unnatural and perverted social conditions.

The Socialist Democrats are not content with Brentano's view, that humanity's goal is perfection, nor with Held's, that the ideal end of progress is the highest perfectionment of all. Hence they have enlarged the idea of perfectionment by making it include a reduction of labor and an increase of bodily and intellectual enjoyment. That, in the event of a refusal to recognize the justice of this demand, force would be employed to complete its demonstration, is expressed without ambiguity. But it is easy to see how welcome it would be to the leaders of the movement could biology be ranged on their side, and could Darwinism, "carried out to its logical conclusions," be inscribed on their banner.

If the Socialist-Democratic doctrines had any organic relationship to the anthropological side of Darwinism, science would find its way into it. It would ill befit Science to complain of this. In fact, the Socialist Democrats believe that such alliance has already been effected between science and their philosophy, and it will do no harm to consider the situation, though as conceived by the Socialist Democrats it implies a fundamental misunderstanding.

In the "Volksstaat" (*ubi supra*) we read:

> The Darwinian theory is an important support for Socialism: it is, so to speak, unconsciously its sanction on the part of natural science. For, after all, what is the principal result or the practical meaning of the Darwinian doctrine? Surely, along with a profound insight into the workings of organic nature, it means *the explicit recognition of the doctrine of equality between all men*. If, etc. . . . then surely we may well preach Socialism, inasmuch as every one knows that each individual is a product evolved by nature, and hence having the same claims on nature.

Then the conclusion is drawn that, inasmuch as the reactionaries will not accept the descent of man, they do all they can to prevent the recognition of Darwinism as a support of Socialist Democracy, and to check its diffusion among the people.

How the Socialist Democrats picture to themselves the equality of all men, and first of all the equal natural condition of all men, we see in Jacoby: "Man is good from the beginning"; "The brain of each individual man is capable of being developed so that it shall of itself do all thinking, just as the hand of each individual man is capable of being developed so that it shall do everything with the aid of machinery." That hitherto we have seen only *capacity* for equal development, while *in fact* there exists great inequality of development, is due to the fact that only those who enjoy unnatural privileges have the time requisite for the development of the consciousness. When men shall once have been properly brought up to equality in the Socialistic state, then equal development, with bias toward the good, will come of itself, for "*knowledge of nature compels us to regard all men as beings capable of development in precisely the same measure.*"

But by the term "all men" is to be understood the male portion of the race, for many Socialist Democrats agree with high authorities in holding that woman, by reason of her abnormal brain-structure, must, in the state of the future, act a subordinate part: judgment, action, are not for her, but only feeling, and the faculty of order.

In all this it were difficult to find a single trait that can be referred to the Darwinian anthropology. The Socialist's "aspiration toward perfection" is associated with his ideal of the equality of mankind. Now, *this illusion Darwinism utterly demolishes.* The very principle of development negatives the principle of equality. So far does Darwinism go in denying equality, that even where in idea we should have equality, Darwinism pronounces its realization an impossibility. *Darwinism is the scientific establishment of inequality*, and hence the assertion that the Darwinian doctrine is above all a recognition of the doctrine of the equality of all men needs no refutation from our side: it has no foundation in fact.

Again, nowhere in the literature of Darwinism do we find the axioms that "every man is from the beginning good," or that "all men are equal in their capacity for development."

As to what the Darwinians think, let me quote from my book, "Darwinism and the Doctrine of Descent": "The grade to which this (intellectual) development rises is generally dependent on the preceding generations. The psychical capacities of each individual bear the family type, and are determined by the laws of heredity. *For it is simply untrue that, independently of color and descent, each man, under conditions otherwise alike, may attain a like pitch of mental development*" (page 296).

Had it not been that we are held answerable for these ideas of the Socialist Democrats, we should never have esteemed them worthy of notice.

The Socialist Democrats anticipate, when their state shall have been founded, the universal contentment of all men, who shall labor partly out of personal inclination, partly by state ordinance. For this, good men will surely be needed, for one year after the proclamation of equality, the "Volksstaat" (1874, No. 30) demands that "the strong and the weak, the bright and the dull, force of mind and force of body, in so far as they are human, shall in a partnership such as befits human beings be associated in labor, and associated in the enjoyment of its fruits."

Here we must consider that fraction of the Socialist Democrats who with Engels (*ubi supra*, pages 223 *sqq*., and especially page 235) imagine that the inequality which man inherits from his brute origin, *an inequality that can not be done away*, will be paralyzed under the new social order. As we have seen, some of the Socialists deduce the inequality of human individuals from the unnaturalness of the old form of social organization; they not only maintain a vague idea of equality, but they also expect to see an equal development of individuals, though strong and weak, bright and dull, still continue.

On the other hand, Engels calls the advanced advocates of equality "ghosts," and the demand of the proletariat for an equality beyond the abolition of class, an "absurdity" (page 84); at the same time he is confident that the struggle for existence will cease on the abolition of class distinctions, and will give place to universal mutual good will. This would require that individuals should disregard all actually existent inequalities, whether mental or bodily. Plainly there is no Darwinism here either, and we leave it to others to contest this conversion of the doctrine of development, a doctrine grounded on observation, into a fiction of the imagination, for we have to do, not with Socialist Democracy as such, but with its relation to Darwinism.

The result of our investigation is, that Socialist Democracy, wherever it appeals to Darwinism, has failed to understand that hypothesis; that, if it has understood it, it knows not how to draw from it any advantage for itself; and that it must deny the unalterable principle of Darwinism, namely, competition.

Such is an account of our relations to Socialist Democracy, a movement whose gravity we look on as a sign of a diseased social state that calls for help and salvation.

It remains to define our position with respect to the views of a few of the friends and counselors of the Socialistic movement who, approaching more nearly to the Darwinian point of view, look for the best results for human progress to result from natural selection *after* present social ills have been cured. I refer in particular to Albert Lange. That so eminent a student of human life should estimate at its true value the struggle for existence which has come down to humanity from the unconscious animal world, was to have been expected. He well knew how little warrant there is for the expectation that the "struggle for the more desirable position" will ever cease. But he based his hope on the idea of liberty and equality, an idea that is slowly developed with the developing reason, and which brings men together, in spite of differences of race, or talents, or station. He hoped that the laws of the conscious intelligence would, as time went on, gain the mastery. He hoped for a deliverance to come in the very remote future from a current of thought and feeling which would arise in the developed human mind, and which would run counter to the natural process of differentiation and division. He hoped for a spiritualization of the physical struggle into a peaceful contest, having for its object the good of the race. It is therefore nothing new if in these days like views are put forth by Socialist Democrats.

In his work "The Labor Question," Lange has intimated that certain social evils are the results of artificial selection, and that these might be corrected by a return to simpler natural conditions. If we were to spin out this thought, as is done, for instance, by Dodel in his "Neuere Schöpfungsgeschichte" (1875, pages 145, 147), we might readily persuade ourselves that under the social conditions now existing the principle of natural selection, indeed any purely natural development, "comes into operation either not at all, or only to a limited extent." Then it seems to be an infraction of the natural order, that they who are born to station, so often, without personal worth or talents, monopolize, in virtue of their inherited wealth, the pleasures and enjoyments of life, leaving for their descendants the same even path. Do these favorites of Fortune, it is asked, who take no part in the struggle for existence, constitute the portion of society which is most highly endowed by nature? Ought we to foster such a class for generations to come? In fact, can it be that its continued and prosperous existence has the highest justification?

Out of this preposterous condition of things, where the salutary principle of natural selection is borne down by artificial selection, our one hope of deliverance, we are told, is in the coming of a time "when all the millions who day by day come into existence shall enjoy equal rights of development, so that each individual favored of Fortune, be his birthplace a hovel or a palace, each one endowed with talent or genius, shall find ready prepared for him all the means requisite for developing his natural powers in proportion to their value, and for afterward employing the same for the common good."

I can not accept as correct this explanation of natural and artificial selection. Each individual has, throughout the whole course of historical development, fortified his existence by all the means at his command, with property, with inherited station, by putting forth all his powers of body or of mind, inherited and personal. Artificial selection has a definite end in view: it aims at transforming for a special purpose that which is offered by nature, and then maintaining the new form for the same end. When the nobility, the great landholder class, maintains its position and becomes rooted, we have not an instance of artificial selection in the Darwinian sense, but it is the *natural course of things*, however unnatural the result may seem to be. If this be not admitted, then the whole education of mankind, and every arrangement in the state or in society made consciously and with the object of adding to man's happiness or developing his powers, must be accounted instruments of artificial selection. And among the most artificial of them all would be a regulation of the state which should insure unlimited freedom of development to each individual's talents.

At our point of view we are ever and again reminded that the idea of the natural struggle for existence does not imply that the victorious one is always physiologically, or, in the case of man, morally, the most deserving. We might, but we can not, imagine an ideal state wherein the most deserving shall always gain the victory, and thus we may represent to ourselves a universal perfectionment as the end of development. Hence we are not in the least pessimists; but, on the other hand, the innumerable evidences of progress which we see in nature, both animate and inanimate, do not suffice to make our idea of the universe purely optimistic. Progress is an asymptote of the ideal of perfectionment, and in recognizing this we give free play to the tendency perfectionward, without attempting on our own part to interfere.

With all the certainty that is attainable by inductive proof, the doctrine of development teaches the brute origin of man. Whether Pfeffel says aright

> Far better for the welfare of the world
> Is pious error, which sustains,
> Than cold truth, which destroys.

may perhaps be open to question in certain cases. In the present case it is worth while to reflect that oftentimes men who awaken from a long-cherished though pious error betray their kinship to the beasts; while truth, sedulously handed down from generation to generation, and advancing enlightenment, make men more human.

Would that we could diffuse abroad a conception of the full truth of the Darwinian doctrine of development, to the end that every thinking man who has not already been caught by the counter-current might know what it comprises and what consequences it does *not* warrant!

6 Energetics: A New Religion of Science and a Scientific Philosophy of History

The dramatic analogies imported from Darwinian science into social, political, and religious concerns during the mid-nineteenth century set a precedent for the adoption of other scientific analogies; and the major changes which society was undergoing as a result of the development of steam and electrical power made energetics or thermodynamics—the science of the transformation and transfer of power—a natural source of inspiration. Thermodynamics seemed additionally apt to prove applicable to man and society because of its great generality and wide-ranging applicability within the sciences. Chemistry, mechanics, cosmology, geology, electricity and magnetism, ideas of heat, and even certain biological phenomena were embraced and enriched by thermodynamic considerations. Who could doubt that these principles might be fruitfully extended to such domains as psychology, social theory, and religion, as well?

Erwin Hiebert's essay "The Uses and Abuses of Thermodynamics in Religion" begins with a short synopsis of some of the fundamental concepts of thermodynamics and an indication of how thermodynamics achieved its widespread applicability within the sciences. Next, it investigates the basis in energetic ideas for a new monistic, humanistic religion. So, to avoid repetition, I will limit my comments here to providing a setting for the selection from Henry Adams' "The Rule of Phase Applied to History."

Henry Adams, son of the American diplomat Charles Francis Adams, grandson of John Quincy Adams, and one of the premiere American historians, illustrates the great impact of scientific thought on humanistically oriented intellectuals of the late 19th century. Appalled by the materialism of many contemporary scientific theories, he was nonetheless convinced that the power—both material and spiritual—released by modern science was the single dominant fact of his era. Furthermore he was convinced that a new science of history could be based on the biologists' evolutionary ideas and the physicists' considerations of energy and entropy. Adams read widely in scientific literature, and toward the end of his long career he wrote three important philosophical essays which were collected and published posthumously. The first and third of these essays based a very pessimistic vision of the future on an interpretation of the second law of thermodynamics, arguing that the dissipation or degradation of energy implied by the second law manifests itself in biological and social degeneration. The second essay, written in 1909, tried to adapt the phase rule of J. W. Gibbs to explain the major turning points in human history.

Adams' essays are symptomatic of a qualitatively new and different use of scientific metaphors for social and humanistic thought—a use which almost consciously avoids a clear understanding of the scientific theory which provides the metaphor and which dismisses the need to critically analyze the logical implications of the analogies employed. Adams cheerily admitted that he was not fully aware of the science he claimed to be using as his model and held that "the most trifling popular science is enough for popular teachers like ourselves." Yet he clearly wanted to establish history on a scientific basis; that is, he wanted to discover exact quantitative historical laws analogous to the inverse square law of gravitational attraction and the second law of thermodynamics. Adams set an important and impressive precedent for the grossest misapplication of scientific theories, for his expositions of the scientific bases of his analogies are hopelessly garbled. Just in "The Rule of Phase Applied to History," for example, he confuses velocity and acceleration, phase and change, volume and concentration; and in one sentence he equates the radically disparate terms solvent, motion, equilibrium, and current. Even the simple arithmetic calculations Adams made were mistaken. In spite of his failings, however, Adams' technique of appropriating the terms and concepts of science appealed to a large class of readers who could not evaluate his fidelity to the science from which he borrowed. His urgings to heed the ideas of the scientists—the new prophets of civilization—encouraged such coming trends as the uncritical appropriation of the physicist's notion of relativity into literature, art, and history.

Adams' patent mishandling of scientific theory should not be too bitterly criticized, however; for metaphors, unlike rigorous declarative statements, are not simply true or false. They may be useful, effective, and fruitful or not; and only to a very small degree does this depend upon the complete understanding of both terms in the implicit analogy. Adams was stimulated by his metaphors to reach several interesting historical conclusions and, as he argued, "Mathematicians assume the right to choose, within the limits of logical contradiction, what path they please in reaching their results, provided that when they come to the end of their process, they consent to test their result by the facts of experience. More than this cannot fairly be asked of historians." Of course, Adams had once more confused the scientists' procedures, equating the divergent fields of mathematics and physics. But the sentiment he expressed may still have some validity. No perfect analogies exist, and even very imperfect ones may be fruitful in one field while they misrepresent another.

The immediate spur to "The Rule of Phase Applied to History" was Josiah Willard Gibbs' establishment of a very general thermodynamic principle which allowed scientists to determine the number of physically distinct, homogeneous portions of a system that could exist in equilibrium with one another when the number of chemically distinct components and a certain number of other variables such as temperature, pressure, and volume were known. The phase rule was an elegantly simple rule that had immense consequences for both theoretical and commercial chemistry and physics, and it was widely acclaimed after its proposal in 1886. Henry Adams was particularly impressed by the application of the rule to the system of water, steam, and ice, and he saw in this application a guide to the understanding of important historical trends.

Thermodynamics and Religion

Erwin Hiebert

. . . While there are a great many general works that employ a broad historical approach to the interaction of science and religion, there are few that discuss the use to which science has been put in answering questions of a fundamentally religious cast. Yet the influence that thermodynamics, for example, has exerted upon religious thought over the past century testifies to such use. The first and second laws of thermodynamics have been used, affirmed, rejected, manipulated, exploited, and criticized in order both to further and to censure religion.

If the history of science is to be more than an internal analysis of the manner in which science came to be what it is today, or at any other time, the historian of science may have to venture into new territory. This new territory stretches beyond the confines of the traditional study of the major documents that have contributed to the experimental and theoretical progress of science. Indeed, if history is a record of man's ideas and behavior in place and time, then it may be quite as legitimate to study the interaction between thermodynamic thought and religion as it is to study the internal history of thermodynamics *per se*—although the documents for the two kinds of historical study are quite different.

In examining the use of thermodynamic principles in religious (and perhaps pseudo-religious) thought, the literature that is encountered may

Source: Erwin N. Hiebert, "The Uses and Abuses of Thermodynamics in Religion," *Daedalus* 95 (1966), 1046-1080.

not always be what, with propriety, may be called first-class. The criterion for considering a given author's treatment of a subject must necessarily be its relevance to history, and not its intrinsic merit as a work of enduring scientific or religious import. This is not to defend the study of *any* crack-pot thermodynamicians, philosophers, or theologians; but, rather, to suggest the pertinence to this historical study of literary works, crack-pot or not, provided they exerted some noteworthy effect upon society. The works, in other words, must mirror something of the feelings and the attitudes of the age, irrespective of whether they would have stood the test of excellence in the eyes of the elite in science or religion. The essential substance of this phase of the history of thermodynamics or religion is, therefore, not what happened but what was thought or said about what happened.

The first law of thermodynamics, the principle of conservation of energy, was enunciated in the 1840's by Mayer and Helmholtz in Germany, Joule in England, and Colding in Denmark. There were a number of other persons who entertained energy-conservation ideas before and after the 1840's, but in one way or another they provided less satisfactory formulations of the principle than the persons mentioned. Suffice it to say that conservation of energy was an independent, multiple discovery that burst forth among various European scientific investigators who were more closely tied to civil and military engineering, medicine, physiology, and brewing than to anything going on at the academic centers in the physical sciences.

In Julius Robert Mayer, the law takes the form of an assertion that various forms of energy are qualitatively transformable, quantitatively indestructible, imponderable objects. In the work of Mayer, Joule, and Colding, there are also theoretical calculations and experimental investigations expressly designed to give the numerical equivalence of various forms of energy. Except for a rider to the first law, which Einstein attached in 1905 for mass-energy conversions, there are no clear-cut natural processes known to violate the principle of conservation of energy. In other words, it is impossible, by mechanical, chemical, thermal, or any other devices, to construct any kind of perpetual-motion machine that creates energy from nothing.

The second law of thermodynamics (an extension of Sadi Carnot's principle of 1824, and Clapeyron's algebraic and graphical representation of the same in 1834) was formulated in various ways by Clausius, Kelvin, Boltzmann, and others after 1850. This law extends beyond the principle of energy conservation by dealing with the direction in which a process can take place in nature; that is, energy conservation does not suffice for a unique determination of natural processes. For example, the second law stipulates how an exchange of heat by conduction can take place between two bodies of different temperature. The principle, therefore, tells something about the quantity of heat that can be converted into mechanical energy in an ideal heat engine operating at a given temperature differential between the combustion chamber and the exhaust.

This means that available and unavailable energy can be distinguished and meaningfully discussed. In fact, the capacity factor for unavailable energy is an extremely versatile thermodynamic function called entropy. In all systems in which boundary conditions are defined in such a way that no energy transfers can take place across the boundaries, the entropy increases for all spontaneous processes within the system. Entropy increase corresponds to a decrease in the available energy. The net result is that in all natural processes some energy ends up being unavailable. An equivalent statement would be to say that systems in nature move spontaneously from order to disorder, from lesser to greater randomness, or toward a state of maximum probability.

In 1865 Clausius put the first and second laws of thermodynamics into the following simple verbal form: "The energy of the universe is constant. The entropy of the universe tends towards a maximum." These two statements, along with various logical and illogical extensions of the energy and entropy principle, constitute the background to the following discussion of the interaction of science and religion.

The acceptance of the laws of thermodynamics, slow at first, especially among physicists and chemists, was followed by a period of rapid exploitation that revolutionized and unified the study of chemistry, heat theory, heat engines, radiation, electricity, and magnetism. Less respectable, but nevertheless real, were the far-reaching deductions, discussions, and speculations regarding the significance of thermodynamic concepts in cosmological works dealing with the source of the sun's energy, the origin of the solar system, and views on the expanding and contracting universe. On the strength of the many far-reaching inferences that had been derived from a small number of axioms, thermodynamics, by the end of the nineteenth century, had taken its place alongside mechanics and electromagnetics as one of the main theoretical pillars of classical physics.

It was virtually inevitable that some of the thermodynamic excitement of the nineteenth century should carry over into speculations concerning the problems of physiology, biological vitalism, and life in general. In truth, thermodynamics not only put its stamp on scientific thought but influenced, as well, social and political thought, psychology, literature, history, philosophy, and religion.

Already in 1863 Sir William G. Armstrong in his Presidential Address to the British Association for the Advancement of Science remarked that the dynamical theory of heat and the new views on energy (or "force," as it was loosely referred to then) probably constituted the most important discovery of the century. By that time, the principle of conservation of energy had been applied not only to motion, heat, light, electricity, magnetism, and chemical affinity, but also to problems encountered in the world of organic life: animal and vegetable heat, digestion, respiration, muscular force, nervous agency, vital power, the development of organized animal and vegetable structure from dormant, primordial germ cells, and the effects of plant and animal sensation and consciousness.

One of the most widely-read authors on this broad "philosophical" subject was the English lawyer and physicist William R. Grove, whose famous *Correlation of Physical Forces*, first published in 1846, went through numerous English and American editions and kept the public informed on how the relentless march of the knowledge of various "forces" was solving the new problems of science. Grove remarked that it was highly probable "that when discovered, and their modes of action fully traced out," these forces would "be found to be related *inter se*, and to these forces as these are to each other." This he believed "to be as far certain as certainty can be predicted of any future event." Thus he concluded:

> Many existing phenomena, hitherto believed distinct, will be connected and explained: explanation is, indeed, only relation to something more familiar, not more known—i.e., known as to causative or creative agencies. In all phenomena the more closely they are investigated the more are we convinced that, humanly speaking, neither matter nor force can be created or annihilated, and that an essential cause is unattainable—Causation is the will, Creation the act, of God.

William B. Carpenter, an eminent and unflagging English investigator in the sciences of zoology, botany, and mental physiology, wrote on the correlation of physical and vital forces with a gasconade that surely must have embarrassed the founders of the principle of conservation of energy. In a paper of 1850, Carpenter took up the problem of the mutual relation (metamorphosis and conversion) of the physical forces to the vital processes of plant and animal growth, development, reproduction, and the evolution of complex heterogeneous structures from homogeneous germinal masses. By applying the principle of correlation (conservation) of forces to vital phenomena, he hoped to open new lines of inquiry that would invest the physiological sciences with the same dynamic aspects as the physical sciences exhibited.

The particular *modus operandi* of "Cell-force" in the cell-formation of plants was compared to "Engine-power," which term, Carpenter remarked, was used "knowing that the steam-engine possesses no power itself, but that it is simply that instrument most commonly employed, because the most convenient and advantageous yet devised, for the application of the expansive force of steam, generated by the application of heat, to the production of mechanical motion." Carpenter found the forces in the growth, development, and movement of animals to be essentially the same as "Cell-force" in plants, except for an additional "Nervous agency" related to the conscious mind, which communicated impressions derived from the external world and was related to the working of contractile tissues in obedience to mental impulses. He wrote: "For, just as electricity developed by chemical change may operate (by its correlation with chemical affinity) in producing other chemical changes elsewhere, so may nerve-force, which has its origin in cell-formation, excite or modify the process of cell-formation in other parts, and thus influence all the vital manifestations of the several tissues, whatever may be their own individual characters." He thus hoped to have established "the general proposition, that so close a mutual relation exists between all the vital forces, that they may be legitimately regarded as *modes* of one and the same force."

As for the relations of the vital to the physical forces, Carpenter proceeded to demonstrate to his satisfaction that "Nervous Force" ("the *highest* of all the forms of vital force, both in its relations to mental action, and in its dominant power over organic processes of every kind") is perfectly correlated and mutually convertible with electricity, heat, light, magnetism, motion, and chemical affinity.

Notwithstanding these physical comparisons, Carpenter retained, as Liebig had, the pre-existence of a living organism necessary for the development of an "organized structure of even the simplest kind." He wrote:

> It is the *speciality* of the material substratum thus furnishing the medium or instrument of the metamorphosis which . . . establishes and must ever maintain a well-marked boundary-line between the Physical and the Vital forces. Starting with the abstract notion of Force, as emanating at once from the Divine Will, we might say that this force, operating through inorganic matter, manifests itself in electricity, magnetism, light, heat, chemical affinity, and mechanical motion; but that, when directed through organized structures, it effects the operations of growth, development, chemico-vital transformation, and the like; and is further metamorphosed, through the instrumentality of the structures thus generated, into nervous agency and muscular power.

Thus, for Carpenter, "all *force* which does not emanate from the will of created sentient beings, directly and immediately proceeds from the Will of the Omnipotent and Omnipresent Creator." What he called "physical forces" were just "so many *modi operandi* of one and the same agency, the creative and sustaining will of the Deity." Carpenter maintained reservations concerning the extension of the doctrine of evolution to man's intellectual and spiritual nature, even though he admitted that man had moved by "grades of organic ascent" from an analysis of physical forces such as heat and light to an analysis of nervous agency and muscular power.

The energy-conservation principle was used in some fairly untamed, speculative, and theologically grounded arguments in the works of the physicist Grove and the physiologist Carpenter. But in none of the early expositions of the energy principle did *philosophy* run riot quite so wildly as in the writings of Herbert Spencer. Spencer was a representative of a widely disseminated naturalistic interpretation of phenomena that rested, in his case, upon a strange synthesis of ideas drawn from evolutionary theory, the principle of conservation of energy, and a humanitarian, religious metaphysics. He was described by one of his friends as "radical all over."

In the *First Principles* of 1862, Spencer attempted to reconcile science and religion on the broad postulate of belief in the *Unknowable* as the cause of all phenomenal existence. The deepest, widest, and most certain of all facts, for Spencer, was that the *Power* which the universe manifests to man is utterly inscrutable. Thus, "ultimate scientific ideas" like force, space, and time were all representative of realities that could not be comprehended; they passed all understanding.

For Spencer, the scientific idea of "force," rooted in primordial experiences, was the ultimate of ultimates. One did not know, he said, what force as an ultimate was; the law of its manifestations could be inferred from experience, but never derived inductively. This law was the law of conservation of energy. "Persistence of force" was the expression that Huxley had suggested and Spencer preferred.

In his chapter on "the correlation and equivalence of forces," Spencer added "mental forces" to the same generalization (the same law of metamorphosis) that he recognized as having been enunciated for physical forces on the basis of the experimental investigations of Mayer, Joule, Helmholtz, and Grove.

> Those modes of the Unknowable which we call motion, heat, light, chemical affinity, etc. are alike transformable into each other, and [also] into those modes of the Unknowable which we distinguish as sensation, emotion, thought: these, in their turns, being directly or indirectly transformable into the original shapes.

He then added:

> Of course if the law of correlation and equivalence holds of the forces we class as vital and mental, it must hold also of those we class as social . . . [and] if we ask whence come these physical forces from which, through the intermediation of the vital forces, the social forces arise, the reply is of course as heretofore—the solar radiations.

Spencer concluded by saying "the deepest truths we can reach are simply statements of the widest uniformities in our experience of the relations of Matter, Motion, and Force; and Matter, Motion, and Force are but symbols of the Unknown Reality." Science, as in the unknowable ultimate of ultimates, namely *force*, coalesced with religion in the "consciousness of an Incomprehensible Omnipotent Power"—an Absolute Being. Religion and science were, nevertheless, both rooted in the common datum of all human thought, in the law of the "persistence of force." But the establishment of the correlation and equivalence between the forces of the outer (matter) and the inner (spirit) world, for Spencer, was a matter of the assimilation of either to the other, "according as we set out with one or the other term."

Related views on force were spelled out by Spencer in the section on "Ecclesiastical Institutions" in the third volume (Principles of Sociology) of his *Synthetic Philosophy* of 1885. Here he was concerned with the history of religion and religious institutions and the elements of *power* that entered into all primitive religions. Acknowledging elements of power beyond consciousness, as evidenced in muscular power, ghost force, sun-worship, the powers of medicine men and priests, Spencer concluded that all religions had a natural genesis that led upward from the most primitive anthropomorphisms to gods, inscrutable, unknowable, and omnipotent. The divinity that was synonymous with superiority eventually became non-anthropomorphic in the idea of force. This idea reached its extreme form in the man of science who could properly interpret the idea of the "persistence of force" in terms of all possible kinds of physical, biological, mental, and nervous phenomena.

Thus, higher faculty and deeper insight raised the sentiments of the man of science to a vision, though dim and incomplete, of ultimate existence.

> And this feeling is not likely to be decreased but to be increased by that analysis of knowledge which, while forcing him to agnosticism, yet continually prompts him to imagine some solution of the Great Enigma which he knows cannot be solved. . . . But one truth must grow ever clearer—the truth that there is an Inscrutable Existence everywhere manifested, to which [man] . . . can neither find nor conceive either beginning or end. Amid the mysteries which become the more mysterious the more they are thought about, there will remain the one absolute certainty, that he is ever in presence of an Infinite and Eternal Energy, from which all things proceed.

Ernst Haeckel, a biologist of enormous literary output and Professor of Zoology at the University of Jena, was a most enthusiastic apostle of Darwinism in Germany. He applied the principle of conservation of energy (in combination with biological evolution) to some of the oldest problems in philosophy and religion. His *Die Welträthsel* of 1899, which appeared in English in 1900 as *The Riddle of the Universe*, adopted an inflexible, materialistic, monistic position that laid down the intrinsic unity of organic and inorganic nature and propounded the evolution of the highest level of the human faculties from unicellular protozoa. While Haeckel rejected the immortality of the soul, freedom of the will, and the existence of a personal deity, he did not, he claimed, reject religion *per se.* His "monistic religion" with its "monistic church" was, he said, a monism connecting religion and science but freed from the dead and dried-up superstitions of the traditional church religions.

In *Der Monismus als Band zwischen Religion und Wissenschaft*, written in 1892, Haeckel proposed that the most important general consequence of the spiritual conquest of modern science belonged to the law of substance (*Substanz-Gesetz*)—and that this was to be designated as the first paragraph of the "monistic religion of reason" (i.e., *die monistische Vernunftsreligion*). In a remarkable essay of 1895, Haeckel wrote concerning this law:

> This supreme basic law of the cosmos actually consists of two intimately related laws: of the "law of the conservation of matter" for which we are indebted to the great French chemist Lavoisier, and of the "law of the conservation of energy" whose founding is shared by two German intellectual heroes —the South-German Robert Mayer and the North-German Hermann Helmholtz. As "matter and energy" are inseparably combined in every thing, so also these two basic "conservation laws" hang together in one law of substance. For the religion of reason of the science of today this law of substance is just as much the immovable foundation stone as the dogma of the "infallibility of the Pope" is for the Catholic church of today—the rudest slap in the face for reason.

Professor Haeckel then ridiculed the behavior of the mourners of the recently deceased Helmholtz. They behaved no better than Darwin's mourners had thirteen years earlier in the pantheon of Westminster Abbey. Haeckel asked whether the highly-respected gentlemen who heard the church bells ring in Berlin for Helmholtz realized that they were honoring "a free-thinker who ought in their eyes to have been a revolutionary, mangy, heretic of the first order." Did none of them know that the greatest contribution of Helmholtz, the "law of substance," was the first paragraph of the "monistic religion," and that it was connected inseparably with Darwin's infamous "materialistic" theory of evolution? . . .

A non-theistic, monistic, and humanitarian exploitation of the philosophy of energeticism as a way of life was expounded in scores of books, pamphlets, and sermons by the renowned German physical chemist and Nobel Prize winner of 1909—Wilhelm Ostwald. In 1913, as President of the German Confederation of Monists, Ostwald delivered a lecture in Vienna that merits attention. This lecture admirably demonstrated the anti-religious views of Ostwald with special reference to thermodynamics. Similar views were presented by Paul Carus, philosopher and editor of *Open Court* and *The Monist* in Chicago, and by Svante Arrhenius, who received the Nobel Prize in 1903 for his electrolytic dissociation theory.

Briefly, Ostwald characterized his brand of monism in this lecture of 1913 as a doctrine "which excludes all double-entry bookkeeping, which removes all barriers, hitherto regarded as insurmountable, between inner and outer life, between the life of the present and that of the future, between the existence of the body and that of the soul, and which comprehends all these things in a single unity that extends everywhere and leaves nothing outside its scope."

Ostwald maintained that the record of man's past, since the pre-Socratic and Greek doctrines of the basic stuff of the universe, revealed man's attempt to struggle toward a unitary, monistic representation of the world. Thus, the history of philosophy was strewed with rejected monistic systems, including one that Ostwald also rejected, namely, the monism of a single supreme God. Ostwald argued that a unified self-consistent world-view was unattainable by means of any *a priori* monism. This included the monism of energy—as a principle. The only monism that Ostwald recognized was an *a posteriori* monism "which proceeds from the diversity of the world as from the data offered by experience." Thus, an *a posteriori* monism, as the future ideal of unity, was science; it was a monism of scientific thought and scientific method.

Ostwald's program for scientific monism rested upon the completion of a process that he found to be already well under way, namely, the abolition of the barrier or dividing line that separated the provinces of science and religion. This extremely movable barrier, Ostwald indicated, shifted invariably in the direction of enlarging and expanding the province of science at the expense of the province of religion. Ostwald predicted that the province of religion would become narrower and narrower until, reduced to almost nothing, religion would be forced to surrender. Drawing from the work of the "pathfinding psychologist" William James of Harvard University, Ostwald argued that the struggle of the priests against science was one that "always allows us to discern the incessant and irresistible advance of science into the departments hitherto occupied by religion."

As a "specimen of the practical Monism" that man should endorse, Ostwald mentioned "the irresistible flow toward the international organization of human affairs." Only the movement toward greater unity in all of science, since the introduction and utilization of energy conservation in science, had made this flow toward international organization possible.

This was Ostwald's so-called doctrine of the "energetic imperative." He postulated: "There must evidently be an imperative and decisive moment in the existence of thought, which gives to the principle of unity . . . so great an impulse that the whole tendency of human intellectual evolution surrenders to its grasp." This movement "out of multifariousness to unity, especially out of the many dualisms to the one Monism" was, for Ostwald, symbolized by the "postulate of economy of energy." In this manner, Ostwald moved from scientific monism to the principle of energy conservation as the grand unifying principle of science, and finally to an energeticism directive for the practical affairs of man in society.

> Thus: We know that we all live by virtue of the free energy that streams from the sun to the earth like a broad and powerful flood. [Our energetic

> imperative must avowedly be]: Waste no energy; turn it all to account. . . . Here, therefore, we have a genuine and far reaching Monism [which reaches from] the simplest technical trade, yes from the daily acts of our half-animal life, to the very highest sociological and ethical problems.

"The whole culture-evolution," said Ostwald, "which has led us from the invention of the slingstone, the lever, [and] fire . . . to a modern giant steamer . . . signifies nothing further than an always finer and more multiform manifestation of the energetic imperative. The same may be said of our moral evolution." And thus, for Ostwald, the whole prophetic quality of science turned on the great synthetic unity of man's expenditure of his available energy.

> Accordingly, at the present time, the principal task of scientific or Monistic thought and labor is manifestly to free the final science in the succession of sciences, sociology, from the hitherto existing influence of the priesthood, and to establish in place of the traditional ethics dependent on revelation a rational scientific ethics, based on facts.

Ostwald concluded:

> Proceeding along this pathway of thought, we have come to recognize that the highest values of Christianity, the kindness and love of the individual towards his fellowman, do not yet represent the highest ethical ideal which mankind can attain. Monism leads far more to the perception that the individual is more and more a mere cell in the collective organism of humanity. Accordingly, the evolution of kindness and love, the evolution of the spirit of self-sacrifice and devotion to the great whole of humanity, becomes more and more a demand of the energetic imperative, therefore an immanent demand of our whole Monistically ordered life. Only through the fact that we have come to recognize kindness and love as a necessity for community life, for the social organization of mankind, has there also been gained by the individual the sole sure and immovable foundation. That we practise mutual kindness and love is no longer the demand of a Godhead standing outside of ourselves, which has once for all transmitted it to us by an unverifiable revelation; but it is a demand of scientific intelligence. To it, of course, only those can belong who dedicate themselves to Monism unreservedly and without any remnants of dualistic thinking and feeling. With the increased broadening and deepening of this intelligence we see arrive the Monistic century, which will not remain the only one of its kind. But it will inaugurate a new epoch for humanity, just as two thousand years ago the preaching of the general love for humanity had inaugurated an epoch. . . .

The second law of thermodynamics raised theological issues of a different character. If the principle of the increase of entropy of any operating system was extended to the world or the universe as a system, then the universe was moving toward a state of maximum probability, a configuration of maximum randomness, a condition of minimum availability of energy. This irreversible process, this unidirectional movement toward disorganization in history (time's arrow), led to a degradation of the energy sources and ultimately to that pessimistic state of affairs, that boredom or Nirvana, that was called the "heat-death" of the universe. Thermodynamics accordingly predicted an end to everything as a function of time.

Such a theory, which injected into the minds of men a majestic dysteleology that would eventuate in the consummation of the world, was bound to unleash polemics from religion. Ends and beginnings—creation's end and origin—were matters that theologians could hardly side-step. What would come to an end? And what constituted a beginning or a creation?

The controversial published correspondence that took place in 1933 and 1934 between Arnold Lunn and J. B. S. Haldane was representative of criticisms of the theological uses of the entropy principle. Lunn, a Madras-born novelist, was converted to Catholicism during the course of his correspondence with the evolutionist Haldane, who was then Professor of Genetics at University College in London. In his discussions on "proofs of God's existence," Haldane chose as his target for criticism the first proof of St. Thomas in the *Summa Theologica*, the argument of the unmoved mover. St. Thomas's philosophy, he concluded, was based on antiquated science and faulty mathematics.

For various "physical reasons," Haldane found invalid the second law, which purported to demonstrate that the world could not have existed for all times and which, consequently, might have given new support for the theory of an unmoved mover. According to Haldane, the universe might be spatially infinite, making any argument for finite systems possibly fallacious. Even if finite, the universe might fluctuate between increasing and diminishing entropies that corresponded to the running-down and the building-up of the available energy; some apparently irreversible events in a space-expanding universe (such as the radiation of heat to cold bodies) might be reversed in a space-contracting universe. All of these physically feasible alternatives excluded the heat-death of the universe. Since they also did not support *processes* or *beginnings* and *creations*, Haldane concluded that the proof of God as first cause or unmoved mover failed by implication.

Lunn's reaction to the argument for God from entropy considerations supported Haldane's criticism—but for a different reason. Lunn responded: "I am reluctant to hitch the wagon of faith to the shooting star of scientific fashion. For all we know, relativity and the quantum theory and entropy will one day join their predecessors in the limbo of discarded scientific fads." Thus, Haldane had no use for supernatural religion, while Lunn had no use for anything but rationalistic religion—which he used to defend miracles and the supernatural, and to combat naturalism, materialism, evolution, Marxism, and Protestantism. Both seemingly agreed on the irrelevance of entropy considerations to the proof of God's existence: Haldane because he could not, at heart, believe in God; Lunn because he did not want his belief in God to be tied to the perennially obsolete fashions of science.

Compare this with the view of Sir Arthur Eddington, astronomer-mathematician at Cambridge University. Eddington's conviction concerning what religion arises from, while not grounded in a rejection of reasoning of the kind Lunn championed, nevertheless rested upon the belief that the ultimate data for reasoning were given through a non-reasoning process. According to Eddington, this process was founded upon a state of self-knowledge or awareness of what was given in consciousness. For Eddington, this was at least as significant as what was given in sensation.

Eddington repudiated, on the one hand, the "idea of proving the distinctive beliefs of religion either from the data of physical science or by the methods of physical science." On the other hand, he did not have so low an opinion of the entropy principle as Lunn had. In his analysis of the relation of entropy to chance coincidences and random elements, Eddington wrote:

> Entropy continually increases. We can, by isolating parts of the world and postulating rather idealized conditions in our problems, arrest the increase, but we cannot turn it into a decrease. That would involve something much worse than a violation of an ordinary law of Nature, namely an improbable coincidence. The law that entropy always increases—the second law of thermodynamics—holds, I think, the supreme position among the laws of Nature. If someone points out to you that your pet theory of the universe is in disagreement with Maxwell's equations—then so much the worse for Maxwell's equations. If it is found to be contradicted by observation—well, these experimentalists do bungle things sometimes. But if your theory is found to be against the second law of thermodynamics I can give you no hope; there is nothing for it but to collapse in deepest humiliation. This exaltation of the second law is not unreasonable . . . ; the chance against a breach of the second law (i.e., against a decrease of the random element) can be stated in figures which are overwhelming.

William Ralph Inge, Dean of St. Paul's in London, approached the second-law problem in a manner radically different from that of Haldane, Lunn, or Eddington. To Inge, an arch reactionary, Christian Neo-Platonist, the myths of progress and evolution were anathemas. Just as Lunn had been convinced that Martin Luther was the father of the modern revolt against reason, Inge, sometimes called "the gloomy dean," was convinced that the fatal error of Catholic theology had been to find a rationalistic foundation for faith. In *God and the Astronomers*, Inge undertook to demonstrate that any philosopher or theologian who wished to write on cosmology ("the relation of God to the universe") had to be informed on modern astronomy and physics. The research in these fields, he argued, could no longer be brushed aside as irrelevant to metaphysics or theology. "Science and philosophy," he wrote, "cannot be kept in watertight compartments."

> God is revealing Himself to our age mainly through the book of nature. . . . This knowledge is given to us for a purpose. . . . Science has been called . . . the *purgatory* of religion. The study of nature . . . purifies our ideas about God and reality. It makes us ashamed of our petty interpretations of the world—ashamed of thinking that the universe was made solely for our benefit.

Thus, Inge suggested that "the man of science worships a greater God than the average church-goer."

Essentially, Dean Inge warned the theologians not to ignore the scientists. What they were doing was directly relevant to theological doctrines. He felt that the religious and philosophical significance of the second law of thermodynamics—that new "Götterdämmerung," as Spengler called it—had something to say about the ultimate fate of the world. Inge argued that if the universe was running down like a clock, the clock must have been wound up at some specific time. If the second law predicted an end in time for the world, then the world must have had a beginning in time. Thus, Inge could ask, "Is science itself driving us back to the traditional Christian doctrine that God created the world out of nothing?"

If the second law led to perdition, Inge was saying, then man must accept it. This was, after all, a very Christian idea. The belief that the species would inherit the earth was hardly a Biblical doctrine. The Bible did not forecast unending temporal progress, nor an evolving God, but rather the blessed hope of everlasting life. "Modern philosophy [with Time itself an absolute value, and progress a cosmic principle] is, as I maintain, wrecked on the Second Law of Thermodynamics; it is no wonder that it finds the situation intolerable, and wriggles piteously to escape from its toils."

Dean Inge found the second law to be at odds with the theory of evolution. Since the universe was moving toward thermal equilibrium—its heat-death—both biological evolution and the idea of progress were illusions. The human species had on earth "no continuing city," and it should therefore "ascend with heart and mind" to its "citizenship . . . in heaven."

The concept of an "end to man's history" raised a host of questions that occupied religious scholars for decades. Was it consistent with God's goodness to annihilate creation through the heat-death? What meaning could be attached to any values—in the mind of God—when man's history on earth was terminated? For whom, and in what way, would these values be preserved and manifested? Would real existence reside solely in God when only dead brute matter remained in the world? If so, how could God's will and purpose be exercised at all? What, if anything, would then become the appropriate vehicle for the exercise of God's will and his creative powers? If creation was one of the symbols of the Creator, then in what way might an eternity of pessimism and doom represent an appropriate image of creation? Did "God," indeed, have any meaning in a universe that was not an abode for conscious life? And why would God create in order to destroy?

The second law seemed to be fatal to any meaningful view of the ultimate relation between the world and intelligent beings—except, of course, within a metaphysical system similar to the one provided by the Dean of St. Paul's. Inge could simply reply that God was not limited to forms of expression that fell within the framework of man's knowledge. God could exist, if he so willed, without man; the world was not so necessary to God as God was to the world. "As a pleasure garden this world may be a failure; but it was never meant to be a pleasure garden..."

The knowledge of the second law of thermodynamics was in the general scientific domain while some of the enthusiastic supporters of biological evolution were indulging in the nineteenth century's most jubilant songs of progress. In fact, the mood of sympathy for movements toward greater perfection, which was often read into evolution, infected most of the religiously enlightened persons during the Victorian reign—not to speak of its spread into the literature of philosophy, aesthetics, and the political, social, and historical sciences. Evolutionary theory was equated with hope, with the law of progress, the lay creed of men of science; entropy considerations prophesied a pessimism of despair, a future of disillusionment.

Spencer could feel empathetic with the energetic concepts and simultaneously identify progress with the increasing complexity offered through evolution. Energeticists like Ostwald stressed those aspects of the second law of thermodynamics that dealt with the spontaneous irreversibility of time and what it signified for man's progress. Ostwald had in mind, naturally, the element of irreversibility and progress in the development of culture, aesthetics, morals, and man's store of knowledge. This was the human race's inheritance; it resulted from man's accumulative and irreversible progress. This was an exploitation of time's arrow that did not dwell on the final, dismal outcome of things. Thus, Ostwald played down the gloomier aspects of the second law by ignoring them completely.

References to thermodynamics and religion—some oblique, some not so oblique, and some quite blunt—are to be found in the writings of many scientists. Mayer, Joule, and Colding included theological comments along with their formulations of the first law. John Tyndall, the Irish physicist, attributed a wider grasp and a more radical significance to the doctrine of the conservation of energy than to the theory of the origin of the species. Energy conservation, he thought, "bound nature fast in faith" by bringing vital as well as physical phenomena under the dominion of causal connection. He wrote: "If our spiritual authorities could only devise a form in which the heart might express itself without putting the intellect to shame, they might utilize a power which they now waste, and make prayer, instead of a butt to the scorner, a potent supplement of noble outward life."

Lord Kelvin excluded living organisms in his statement of the second law. Svante Arrhenius, while admitting the universality of the second law, thought that rare exceptions might occur leading to the rebuilding of worlds

—perhaps by some conscious Maxwellian demon. Emile Meyerson concluded that the second law was an obstacle fatal to the mechanical explanation of the universe.

The philosophies of Nietzsche, Whitehead, and Russell were not inconsequent on these problems. Henri Bergson called the second law the most metaphysical of all the principles of nature because of its implication of a beginning in time. Freud and Jung took a new look at religious experience with the help of wild ideas and brand-new expressions: the energy-laden world waiting to react on man, traumatic neuroses in relation to bound "psychic energy," the *Ich-Energie* as narcissistic libido or desexualized eros, cathectic energy, entropy and the death wish. William James wrote: "Though the ultimate state of the universe may be its vital and physical extinction, there is nothing in physics to interfere with the hypothesis that the *penultimate* might be the millennium. The last expiring pulsation of the universe might be—I am so happy and perfect that I can stand it no longer." Henry Adams thought the universe had been so terribly narrowed by thermodynamics that history and sociology had to "gasp for breath." The worship of the dynamo had replaced the worship of the Virgin of Chartres.

The Rule of Phase Applied to History

Henry Adams

. . .The question how order could have got into the universe at all was the chief object of human thought since thought existed; and order,—to use the expressive figure of Rudolph Goldscheid,—was but Direction regarded as stationary, like a frozen waterfall. The sum of motion without direction is zero, as in the motion of a kinetic gas where only Clerk Maxwell's demon of thought could create a value. Possibly, in the chances of infinite time and space, the law of probabilities might assert that, sooner or later, some volume of kinetic motion must end in the accident of Direction, but no such accident has yet affected the gases, or imposed a general law on the visible universe. Down to our day Vibration and Direction remain as different as Matter and Mind. Lines of force go on vibrating, rotating, moving in waves, up and down, forward and back, indifferent to control and pure waste of energy,—forms of repulsion,—until their motion becomes guided by motive, as an electric current is induced by a dynamo.

Source: Henry Adams, *The Degradation of Democratic Dogma* (New York: The Macmillan Company, 1919), pp. 279-311.

History, so far as it recounts progress, deals only with such induction or direction, and therefore in history only the attractive or inductive mass, as Thought, helps to construct. Only attractive forces have a positive, permanent value for the advance of society on the path it has actually pursued. The processes of History being irreversible, the action of Pressure can be exerted only in one direction, and therefore the variable called Pressure in physics has its equivalent in the Attraction, which, in the historical rule of phase, gives to human society its forward movement. Thus in the historical formula, Attraction is equivalent to Pressure, and takes its place.

In physics, the second important variable is Temperature. Always a certain temperature must coincide with a certain pressure before the critical point of change in phase can be reached. In history, and possibly wherever the movement is one of translation in a medium, the Temperature is a result of acceleration, or its equivalent, and in the Rule of historical phase Acceleration takes its place.

The third important variable in the physico-chemical phase is Volume, and it reappears in the historical phase unchanged. Under the Rule of Phase, therefore, man's Thought, considered as a single substance passing through a series of historical phases, is assumed to follow the analogy of water, and to pass from one phase to another through a series of critical points which are determined by the three factors Attraction, Acceleration, and Volume, for each change of equilibrium. Among the score of figures that might be used to illustrate the idea, that of a current is perhaps the nearest; but whether the current be conceived as a fluid, a gas, or as electricity,—whether it is drawn on by gravitation or induction,—whether it be governed by the laws of astronomical or electric mass,—it must always be conceived as a solvent, acting like heat or electricity, and increasing in volume by the law of squares.

This solvent, then,—this ultimate motion which absorbs all other forms of motion is an ultimate equilibrium,—this ethereal current of Thought,—is conceived as existing, like ice on a mountain range, and trickling from every pore of rock, in innumerable rills, uniting always into larger channels, and always dissolving whatever it meets, until at last it reaches equilibrium in the ocean of ultimate solution. Historically the current can be watched for only a brief time, at most ten thousand years. Inferentially it can be divined for perhaps a hundred thousand. Geologically it can be followed back perhaps a hundred million years, but however long the time, the origin of consciousness is lost in the rocks before we can reach more than a fraction of its career.

In this long and—for our purposes—infinite stretch of time, the substance called Thought has,—like the substance called water or gas,—passed through a variety of phases, or changes, or states of equilibrium, with which we are all, more or less, familiar. We live in a world of phases, so much more astonishing than the explosion of rockets, that we cannot, unless we are Gibbs or Watts, stop every moment to ask what becomes of the salt we put in our soup, or the water we boil in our teapot, and we are apt to remain stupidly stolid when a bulb bursts into a tulip, or a worm turns into a butterfly. No phase compares in wonder with the mere fact of our own existence, and this wonder has so completely exhausted the powers of Thought that mankind, except in a few laboratories, has ceased to wonder, or even to think. The Egyptians had infinite reason to bow down before a beetle; we have as much reason as they, for we know no more about it; but we have learned to accept our beetle Phase, and to recognize that everything, animate or inanimate, spiritual or material, exists in Phase; that all is equilibrium more or less unstable, and that our whole vision is limited to the bare possibility of calculating in mathematical form the degree of a given instability.

Thus results the plain assurance that the future of Thought, and therefore of History, lies in the hands of the physicists, and that the future historian must seek his education in the world of mathematical physics. Nothing can be expected from further study on the old lines. A new generation must be brought up to think by new methods, and if our historical department in the Universities cannot enter this next Phase, the physical department will have to assume the task alone.

Meanwhile, though quite without the necessary education, the historical inquirer or experimenter may be permitted to guess for a moment,—merely for the amusement of guessing,—what may perhaps turn out to be a possible term of the problem as the physicist will take it up. He may assume, as his starting-point, that Thought is a historical substance, analogous to an electric current, which has obeyed the laws,—whatever they are,—of Phase. The hypothesis is not extravagant. As a fact, we know only too well that our historical Thought has obeyed, and still obeys, some law of Inertia, since it has habitually and obstinately resisted deflection by new forces or motives; we know even that it acts as though it felt friction from resistance, since it is constantly stopped by all sorts of obstacles; we can apply to it, letter for letter, one of the capital laws of physical chemistry, that, where an equilibrium is subjected to conditions which tend to change, it reacts internally in ways that tend to resist the external constraint, and to preserve its established balance; often it is visibly set in motion by sympathetic forces which act upon it as a magnet acts on soft iron, by induction; the commonest school-history takes for granted that it has shown periods of unquestioned acceleration. If, then, society has in so many ways obeyed the ordinary laws of attraction and inertia, nothing can be more natural than to inquire whether it obeys them in all respects, and whether the rules that have been applied to fluids and gases in general, apply also to society as a current of Thought. Such a speculative inquiry is the source of almost all that is known of magnetism, electricity and ether, and all other possible immaterial substances, but in history the inquiry has the vast advantage that a Law of Phase has been long established for the stages of human thought.

No student of history is so ignorant as not to know that fully fifty years before the chemists took up the study of Phases, Auguste Comte laid down in sufficiently precise terms a law of phase for history which received the warm adhesion of two authorities,—the most eminent of that day,—Émile Littré and John Stuart Mill. Nearly a hundred and fifty years before Willard Gibbs announced his mathematical formulas of phase to the physicists and chemists, Turgot stated the Rule of historical Phase as clearly as Franklin stated the law of electricity. As far as concerns theory, we are not much further advanced now than in 1750, and know little better what electricity or thought is, as substance, than Franklin and Turgot knew it; but this failure to penetrate the ultimate synthesis of nature is no excuse for professors of history to abandon the field which is theirs by prior right, and still less can they plead their ignorance of the training in mathematics and physics which it was their duty to seek. The theory of history is a much easier study than the theory of light.

It was about 1830 that Comte began to teach the law that the human mind, as studied in the current of human thought, had passed through three stages or phases:—theological, metaphysical, and what he called positive as developed in his own teaching; and that this was the first principle of social dynamics. His critics tacitly accepted in principle the possibility of some such division, but they fell to disputing Comte's succession of phases as though this were essential to the Law. Comte's idea of applying the rule had nothing to do with the validity of the rule itself. Once it was admitted that human thought had passed through three known phases,—analogous to the chemical phases of solid, liquid, and gaseous,—the standard of measurement which was to be applied might vary with every experimenter until the most convenient should be agreed upon. The commonest objection to Comte's rule,—the objection that the three phases had always existed and still exist, together,—had still less to do with the validity of the law. The residuum of every distillate contains all the original elements in equilibrium with the whole series, if the process is not carried too far. The three phases always exist together in equilibrium; but their limits on either side are fixed by changes of temperature and pressure, manifesting themselves in changes of Direction or Form.

Discarding, then, as unessential, the divisions of history suggested by Comte, the physicist-historian would assume that a change of phase was to be recognized by a change of Form; that is, by a change of Direction; and that it was caused by Acceleration, and increase of Volume or Concentration. In this sense the experimenter is restricted rigidly to the search for changes of Direction or Form of thought, but has no concern in its acceleration except as one of the three variables to which he has to assign mathematical values in order to fix the critical point of change. The first step in experiment is to decide upon some particular and unquestioned change of Direction or Form in human thought.

By common consent, one period of history has always been regarded, even by itself, as a Renaissance, and has boasted of its singular triumph in breaking the continuity of Thought. The exact date of this revolution varies within a margin of two hundred years or more, according as the student fancies the chief factor to have been the introduction of printing, the discovery of America, the invention of the telescope, the writings of Galileo, Descartes, and Bacon, or the mechanical laws perfected by Newton, Huyghens, and the mathematicians as late as 1700; but no one has ever doubted the fact of a distinct change in direction and form of thought during that period; which furnishes the necessary starting-point for any experimental study of historical Phase.

Any one who reads half a dozen pages of Descartes or Bacon sees that these great reformers expressly aimed at changing the Form of thought; that they had no idea but to give it new direction, as Columbus and Galileo had expressly intended to affect direction in space; and even had they all been unconscious of intent, the Church of Thought. The maximum movement possible in the old channel was exceeded; the acceleration and concentration, or volume, reached the point of sudden expansion, and the new phase began.

The history of the new phase has no direct relation with that which preceded it. The gap between theology and mathematics was so sharp in its rapid separation that history is much perplexed to maintain the connection. The earlier signs of the coming change,—before 1500—were mostly small additions to the commoner mechanical resources of society; but when, after 1500, these additions assumed larger scope and higher aim, they still retained mechanical figure and form even in expanding the law of gravitation into astronomical space. If a direct connection between the two phases is more evident on one line than on another, it is in the curious point of view that society seemed to take of Newton's extension of the law of gravitation to include astronomical mass, which, for two hundred years, resembled an attribute of divinity, and grew into a mechanical theory of the universe amounting to a religion. The connection of thought lay in the human reflection of itself in the universe; yet the acceleration of the seventeenth century, as compared with that of any previous age, was rapid, and that of the eighteenth was startling. The acceleration became even measurable, for it took the form of utilizing heat as force, through the steam-engine, and this addition of power was measurable in the coal output. Society followed the same lines of attraction with little change, down to 1840, when the new chemical energy of electricity began to deflect the thought of society again, and Faraday rivalled Newton in the vigor with which he marked out the path of changed attractions, but the purely mechanical theory of the universe typified by Newton and Dalton held its own, and reached its highest authority towards 1870, or about the time when the dynamo came into use.

Throughout these three hundred years, and especially in the nineteenth century, the acceleration suggests at once the old, familiar law of squares. The curve resembles that of the vaporization of water. The resemblance is too close to be disregarded, for nature loves the logarithm, and perpetually recurs to her inverse square. For convenience, if only as a momentary refuge, the physicist-historian will probably have to try the experiment of taking the law of inverse squares as his standard of social acceleration for the nineteenth century, and consequently for the whole phase, which obliges him to accept it experimentally as a general law of history. Nature is rarely so simple as to act rigorously on the square, but History, like Mathematics, is obliged to assume that eccentricities more or less balance each other, so that something remains constant at last, and it is compelled to approach its problems by means of some fiction—some infinitesimal calculus—which may be left as general and undetermined as the formulas of our greatest master, Willard Gibbs, but which gives a hypothetical movement for an ideal substance that can be used for relation. Some experimental starting-point must always be assumed, and the mathematical historian will be at liberty to assume the most convenient, which is likely to be the rule of geometrical progression.

Thus the first step towards a Rule of Phase for history may be conceived as possible. In fact the Phase may be taken as admitted by all society and every authority since the condemnation of Galileo in 1633; it is only the law, or rule, that the mathematician and physicist would aim at establishing. Supposing, then, that he were to begin by the Phase of 1600-1900, which he might call the Mechanical Phase, and supposing that he assumes for the whole of it the observed acceleration of the nineteenth century, the law of squares, his next step would lead him backward to the far more difficult problem of fixing the limits of the Phase that preceded 1600. . . .

No question in the series is so vital as that of fixing the limits of the Mechanical Phase. Assuming, as has been done, the year 1600 for its beginning, the question remains to decide the probable date of its close. Perhaps the physicist might regard it as already closed. He might say that the highest authority of the mechanical universe was reached about 1870, and that, just then, the invention of the dynamo turned society sharply into a new channel of electric thought as different from the mechanical as electric mass is different from astronomical mass. He might assert that Faraday, Clerk Maxwell, Hertz, Helmholz, and the whole electro-magnetic school, thought in terms quite unintelligible to the old chemists and mechanists. The average man, in 1850, could understand what Davy or Darwin had to say; he could not understand what Clerk Maxwell meant. The later terms were not translatable into the earlier; even the mathematics became hyper-mathematical. Possibly a physicist might go so far as to hold that the most arduous intellectual effort ever made by man with a distinct consciousness of needing new mental powers, was made after 1870 in the general effort to acquire habits of electro-magnetic thought,—the familiar use of formulas carrying indefinite self-contradiction into the conception of force. The physicist knows best his own difficulties, and perhaps to him the process of evolution may seem easy, but to the mere by-stander the gap between electric and astronomic mass seems greater than that between Descartes and St. Augustine, or Lord Bacon and Thomas Aquinas. The older ideas, though hostile, were intelligible; the idea of electro-magnetic-ether is not.

Thus it seems possible that another generation, trained after 1900 in the ideas and terms of electro-magnetism and radiant matter, may regard that date as marking the sharpest change of direction, taken at the highest rate of speed, ever effected by the human mind; a change from the material to the immaterial,—from the law of gravitation to the law of squares. The Phases were real: the change of direction was measured by the consternation of physicists and chemists at the discovery of radium which was quite as notorious as the consternation of the Church at the discovery of Galileo; but it is the affair of science, not of historians, to give it a mathematical value.

Should the physicist reject the division, and insist on the experience of another fifty or a hundred years, the consequence would still be trifling for the fourth term of the series. Supposing the Mechanical Phase to have lasted 300 years, from 1600 to 1900, the next or Electric Phase would have a life equal to $\sqrt{300}$, or about seventeen years and a half, when—that is, in 1917—it would pass into another or Ethereal Phase, which, for half a century, science has been promising, and which would last only $\sqrt{17.5}$, or about four years, and bring Thought to the limit of its possibilities in the year 1921. It may well be! Nothing whatever is beyond the range of possibility; but even if the life of the previous phase, 1600-1900, were extended another hundred years, the difference to the last term of the series would be negligible. In that case, the Ethereal Phase would last till about 2025.

The mere fact that society should think in terms of Ether or the higher mathematics might mean little or much. According to the Phase Rule, it lived from remote ages in terms of fetish force, and passed from that into terms of mechanical force, which again led to terms of electric force, without fairly realizing what had happened except in slow social and political revolutions. Thought in terms of Ether means only Thought in terms of itself, or, in other words, pure Mathematics and Metaphysics, a stage often reached by individuals. At the utmost it could mean only the subsidence of the current into an ocean of potential thought, or mere consciousness, which is also possible, like static electricity. The only consequence might be an indefinitely long stationary period, such as John Stuart Mill foresaw. In that case, the current would merely cease to flow.

But if, in the prodigiously rapid vibration of its last phases, Thought should continue to act as the universal solvent which it is, and should reduce the forces of the molecule, the atom, and the electron to that costless servitude to which it has reduced the old elements of earth and air, fire and water; if man should continue to set free the infinite forces of nature, and attain the control of cosmic forces on a cosmic scale, the consequences may be as surprising as the change of water to vapor, of the worm to the butterfly, of radium to electrons. At a given volume and velocity, the forces that are concentrated on his head must act.

Such seems to be, more or less probably, the lines on which any physical theory of the universe would affect the study of history, according to the latest direction of physics. Comte's Phases adapt themselves easily to some such treatment, and nothing in philosophy or metaphysics forbids it. The figure used for illustration is immaterial except so far as it limits the nature of the attractive force. In any case the theory will have to assume that the mind has always figured its motives as reflections of itself, and that this is as true in its conception of electricity as in its instinctive imitation of a God. Always and everywhere the mind creates its own universe, and pursues its own phantoms; but the force behind the image is always a reality,—the attractions of occult power. If values can be given to these attractions, a physical theory of history is a mere matter of physical formula, no more complicated than the formulas of Willard Gibbs or Clerk Maxwell; but the task of framing the formula and assigning the values belongs to the physicist, not to the historian; and if one such arrangement fails to accord with the facts, it is for him to try another, to assign new values to his variables, and to verify the results. The variables themselves can hardly suffer much change.

If the physicist-historian is satisfied with neither of the known laws of mass,—astronomical or electric,—and cannot arrange his variables in any combination that will conform with a phase-sequence, no resource seems to remain but that of waiting until his physical problems shall be solved, and he shall be able to explain what Force is. As yet he knows almost as little of material as of immaterial substance. He is as perplexed before the phenomena of Heat, Light, Magnetism, Electricity, Gravitation, Attraction, Repulsion, Pressure, and the whole schedule of names used to indicate unknown elements, as before the common, infinitely familiar fluctuations of his own Thought whose action is so astounding on the direction of his energies. Probably the solution of any one of the problems will give the solution for them all.

7 Freud: A New Emphasis on the Animality of Man

Both the creation of Sigmund Freud's psychoanalytic theory and its extension to societies as well as individuals testify to the immense power and appeal of scientific analogies and metaphors. The growth of psychoanalytic theory involved an essential and conscious attempt to use analogies from evolutionary thought and from the physical science of energetics in explaining mental phenomena. And Freud's expansion of psychoanalytic notions into cultural concepts depended on a similar conscious emphasis on the likeness of societal development to the growth of an individual human personality.

Freud's theories, moreover, have significantly changed the way that we think and talk about ourselves and our society. In America today there are few educated men or women who have not read and reacted to the writings of such neo-Freudian authors as Erich Fromm, Norman O. Brown, and Herbert Marcuse or to sociologists like David Riesman, anthropologists like Margaret Mead, and pediatricians like Benjamin Spock. None of the above authors would agree with all aspects of psychoanalytic theory; but all have formulated their ideas in terms of theories of personality stimulated by Freud's work. Nor is Freud's influence confined to the well-read. Almost all of us speak of aggression, frustration, repression, wish fulfillments, unconscious impulses, anxieties, neuroses, and defense mechanisms; and all of these terms as they are used today carry important connotations given to them by Freud. Finally, psychoanalytic ideas have led in a very important way toward a new interest in sensuality and sexuality and toward a general celebration of self-fulfillment and happiness as the major goals in life—a celebration which challenges the immensely powerful work-oriented Puritan ethic, and which manifests itself in much modern art, modern literature, modern drama, and, above all, in the daily behavior of increasing numbers of people.

The fact that Freud was a prolific author, producing more than twenty major works over a period of nearly fifty years, and that he continued to develop and revise his ideas on even some of the most basic elements of psychoanalytic theory and its societal implications throughout his long life, makes it particularly difficult to catch the essence of Freud's ideas in any short series of selections. Moreover, the fact that much of his greatest impact on Western culture came through the work of revisionists who viorently disagreed with Freud and with each other adds to the difficulty of characterizing his influence in a simple and coherent way. All I can hope to do is to provide hints of some of the directions taken by Freud's thought and by those who were stimulated by it.

The first selection, "Freud's Theory of Mind," from Ernest Jones' *The Life and Works of Sigmund Freud*, traces some of the sources of the fundamental concepts of psychoanalysis. Unlike Descartes, whose analogies between machines and biological organisms led him to equate the two, Freud was profoundly aware that the parallels he saw between physical and mental phenomena did not imply their identity or causal connection. Nonetheless, he depended on physics and physiology to suggest the nature of mental mechanisms. The law of the conservation of energy led Freud to propose an analogous law of constancy of some kind of mental stuff that was later called psychic energy. Similarly, the gradual buildup of electrical charge in the nerve synapses with its subsequent instantaneous release led Freud to postulate the buildup of some kind of psychic tension with mechanisms for rapid release. Since virtually all of Freud's later ideas depend on the principles that pleasure is achieved by the release of psychic tension or energy, and that such energy demands release in one way or another, finding secondary outlets if primary ones are blocked, this analogy is absolutely central to psychoanalysis. The "Project for a Scientific Psychology" in which many of Freud's physicalistic ideas were developed was, as Jones points out, renounced by Freud shortly after he finished it. But in many important ways Freud's subsequent career can be seen as an attempt to transform the physiological notions and laws of the "Project" into directly analogous psychological concepts and statements.

The second selection, written in 1923 by Jacques Rivière, editor of the influential *Nouvelle Revue Française*, provides a clear idea of the impact of Freud's psychoanalytic theory on the literary community of the early twentieth century. Rivière provides an exposition of certain key psychoanalytic insights and draws a number of important and unsettling inferences from them—that, "all of our feelings are the strict equivalent of neurotic symptoms, . . . that all human beings are driven to hypocrisy," and that, "the source of all spiritual creation is carnal." Such inferences from Freudian theory did much to alter the focus of modern literature, encouraging an ethical nihilism which argued that all values are irrational and that men's inner lives are mere collections of senseless compulsions. Rivière retreats from this radical interpretation of psychoanalysis, emphasizing the fundamental determinism and rationalism of Freud's works, but many could not.

Up to 1910, Freud concentrated on biological factors in the creation of human personality; but from then on he tended to balance this initial emphasis by considering more and more the individuals' relations to society and to the development of cultural and moral systems. He began to speak of basic aggressive tendencies in man as well as sexually oriented drives toward pleasure; and he began to postulate a system of culturally generated moral repressive control—the superego—that goes beyond the value-neutral ego in redirecting the basic biological drives in order to protect society from the aggressive tendencies of individuals.

In 1913, Freud made his first attempt to explain cultural development through psychoanalysis. In *Totem and Taboo* he argued that the earliest human society was a "primal horde" like that described by Charles Darwin. The dominant male, the father of this horde, subjected all younger males to his rule and kept all females for his own use. Eventually the young males banded together, killed the father, and cannibalized his body. But soon after this act the males felt a need for atonement; they forbade the killing of the totem animal which symbolized the father and instituted a ceremonial feast in which the original crime was ritually reenacted. This explains the widespread totem worship among primitive tribes. Furthermore, since the women had been the original cause of the patricide, the men forbade marriage with the women within the tribe and created a taboo against killing within the group in order to avoid a repetition of the disruptive act; that is, they created social regulations to curb men's sexual drives. J. A. C. Brown describes the results of Freud's analysis as follows: "In a single hypothesis, he explains the origin of society, of religion and law, of totemism, of the incest taboo and exogamy, and of rituals and myths. Law curbs the sexual and aggressive drives, religion, myth, and ritual commemorate the crime and assuage guilt, and society is the overall mechanism of control."[1]

The basic argument of *Totem and Taboo* was clearly not accepted by ethnographers and anthropologists in general—but the notion that psychoanalytic principles could be a guide to the analysis of culture was widely adopted. Freud himself continued in this vein, writing the *Future of an Illusion* and *Civilization and Its Discontents* in 1930 and *Moses and Monotheism* in 1939; and during the 1940's and 50's there was a wider movement to base the analysis of culture on psychological principles. Margaret Mead's *Balinese Character* (1942), Ruth Benedict's *The Chrysanthemum and the Sword* (1946), and Abraham Kardiner's *The Individual and His Society* (1939) are outstanding examples of the trend. One of the problems of this movement is that it is similar in many ways to social Darwinism; the extension of psychoanalytic ideas beyond individual development has been used to justify, condemn, and call for a wide range of cultural adjustments. Herbert Marcuse in *Eros and Civilization*, for example, calls on Freud to sanction a totally nonrepressive society, while Abraham Kardiner seeks a new author-

---

[1]J. A. C. Brown, *Freud and the Post-Freudians* (Baltimore: Penguin Books, 1961), pp. 116-117.

ity—a new, more effective system of repressions—to avoid the distressing ambiguities of modern life. Like Darwinism, Freudian theory does not seem to provide sufficiently unambiguous analogies to force agreement on its societal implications. In his "Freud, the Revisionists, and Social Reality," Will Herberg analyzes the impact of Freudian ideas on social thought. He emphasizes Freud's pessimistic belief in the unavoidable conflict between individuals' biological instincts and cultural demands and contrasts that with the confident belief in the harmony of men in society which characterizes the work of Erich Fromm, one of the most influential of neo-Freudian authors.

Freud's Theory of Mind (1900)

Ernest Jones

Freud came from his early training deeply imbued with the belief in the universality of natural law and with a disbelief in the occurrence of miracles or spontaneous or uncaused acts. Scientific investigation would indeed be pointless if the order it strove to ascertain did not exist. He would certainly have subscribed to the closing words of his teacher Meynert's address to the 54th *Versammlung Deutscher Naturforscher und Ärzte*: "On the Lawfulness of Human Thought and Behavior": "I should like to say in general that all philosophy, all human acceptance of wisdom so far as history spans, has really brought to light only two conclusions in which the outlook of those who have made use of the thought of all mankind differs from that of the common man. One is that everything in the world is only appearance and the appearance is not identical with the essence of things; the second is that even the freedom we feel in ourselves is only apparent."

The reason Meynert gives for the illusion of free will is that we are not yet able to follow in the finest details the regular processes in the life of the brain. Nevertheless, the apparent freedom is really based on law, therefore on necessity.

Herbart, to whom many of Meynert's and Freud's ideas can be ultimately traced, had also in 1824 protested against "this false doctrine of a free will that has raised its head in recent years,"—referring here to his own teacher Fichte, about whose idealistic philosophy he felt equally strongly.

Source: Ernest Jones, M. D., *The Life and Works of Sigmund Freud*, vol. I (New York: Basic Books, Inc., 1953), Chapter 17, pp. 265-399.

Writing on this subject in 1904, Freud gave the reason for our unshakable conviction of freedom of choice. He remarked that it is far stronger with trivial decisions than with weighty ones; with the latter we commonly feel that our inner nature compels us, that we really have no alternative. With the former, however, for example the arbitrary choice of a number, we discern no motive and therefore feel it is an uncaused act on the part of our ego. If now we subject the example to a psychoanalysis we discover that the choice has after all been determined, but this time the motive is an unconscious one. We actually leave the matter to be decided by our unconscious mind and then claim the credit for the outcome. If unconscious motivation is taken into account, therefore, the rule of determinism is of general validity.

Freud never wavered in this attitude and all his research into the workings of the mind is entirely based on a belief in a regular chain of mental events. He would have endorsed the view of the great anthropologist Tylor that "the history of mankind is part and parcel of the history of Nature, that our thoughts, wills and actions accord with laws as definite as those which govern the motion of the waves." When enumerating the essential elements of psychoanalytical theory, in 1924, he included "the thoroughgoing meaningfulness and determinism of even the apparently most obscure and arbitrary mental phenomena." He does not appear ever to have expressed any opinion on the general theory of causality, but he presumably held the simple nineteenth-century view of invariable antecedents. . . .

Mind and Brain

In this narrower field there is more to be said. The first statement on it is to be found in Freud's *Aphasia* (1891). There he proclaimed himself an adherent of the doctrine of psychophysical parallelism. "The chain of the physiological processes in the nervous system probably does not stand in any causal relation to the psychical processes. The physiological processes do not cease as soon as the physical ones begin; the physiological chain continues, but from a certain moment onwards there corresponds with each link in it (or several links) a psychical phenomenon. Thus the psychical is a process parallel to the physiological ('a dependent concomitant')." He then proceeds to quote the following passage from Hughlings Jackson: "In all our studies of diseases of the nervous system we must be on our guard against the fallacy that what are physical states in lower centers fine away *into* psychical states in higher centers; that, for example, vibrations of sensory nerves *become* sensations, or that somehow or other an idea produces a movement." . . .

Freud held that not only was the essential nature of both mind and matter quite unknown, but they were so intrinsically different in kind as to make it a logical error to translate a description of processes in the one into terms of the other. Nor was there any clue for elucidating the direct relationship of one to the other. How an excitation of the retina could be followed by a perception of light or form was an unapproachable mystery. Of course, like all doctors and many other people, Freud would often use loose language incompatible with what has just been expounded: bodily, e.g., sexual, changes would *produce* anxiety, or an emotion would *produce* paralysis of a limb. Clearly, however, these are shorthand expressions not meant to be taken as literal exactitude. Psychosomatic medicine, for example, is replete with phraseology of this kind.

Nevertheless Freud believed, much more strongly in his early years but perhaps to some extent always, that the correlation of mental processes with physiological ones hinted at a similarity in the way both worked. As we shall see presently, he cherished the hope for a time that by applying physical and physiological concepts, such as those of energy, tension, discharge, excitation, etc., to mental processes it would be possible to achieve a better understanding of such processes. He even made a valiant, if somewhat forlorn, endeavor to put this into operation and he wrote a brochure (in 1895) describing his effort in detail. . . .

Physics, Physiology, and Psychology

We know that Freud's main aim in life, certainly in the early productive period and perhaps always, was to formulate a theoretical basis for the new discoveries he was making in psychopathology and, with the help of that, to found a theory of the mind that would take into due account the peculiar features of the unconscious; the outcome of his endeavors is called psychoanalysis. It is therefore pertinent to inquire into what foundation he had to work on apart from his own clinical observations. . . .

Much work has been done in tracing the genealogy of the basic ideas that Freud employed in his psychology. The most painstaking example is a slight volume by Maria Dorer, of Darmstadt, one unfortunately somewhat marred by an undue intentness on establishing a particular thesis. Some of her conclusions will presently be mentioned. It was a Polish psychologist, Luise von Karpinska, who first called attention to the resemblance between some of Freud's fundamental ideas and those promulgated by Herbart seventy years previously. We have mentioned above one on freedom of the will. The one Karpinska especially dwells on is Herbart's conception of the unconscious, which was the only dynamic one before Freud's. According to it, unconscious mental processes are dominated by a constant conflict which Herbart describes in terms of ideas of varying intensity—a notion which Freud later replaced by a conflict of affects; with Herbart ideas are always primary to affects, as in the later James-Lange theory. The conflict Herbart describes is partly intrapsychical but more characteristically between those of one person and of another. The latter are treated as disturbing, or aggressive, elements which evoke "self-preservative" efforts on the part of the subject. Mental life is throughout dualistic, as Freud also always conceived it. Herbart actually describes an idea as "*verdrängt*" when it is unable to reach consciousness because of some opposing idea or when it has been driven out of consciousness by one! He conceives of two thresholds in the mind, which correspond topographically with the position of Freud's two censorships. One, the "static threshold," is where an inhibited idea is robbed of its activity and can enter consciousness only when the inhibition is lifted; it is, therefore, not unlike a "suppressed" idea in the preconscious. At another level is what he calls the "mechanistic threshold" where wholly repressed ideas are still in a state of rebellious activity directed against those in consciousness and succeed in producing indirect effects, e.g., "objectless feelings of oppression (*Beklemmung*)." "Science knows more than what is actually experienced [in consciousness] only because what is experienced is unthinkable without examining what is concealed. One must be able to recognize from what is experienced the traces of what is stirring and acting 'behind the curtains'!"

All this is very interesting, but there is more. People vary in the way in which the body responds to affects (Freud's somatic compliance), which Herbart calls the "physiological resonance"; this leads to a "condensation of the affects in the nervous system." Mental processes are characterized by a "striving for equilibrium" (Freud's constancy principle). "Ideas" are indestructible and are never lost. Nor do they ever exist alone, only in chains of ideas that are so interwoven with one another as to form networks. Affects arise only when the equilibrium is disturbed through an excessive quantity of intensity being present in the ideas. Consciousness of self (the ego) comes about when active ideas are inhibited (frustrated?).

Herbart's principal thesis was that mental processes must be capable of being resumed under scientific laws. "Regular order in the human mind is wholly similar to that in the starry sky." The processes must ultimately be measurable in terms of force and quantity. He dreamt of a "mathematical psychology" and drew up a project for one. Some years later Fechner seemed to have made important progress in this direction by generalizing Weber's law into the statement that the strength of our sensations increases proportionally to the logarithm of that of the stimulus. Perhaps these hopes could be traced back to Spinoza's use of geometry in his *Ethics*, but they overlook Pascal's classic remark that "the heart has reasons which the reason knows not of." . . .

Notions such as those just described could have filtered through to Freud from many sources, but those echoes from the past are nonetheless noteworthy. It is not very likely that Freud would ever have had reason to make a study of Herbart's writings, though it is of course possible. We do not even know if Meynert did, but his published works make it certain that he was very familiar with the Herbartian psychology, on which his own was based and of which his was an extension and modification. He must in any case have had access to it through the full exposition of it by Griesinger, of whose writings Meynert thought highly and which Freud probably also read.

After this paragraph was written, Dr. and Mrs. Bernfeld, thus increasing my great debt to them, communicated the remarkably interesting fact that in Freud's last year at the Gymnasium the following textbook was in use: Gustaf Adolf Lindner, ***Lehrbuch der empirischen Psychologie nach genetischer Methode*** (Textbook of Empirical Psychology by Genetic Method) (1858). The author's teacher was Franz Exner, the father of Freud's instructor at the Brucke Institute. Now in the preface to the book the author states categorically that only thinkers of the Herbart school come into consideration, and in fact the book may be described as a compendium of the Herbartian psychology. It contains among other things, this passage: "A result of the fusion of ideas proves that ideas which were once in consciousness and for any reason have been repressed (*verdrängt*) out of it are not lost, but in certain circumstances may return." There is a detailed account of the conflict between stronger and weaker ideas along correct Herbartian lines.

Fechner's psychology is altogether built on Herbart's, whose main principles (except the metaphysical ones) he fully shared; it was strengthened by his endeavor to apply to living organisms the recently discovered principle of the conservation of energy. He went further in maintaining that pleasure-unpleasure phenomena were themselves susceptible of quantitative treatment, and not simply qualitative. The word "threshold" stands at the center of all his writings, and he maintained that whenever certain physiological processes attained a given intensity they would be followed by conscious ones. He did not commit himself on the question of whether unconscious processes could be psychical, but of their importance otherwise he was convinced. "What is below the threshold *carries* the consciousness, since it sustains the physical connection in between." He likened the mind to an iceberg which is nine-tenths under water and whose course is determined not only by the wind that plays over the surface but also by the currents of the deep.

Fechner exercised an important influence on Brucke, who held that "movements in the nervous system give rise to ideas," on Meynert, whom we shall consider presently, and on Breuer, who ranked him next only to Goethe. Freud, who had studied his writings at first hand, also spoke highly of him. He said, "I was always open to the ideas of G. T. Fechner and have followed that thinker upon many important points."

For Meynert mind and brain were so closely connected that they could be spoken of in the same breath, and sometimes interchangeably; the "mechanics of the brain" was a favorite phrase of his. Although he was much influenced by Kant and Schopenhauer—he was altogether well read in philosophy—his psychology was essentially founded on the "association psychology" of Herbart and Fechner. All three repudiated the "faculty psychology" that had such a vogue in England in the early nineteenth century.

A prominent characteristic of Meynert's psychology was his "projection" theory, in which he uses various optical analogies (camera, etc.), as Freud did later. He used this term to indicate the gathering from various sources of impressions that finally reached the cells of the cortex, and then are from there "projected into consciousness."...

Project for a Scientific Psychology

In the days of which we are now writing a great deal was known of both the gross anatomy and the histology of the brain, subjects which Freud had fully mastered, but very little of its physiology. Fritsch and Hitzig in Germany, with Ferrier and Horsley in England, had shown that electrical stimulation of certain areas of the cortex produced movements in the limbs on the opposite side of the body, and it was also known that destruction of certain other areas would lead to aphasia, either motor or sensorial. Beyond this there was very little, and a great deal of the talk about brain physiology and its application to the mind was really little more than using the language of physics—with terms like energy, tension, force, etc.—in another sphere.

Doubtless all psychologists expected, as they still must do, that one day scientific law and order would prevail in the apparent chaos of mental processes as it has in other sections of the universe. The illusion, however, that scientists of those days labored under was that the most promising approach to that desideratum was through brain physiology. In the event brain physiology has turned out to be even more refractory than psychology itself.

Freud seems to have shared this illusion himself for many years, until there was a gradual liberation from it which was complete by 1897. It brought about a very interesting episode in his life, the writing of a long essay which is here called the "Project." We know that his ambition in those years was to advance in knowledge along the avenue signaled by the designations: anatomy of the brain, physiology of the brain, psychopathology, psychology, philosophy. The first two of these proved to be will-o'-the-wisps, and the final goal was only partially attained. But of the advance there is no doubt.

The earlier ambition of proceeding direct from the brain to the mind reached its climax in 1895. On April 27, in the month after he wrote the chapter on psychotherapy for the *Studies in Hysteria*, he wrote to Fliess that "I am so deeply immersed in the 'Psychology for Neurologists' as to be entirely absorbed until I have to break off, really exhausted by overwork. I have never experienced such intense preoccupation. I wonder if anything will come of it? I hope so, but it goes on slowly and with difficulty." A month later (May 25) he described his aims in a passage quoted earlier, and, after giving as one of them the hope of extracting from psychopathology something useful for psychology, he went on: "Actually a satisfactory general understanding of neuropsychotic disturbances is not possible unless one can make a connection with clear presuppositions about normal psychical processes. I have devoted every free minute to this work during the last few weeks, spending the hour from eleven to twelve in phantasy, in translation from one field to the other, and guesswork, and ceasing only when I have reduced some idea to an absurdity or else really and seriously overworked to the extent of losing all interest in my professional activities." (May 25.)

It is not surprising that this obsessive preoccupation found a limit. By August 16 we hear that he was throwing it aside and pretending to himself that it was not interesting. He was on holiday and felt he had been setting himself a crucifying task. "Anyhow, playing at skittles and searching for mushrooms are much healthier occupations." He had found that one topic had led to another, to memory, to sleep, and so on in such an endless fashion as to give the whole project mountainous proportions.

One suspects that an important immediate stimulus to this gigantic undertaking, for which he was assuredly ill-prepared and untrained, had been given by a huge volume that his teacher Exner had published only the year before on the same theme. It was written more diffusely than Freud's "Project," and was eight times as long, but there is a good deal of similarity between the two. Exner held that the degrees in excitation in the nervous system must be subject to quantitative laws. He uses the same phrase as Freud, "summation of excitations," and discusses at length the function of inhibition in the control of simple stimuli. It was he who developed the conception of *Bahnung* (facilitation of the flow of excitation) that plays an important part in Freud's exposition. Like Freud, he covers a wide field, dealing with the topics of perception, judgment, ideation, recognition, thinking processes, and so on. The pleasure-unpleasure principle is a regulating one. His disbelief in free will is based on the same arguments as Freud's, the illusion of it arising from a break in the antecedent ideas in consciousness; where, however, Freud demonstrated the existence in the unconscious of the missing links, Exner could only fall back on a vague remark about the continued activity of the subcortical centers. His fundamental aim was the same as Freud's: to deprive mental processes of their peculiar status among natural phenomena. By way of contrast, however, it must be said that Freud's essay was not only far more concise, but much more closely reasoned than Exner's.

The year before that another of his seniors in the Brücke Institute had published (posthumously) a volume which would provide Freud with any basis he needed for his cerebral physiology.

The respite the holiday gave did not last long, and the tyrannical compulsion to get his ideas set forth was soon again in action. On September 4 Freud went to Berlin to visit Fliess and doubtless to take over the problems concerned. The result of this was so exciting that Freud could not wait to get home but started composing in the train; the first part of his essay was thus written in pencil. The whole essay, which would make a brochure of some hundred pages, was written feverishly in a couple of weeks, with a few necessary intervals. All three parts were dispatched to Fliess on October 8, and fortunately he perserved them; they were published as an Appendix to the letters of the *Aus den Anfängen der Psychoanalyse* (From the Beginnings of Psychoanalysis) (1950).

It is noteworthy that Freud never asked for the return of that interesting manuscript which had cost him so much trouble, nor apparently did he ever want to see it again. He had been relieved of an oppressive burden, and his attitude towards what he had given birth to rapidly changed. In sending it off he said he was keeping back a final section, on the "Psychopathology of Repression," which he was finding extremely difficult; his mood about it was alternately "proud and happy" or "ashamed and miserable." This final section has not come to light; possibly it was never completed. The following sentence, written a week later, probably refers to it: "For two weeks I have been in a fever of writing and believed I had the secret; now I know I haven't and so have put the matter aside." Three weeks later he wrote that he had deposited the manuscript of the "Psychology" (probably again that troublesome section) in a drawer, partly because he ought to begin writing the monograph for Nothnagel—for which he had only another six weeks, but which actually plagued him for another whole year—and doubtless also because he could not find a satisfying solution to the problems.

The elation at his recent accomplishment lasted a little while and was accompanied by the following interesting experience. In a letter of October 20, less than a fortnight after finishing what we possess of the "Project," he wrote: "One evening last week when I was hard at work, tormented with just that amount of pain that seems to be the best state to make my brain function, the barriers were suddenly lifted, the veil drawn aside, and I had a clear vision from the details of the neuroses to the conditions that make consciousness possible. Everything seemed to connect up, the whole worked well together, and one had the impression that the Thing was now really a machine and would soon go by itself. The three systems of neurones, the free and bound state of Quantity, the primary and secondary processes, the main tendency and the compromise tendency of the nervous system, the two biological laws of attention and defense, the indications of Quality, Reality, and Thought, the (particular) position of the psychosexual group, the sexual determinant of repression, and finally the necessary conditions for consciousness as a function of perception: all that was perfectly clear, and still is. Naturally I don't know how to contain myself for pleasure." This passage, with its hint of artistic intuitiveness, may mark the transition from Freud the dogged worker to Freud the imaginative thinker.

The mood of elation, however, could not last, and the excitement soon died down. Only ten days later he was telling Fliess, that on reflection he saw that what he had sent him had in part lost its value and was meant to be only a preliminary draft, but he hoped it might come to something. And a month later it was all over: "I no longer understand the state of mind in which I hatched out the "Psychology," and I can't understand how I came to inflict it on you. I consider you are always too polite; to me it seems pure balderdash."

We may now ask what claims this curious document has on our interest. It has many.

In the first place, it is a magnificent *tour de force*. Nowhere in Freud's published writings do we find such a brilliant example of his capacity for abstruse thought and sustained close reasoning. It reveals a brain of the highest order and makes a mockery of his own complaints about his poor intellect. With its elliptical style and somewhat obscure phraseology, almost unrelieved by concrete examples, it imposes more exacting demands on the reader than any of his published work; there must be very few who can apprehend its full meaning without several perusals. Much of this difficulty, of course, is because the essay was couched in language familiar only to its solitary audience.

Then it is of great value to the student of Freud's psychology because it throws light on many of his later conceptions, some of them rather alien, which he seldom defined or even elucidated. The language of physics and cerebral physiology in the "Project" was Freud's natural one, to which he in great part adhered later even when he was dealing with purely psychological problems. It is true that he then gave the terms he used psychological meanings which take them away from their original context, but all the same they are often terms that no psychologist would have employed to start with. It has not been easy to translate some of them into more familiar terminology or to define their precise signification in Freud's mind. For this the study of the "Project" should prove most helpful.

The richness of the ideas contained in the "Project," and the extraordinarily close relationship subsisting among them, provide the student with a wealth of material for research. There is room for several monographs devoted to the elucidation of this little essay.

For the student of Freud's personality the "Project" has several instructive lessons. It shows what a hold the concrete studies of his youth had obtained over him. In the realm of the visual, of definite neural activities that could be seen under the microscope, he had for many years felt entirely at home; he was as safe there as at the family hearth. To wander away from it and embark on the perilous seas of the world of emotions, where all was unknown and where what was invisible was of far greater consequence than the little that was visible, must have cost him dear. He was called to a high endeavor, which he was now on the brink of undertaking, and we may regard the feverish writing of the "Project" as a last desperate effort to cling to the safety of cerebral anatomy. If only the mind could be described in terms of neurones, their processes and synapses! How fond the thought must have been to him.

Another important consideration is that never again until the last period of his life, and never before so far as we know, did Freud indulge in deductive reasoning as he did here. The great Herbart, it is true, had maintained that in psychology deduction has equal rights with induction, but this metaphysical heresy had been vehemently repudiated by both Griesinger and Meynert, and Freud himself had been drilled in the sacred doctrine that all conclusions were to be founded on experience, and experience alone. Yet in the "Project" there is very little direct reference to any experience. Axioms and assumptions—whether plausible or not is beside the point—are taken as the basis for far-reaching trains of thought and somewhat dogmatic conclusions. It is an essay one would have expected from a philosopher rather than a pathologist.

The word philosopher is suggestive. Perhaps Freud was here for the first time releasing his early, and so thoroughly checked, tendency to philosophize. The feverish obsessiveness with which he wrote the essay hints at some deep underground activity, one of which his quick subsequent repudiation expressed disapproval. If not checked, this tendency might end in empty speculation, an arid intellectualizing of the underground urges. Fortunately the issue was otherwise. Freud returned to the empirical experience of his clinical observations, but he had taken the crucial step of releasing, even if only for a month or two, something vital in him that was soon to become his scientific imagination—a realm in which both sides of his nature were to find free play in a fertile cooperation.

It is indeed remarkable how closely the "Project" already unites and expresses the two opposite sides of Freud's nature, the conservative and the freely imaginative. It was doubtless that combination, once effected, that gave such a powerful urge to its composition. Its relative sterility is to be explained by its divorce from clinical data. He had yet to find more fruitful outlets, which only the courage to explore emotional experiences could provide.

After such a long preamble it is time to consider what is contained in this "Project." Unfortunately it is not only highly abstruse but is already so condensed that any synopsis of it would be totally unintelligible. I must therefore content myself with some general remarks and with providing a list of the topics dealt with, most of which—though not all—Freud expanded in his later writings.

The essay itself had no title, though Freud had spoken of a monograph to be called "Psychology for Neurologists." The editors of the *Anfänge* decided to label it "*Entwurf einer Psychologie*" (Project for a Scientific Psychology).

What is available to us is divided into three parts: (1) General Plan; (2) Psychopathology, particularly of hysteria; (3) Attempt to represent normal psychical processes.

The General Plan is introduced by a statement that the aim of the "Project" is to furnish a psychology which shall be a natural science. This Freud defines as one representing psychical processes as quantitatively determinate states of material elements which can be specified. It contains two main ideas: (1) to conceive in terms of Quantity, which is subject to the general laws of motion, whatever distinguishes activity from rest; (2) to regard the neurones as the material elements in question.

Freud's aim was therefore to combine into a single whole two distinct theories. One was the neurone theory derived from his neuro-histological studies. The other dated farther back to the Helmholtz-Brücke school, although it must have been powerfully reinforced by Meynert and the other authors discussed above. It was to the effect that neurophysiology—and consequently psychology—was governed by the same laws as those of chemistry and physics. . . .

The functioning of the nervous system, according to the "Project," was subject to two general and closely related principles. One was that of "inertia" (*Trägheit*), which stated that neurones tend to get rid of any Quantity they contained; in later years it became the "pleasure-unpleasure principle." Freud maintained that this idea was born in his clinical observations of the psychoneuroses, from the notions of intensive ideas, of stimulation, substitution, discharge, and so on, and that he felt it legitimate to transfer it to the field of neuronic activity (which in its turn was to explain the working of mental processes). Reflex movement, where a sensory excitation is followed by a motor discharge, is the purest form of this principle in animals. Freud traced the neuronic reaction to an origin in the excitability of the surface layer of protoplasm.

This discharge may be called the primary function of the neuronic system; nothing matters but the discharge in any direction. A secondary function comes into play when paths of discharge are chosen that put an end to the stimulation—"Flight from excitation." But what really causes the mechanism to break down, so that it has to be modified, is its inability to deal with internal stimulation (from the body) in the simple way it can with external stimulation. Here both discharge and flight are impossible, and help can be obtained only by specific changes to be brought about in the outer world, e.g., getting food. For this purpose the neurones have to keep in readiness a certain reserve of Quantity, reducing it to nil being no longer feasible.

The other principle, that of "constancy," was more obviously derived from the domain of physics. It was one used extensively by both Breuer and

Freud, although Breuer gave Freud the credit for perceiving its importance in the present connection. They had defined it together in 1892 as follows: "The nervous system endeavors to keep constant something in its functional condition that may be described as the 'sum of excitation'."

On the basis just described Freud was able to construct what James Strachey has truly called "a highly complicated and extraordinarily ingenious working model of the mind as a piece of neurological machinery," one which Freud said "nearly worked by itself."

The first problem to arise was that of *memory*. How was it possible for neurones to be permanently altered from receiving an impression, and storing its traces, and yet to receive fresh ones in just the same manner as before? This riddle Freud solved by postulating two classes of neurones, one for each purpose. One class, to which he gave the designation ϕ, allowed a current of excitation to pass through without effecting any change; this was specially concerned with stimuli from the outer world and could be identified as the nuclei of afferent nerves. The other, which were labeled ψ, retained permanent traces of any stimulation affecting them; they were the nerve cells of the brain proper. These latter are concerned principally with stimuli arising in the body itself, and Freud even speculates—strangely enough—that the brain could have arisen from an enormous complication of sympathetic ganglia.

The correlation between two classes of neurones, easily permeable and less permeable, on the one hand, and the phenomena of perception and memory on the other, was a fundamental feature of Freud's theory, one which was retained later when expressed in psychological language only. In the present context the variation in permeability was supposed to reside in the contact barriers of the neurones (synapses), which would be different not only with different cells but also with the various processes belonging to each neurone. It is a conception he fully shared with Breuer, who may well have originated it; in his theoretical chapter Breuer had phrased it thus: "This perceptual apparatus, including the cortical areas for the senses, must be different from the organ that preserves and reproduces sensorial impressions as memory images. For the fundamental condition for the functioning of the perceptual apparatus is the swiftest *restitutio in statum quo ante*; otherwise no further proper perception could occur. The condition for memory, on the contrary, is that no such restitution takes place, but that every perception produces lasting changes. It is impossible for one and the same organ to fulfill both of these contradictory conditions; the mirror of a reflective telescope cannot be at the same time a photographic plate." Breuer also followed Meynert, as did Freud, in ascribing hallucinations to the reverse process, i.e., a stimulation of the perceptual apparatus emanating from the memory images.

Since no support for this distinction between two kinds of cells can be found in histology, Freud threw out the suggestion that permeability and impermeability depended not on the type of cell but on the Quantity emanating from two different sources; with the ϕ neurones from the outer world, with the ψ neurones from the cells of the body (plus perceptual stimuli passed on to them through ϕ). There is every reason to suppose that the strength of the former stimuli is much greater than that of the latter, so much so that Freud presumed that the arrangement of the peripheral nerve endings act as a screen (an idea he was to elaborate later in his psychology proper); nevertheless they easily overcome the resistance at the ϕ contact barriers. He thus associates his distinction with the fundamental biological functions of the nervous system, of dealing with the stimuli of the outer world and of the body respectively.

There was one thing that could dislocate all the machinery—bodily pain; no contact barriers could hold up painful stimuli. Flight from pain is therefore a primordial tendency of the nervous system, and the memory itself of the object that had evoked the pain causes unpleasure (*Unlust*). This conception of the breaking of barriers Freud later linked both with the importance of the part played in psychopathology by trauma and also with the theory of fright and the warning signals of anxiety.

The problem of consciousness naturally gave great difficulty, since in addition to the physical laws governing variations in Quantity a new factor, Quality, enters. Freud therefore introduced a third set of neurones, which he labeled ω, intermediate between ϕ and ψ, the function of which was somehow to transform Quantity into Quality; the other two operated only in terms of the former (ω). Freud saw consciousness neither as a mere appendage of the physiological-psychical processes nor as the subjective aspect of *all* psychical processes, but as the subjective aspect of a *part* of them, namely, the perceptual processes (ω), so that the apparatus is different when consciousness is absent inasmuch as then no contribution is being made from the ϕ neurones. Nevertheless, the mechanism seemed to creak a little at that point. Freud had to introduce a third dimension into his measurements, that of time: he terms it a "Period."

We meet here the familiar conception that unpleasure (*Unlust*) signifies a heightening of the level of Quantity, pleasure its discharge. It is the basis of the pleasure-unpleasure principle.

It was remarked earlier that the ψ neurones received stimuli from two sources, from ϕ neurones and from the interior of the body. Freud accordingly postulated two sets of neurones: "mantle" neurones and "nuclear" neurones respectively. It is stimulation of the latter by instinctual processes that manifests itself as the *Will*, and supplies the motive power for the whole machine. This necessitates some action in the outer world. With the infant it means crying for help from another person, a complicated method of getting the inner tension discharged. Here Freud added an illuminating formula: "This path of discharge thus acquires the highly important secondary function of *establishing human contact*, and it is the early helplessness of human beings that provides the *original source of all moral motives.*" It anticipates his later account of the part played by human relationships in the transition from the pleasure to the reality principle.

The *ego* is described as an organization of neurones charged with a constant reserve of Quantity, with fairly free communication between themselves. An essential function of it is its capacity to inhibit incoming excitation; this is accomplished through part being diverted to what Freud called "side cathexes," i.e., charged neurones connected through simultaneity with the first neurone stimulated; they are always available.

We come next to a theme of the utmost importance, the distinction Freud established between what he called "primary processes" and "secondary processes." It was perhaps his most fundamental contribution to psychology, one which he defined and extended on several occasions. The only feature he considers in the present context, however, is the difference in the flow of discharge with the two processes. The working of the apparatus can suffer in two ways. With a wish clamoring for satisfaction the excitation may, after stimulating the memory image of the satisfying object, revert to ω, evoking an hallucination that is biologically futile; an indication is needed to distinguish between this and the perception of the real object. A criterion of reality is similarly needed in the second case, that of a "hostile" memory associated with unpleasure; when it is present a defensive flight can occur, whereas otherwise the excitation would evoke the full amount of unpleasure.

The criterion of reality is afforded by the inhibiting powers of the ego. The stimuli proceeding from within the body can stir the qualitative difference characteristically associated with external reality only if they are strong; when subject to the inhibition exercised by the ego they cannot. ϕ excitation (from external stimuli), on the other hand, can provoke the qualitative difference, however weak it may be. What actually gives the ψ neurones the information that a stimulus from the "real" outer world is operative is an excitation from the (motor) discharging mechanism which always then comes into play to some extent; even with thought there is some slight involuntary action of the muscles concerned with speech.

In a word, therefore, the "primary process" is an uninhibited one, with a freely flowing current; the "secondary process," distinguishing between an external and an internal stimulus, is an inhibiting one, the level of whose Quantity is therefore lower. The distinction between "free mobile" energy and "tonic, bound" energy Freud ascribed to Breuer, and he expressed the opinion that it represents our deepest insight into the nature of nervous energy. Incidentally, it may be remarked that this nomenclature is the reverse of the one used in physics, where kinetic energy is "free" as contrasted with potential energy.

Taking the search for satisfaction, "wish-fulfillment," as the fundamental motive power of the machine, Freud analyzed in physiological terms what may be supposed to happen in and between the various neuronic systems during the processes of *judging, recognizing, distinguishing, remembering, and thinking*, all of which are complicated methods of conducting the search. Freud asserted indeed that the search, when wishes arise, is the biological justification of every process of thought.

The beginnings of the theory of dreams that the "Project" contains have been mentioned in the chapter on that topic. It was perhaps in that field that Freud had found the purest example of the "primary process"; during sleep the diminution of bodily needs makes the secondary function of the ego superfluous.

In the *Second Section*, on psychopathology, Freud distinguished between a "primary defense" against pain, on the one hand, and, on the other, "repression," which is an exclusion from consciousness—or, as he says, more strictly an exclusion from the processes of thought—of an idea which would cause unpleasure in the ego; provided always that the idea in question takes its origin in a sexual impulse. The peculiarity of the latter feature is that so often with it the release of affect occurs in connection with the memory rather than with the sexual event itself. Freud attributed this to the lateness of puberty, an explanation that became inadequate after his discovery of infantile sexuality. The enigma of repression, however, was not fully solved. But the conclusion that affects hinder thought—which can only operate with small "testing" amounts of Quantity—by facilitating the passage of "primary processes" remains valid.

In the *Third Section* Freud applied the general principles mentioned above to the working of the normal mind as a whole. This section is so technical, complicated, and closely thought out that it would need a special treatise adequately to expound it. It is mostly concerned with the changes in cathexis, and the flow of energy from one set of neurones to another that is supposed to take place during different kinds of thinking—observing, recognizing, judging, discriminating, "practical" thinking, reflective thinking; and also with the various kinds of errors in the processes of thought.

A prominent idea is that of *Attention*. Freud described how the ego neurones respond to an external stimulus not simply passively, but by actively directing a charge of energy to the ω neurones. This happens particularly when consciousness is present. He thought it easier to explain biologically than mechanically, though he endeavored to do this also. It is a theme that plays little part in his later writings. Nevertheless he here establishes as one of his two "biological laws"—the other is the primary defense against pain—the cathecting by the ego of any perceptual element that has the indication of reality (*Realitätszeichen*), i.e., one signifying a source in the outer world.

Great importance is attached to *Speech Associations*. They have two special features: they are limited in number, and they are exclusive or peculiar. Their motor aspects are valuable in affording one of the tests of reality, and also in furthering the processes of memory. In his later writings speech is regarded as an attribute distinguishing the preconscious from the unconscious proper.

Speech also plays an important part in the early stages of development in the relations with the human environment. Here the inevitable frustrations are the main stimulus to the sense of reality, of the outer world, and the act of crying is the first attempt to replace the original relief through hallucinatory imagination by the circuitous route of bringing about changes in the outer world.

Freud found the origin of the *ego* the most obscure problem of all. He explained it by a particular interaction between the nuclear neurones (those fed by somatic stimuli) and the process of satisfying a desire. When this occurs associations are forged in the "ego" with the perceptual image of the satisfying object on the one hand and on the other hand the information derived from the motor activity that brought that object within reach. This primitive ego has to learn through experience, however, not to cathect these images of movement before certain conditions are fulfilled in respect of the perception, i.e., before tests of reality have been applied. Nor, moreover, must it allow the wish-idea to be cathected beyond a certain degree, lest it regress to the original hallucinatory fulfillment. If it commits either of these mistakes, no satisfaction of a real kind will be attained, and the situation will become one of unpleasure (*Unlust*). This is the threat that leads to the growth and development of the ego: "Unpleasure remains the sole means of education."

The list of topics dealt with in the "Project" is as follows. All except the last three were developed further in Freud's later writings, often thirty years later.

Principles of Inertia and Constancy
Primary and Secondary Processes
Unconscious and Preconscious
Urge towards Wish-Fulfillment
Hallucinatory and real fulfillment of Wishes
Criteria of Reality
Inhibitory function of the ego—Mobile and bound energy
Separation of function between perception and memory
Relation of Memory to contact barriers and facilitations
Three conditions for the arising of consciousness
Significance of Speech
Thought as experimental small-scale action
Traumas and pain as excessive stimuli
Protective screen against them and concentration of cathexes to deal with irruptions
No screen against internal stimuli
Signals of unpleasure instead of full doses
Dreams: wishful, hallucinatory, regressive, distorted—No motility during sleep
Parallelism of dreams and neurotic symptoms
Importance of sexuality in neuroses
Hysteria: defense, repression, displacement, distortion

Significance of Attention
Analysis of intellectual processes, including logical errors, etc.
Connection between repression and retardation of puberty . . .

We have told earlier how Freud soon cast aside this remarkable production as a thing of no value, of which he was almost ashamed. It is not the least curious feature of the story. His revulsion in attitude, from elation to depreciation, did not proceed, as one might have supposed, from insight into the inherent incompatibility of the double task he had attempted to perform. On the contrary, he continued for more than a year longer to bring emendations to his theory in the same terms of brain anatomy and physiology. It was not that he had appreciated the impossibility of the task, merely that he was dissatisfied with his endeavor to carry it out.

One cannot, however, be quite sure on this point. As we saw earlier in connection with his seduction theory, Freud had a way of rather obstinately persisting with an idea even when he was uneasily half aware of being on a wrong track. To have to retrace one's steps is never pleasant. So perhaps, after all, the alternative explanation of dawning insight may be the correct one.

However that may be, we have in a letter to Fliess of January 1, 1896, the continuation of the same train of thought in physiological terminology. There he made several emendations to the earlier exposition which to some extent simplified it. Perhaps the most important was the doctrine that the Quantity reaching the neurones is derived from one source only, the internal organs; they are powerless to excite the ω neurones. Stimuli reaching the brain from the sense organs do not increase the Quantity in the neurones; they merely excite them. They convey, however, their qualitative attribute to the ω section of the ψ neurones, and in this way evoke consciousness. This simplification might seem to make the concept of special ϕ neurones superfluous. Yet it was retained as follows. They receive the stimulation from the sense organs and pass on to the ω neurones the qualitative element; these in their turn stimulate the ψ neurones without passing on to them either Quantity or Quality.

The ψ neurones do not in themselves involve consciousness until they have become connected with speech associations.

The reversal of the excitation in hallucinations is no longer towards the ψ neurones, only to the ω neurones.

He now traced the arising of unpleasure to a conflict between the Quantity in the ψ neurones derived from internal organs (including sexual energy) and the processes of consciousness. In other words, he had perceived that mental conflict and suffering emanate essentially from man's difficulty in coping with his bodily needs and impulses, predominantly the sexual ones.

In a letter of December 6, 1896, Freud showed he had made further progress in some important respects. The most novel feature here was the conception of memory traces not being deposited once and for all, but undergoing several rearrangements as time goes on. He had in his book on Aphasia made a similar suggestion concerning the incoming pathways from the periphery, a point he cites in his letter.

He maintained his view that no memory trace is left in the ω neurones, those associated with consciousness, so that memory and consciousness are "mutually exclusive." The first trace is laid down (evidently in the "primary process") according to the association law of simultaneity. The second transcription, formed in accord with causal connections, is also in the unconscious, like the last one inaccessible to consciousness; it corresponds to conceptual thinking. The third, bound up with verbal imagery, belongs to the preconscious system. The secondary thinking consciousness probably depends on an hallucinatory reanimation of the verbal images, so that once again one sees that consciousness is bound up with the perceptual ω neurones which themselves contain no memory traces.

Freud added that were he only able to furnish a complete account of the psychological characteristics of perception, and the three sets of memory traces, he would be in a position to construct a new psychology.

He then made the interesting suggestion, one he hardly followed up, that the three sets of traces are laid down in different periods of life.

There is a normal defense against the release of unpleasure (*Unlust*) in each of these three phases. Pathological repression, however, is different. That signifies the blocking of the transition of the memory trace from one phase to the next, and this is invariably due to the avoidance of the unpleasure that such a transcription would evoke. It is not, however, a mere question of the *amount* of unpleasure that decides this; it happens only when the awakening of a memory evokes *fresh* unpleasure, not simply the recall of an old one. This state of affairs is peculiar to certain unpleasant sexual experiences, so only sexual impulses are capable of undergoing repression.

One easily sees that, in spite of the neuronic systems being still evoked in the descriptions, Freud was now rapidly moving into the field of pure psychology. It is, in fact, the last we ever hear of brain physiology.

There is a later echo in a letter of September 22, 1898, when he wrote: "I have no inclination at all to keep the domain of the psychological floating, as it were, in the air, without any organic foundation. But I have no knowledge, neither theoretically nor therapeutically, beyond that conviction, so I have to conduct myself as if I had only the psychological before me." He never moved from this position. In 1905, for instance, he wrote: "To avoid any misunderstanding I would add that I am not attempting to proclaim cells and fibers, or the systems of neurones that nowadays have taken their place, as psychical paths, although it should be possible to represent such paths by organic elements of the neurone system in ways that cannot yet be suggested." And, again, in 1917: "Psychoanalysis hopes to discover the common ground on which the coming together of bodily and mental disturbances will become intelligible. To do so it must keep free of any alien preconceptions of an anatomical, chemical or physiological nature, and work throughout with purely psychological auxiliary hypotheses."

We come finally to Freud's published account of his theory of the mind. It is contained in the seventh chapter of *The Interpretation of Dreams*, which served ever after as the basis for subsequent extensions and modifications of his views.

Freud employed here a working model of the mind very similar to the one he had in the "Project" and also a good many of the same fundamental conceptions, but the physiological terminology has almost entirely disappeared. In comparison with the Project it is both simpler and more lucid; one reason for this is that he was writing for a wider and less informed audience.

Freud had of course gathered his knowledge of practical psychololology almost entirely from his clinical experience, and it had long been his ambition to make use of it in formulating a theoretical psychology. It must, therefore, have been very hard for him to renounce this cherished plan in favor of one based on his recently acquired knowledge of dream processes, and yet if he was to make his book on dreams complete there were cogent reasons why he should make that decision. For a while he played with the idea of combining the two sources of knowledge, and presumably it was his sense of artistic congruity that impelled him to make *The Interpretation of Dreams* a book on dream life alone. From that point of view he was undoubtedly right, but we have suffered thereby in being offered a less comprehensive statement. He himself insisted that the view of the mind he there presents is necessarily a partial one, which needed amplifying by studies based on other data than those of dreams alone.

In what follows here the general psychological principles will be abstracted from the application of them to the special problems of dream psychology, with which the chapter in Freud's book is primarily concerned.

Of psychological dicta in Freud's writings preceding *The Interpretation of Dreams*, two in particular deserve to be singled out, both dating from 1894. One is the regulating Principle of Constancy, to which he always adhered. It was evidently derived from Helmholtz's principle of the conservation of energy, according to which the sum of forces remains constant in every isolated system. Freud's own definition of it was given earlier in this chapter. He seems to have felt some proprietary rights in the idea of its being applicable to the nervous system, since he relates in a letter of November 29, 1895, being annoyed at its being appropriated by Heinrich Sachs.

The other is one of his only too rare definitions: "There is to be differentiated in psychical functions something (an amount of affect, a sum of excitations) which has all the attributes of a Quantity—although we possess no means of measuring it—something capable of being increased, diminished, displaced, or discharged, and which extends itself over the memory traces of ideas, rather as an electric charge does over the surface of bodies." The two words in brackets indicate that the property in question can be described either in psychological (*Affektbetrag*) or in physiological (*Erregungssumme*) terms. The conception that it has a certain autonomy, that it can be "displaced" from one idea to another, was a momentous one and one alien to the psychology of the time; the idea of the affect being independent and detachable differentiated it sharply from the old conception of "affective tone." It was a pure gain from the realm of psychopathology.

It is generally held that Freud's greatest contribution to science, the one usually associated with *The Interpretation of Dreams*, was his conception of an *unconscious mind*. It was certainly one of the two that provoked the strongest opposition; the other was of course his libido theory. Two comments are in place here. It is interesting to remember that the idea of unconscious mental processes was much more widely accepted in the last twenty years of the nineteenth century than in the first twenty of the present one, when it was met with a spate of incredulity and ridicule. A couple of examples may be quoted. In 1885 Sir Samuel Wilkes, a distinguished London physician who was later President of the Royal College of Physicians, in reviewing a book by Hack Tuke, agreed that the author shows "that consciousness is not an essential element in all our mental acts," and added, "It is now generally admitted that the higher centers of the brain (i.e., the mind) may be in full operation without consciousness being called up." The other is from Theodor Lipps, Professor of Psychology in Munich, a man whose writings Freud much admired. The following passage Freud had underlined in his copy of a book of Lipps', which he read in 1898 (August 31): "We maintain not merely the existence of unconscious mental processes besides the conscious ones. We postulate further that unconscious processes are the basis of conscious ones and accompany them. Conscious processes rise out of the unconscious when conditions are favorable, and then sink back again into the unconscious."

One would conclude from this observation that the change after 1900 from easy acceptance to bitter opposition was not so much, as it appeared to be, to the idea itself of an unconscious mind as to its content which Freud had revealed.

In the second place, careful students have perceived that Freud's revolutionary contribution to psychology was not so much his demonstrating the existence of an unconscious, and perhaps not even his exploration of its content, as his proposition that there are two fundamentally different kinds of mental processes, which he termed primary and secondary respectively, together with his description of them. The laws applicable to the two groups are so widely different that any description of the earlier one must call up a picture of the more bizarre types of insanity. There reigns in it a quite uninhibited flow towards the imaginary fulfillment of the wish that stirs it—the only thing that can. It is unchecked by any logical contradiction, any causal associations; it has no sense of either time or of external reality. Its goal is either to discharge the excitation through any motor exit, or, if that fails, to establish a perceptual—if necessary, an hallucinatory—identity with the remembered perception of a previous satisfaction.

In the secondary process there occur extensive inhibitions of that freely flowing energy; it is attached, "bound," and allowed to flow only after processes of thought have found a direction in which it is possible to find a "real" satisfaction of the wish through taking into account the facts of the outer world. "It is from the contrast between Reality and Wish-Fulfillment that our psychical life grows."

This division of the mind has, of course, a physiological counterpart, one with which Freud was very familiar. It is that between the simple reflex on the one hand, where the movement of excitation from the sensory to the motor fibers is immediate and unchecked, and on the other hand the various complex reactions to stimuli which may or may not be followed by a motor response and in the course of which inhibition always plays a part. It is therefore not the division itself that is original, but Freud's detailed exploration and description of the two sets of processes, something that had never before been even attempted.

There is, it is true, a rough correlation between this differentiation in the nature of mental processes and the distinction between the unconscious and consciousness, but the latter is the broader conception and needs further description which will presently emerge.

A noteworthy remark of Freud's in this connection is that from certain points of view the contrast between "the ego" and "the repressed" is more instructive than that between consciousness and the unconscious.

Many of the ideas expounded in *The Interpretation of Dreams* have already been considered, in another language, in the account given of the "Project." Some of them have simply to be translated. Thus, instead of a system of neurones we have psychical constellations, instead of the physical concept of Quantity we have a hypothetical "cathexis" of psychical energy, and the physical principle of inertia develops into the well-known pleasure-unpleasure (***Lust-Unlust***) principle.

Freud still uses the word "Apparatus" and the model he provides is constructed on lines very similar to those of the physiological model. But here, in terms of psychical processes, the model comes to life. . . .

An Intelligent Layman's Reaction to Freud

Jacques Riviere

Freud has been accused by some, Jules Romains among them, of a certain scientific levity, *i.e.*, a certain tendency to convert his hypotheses into laws, before he has accumulated sufficient experimental and objective data to warrant him in doing so.

"He does not hesitate," says Jules Romains, "to link up two scientific statements with one of those 'brilliant views,' which certainly show proof of a great activity of thought, which we are at first inclined to rate as genuine discoveries, but which we do not, on reflection, put away in that corner of the mind where we keep good scientific money. They are fiduciary bonds, bound up with the fate of the bank that issues them."

In many passages, however, Freud exhibits a quite remarkable prudence, and even takes the trouble to indicate himself the gaps in his doctrine, and the points where observation has not yet confirmed it. "The reply to this question," he writes in the *Introduction to Psycho-analysis*, "is not, I think, urgent, and, moreover, it is not sufficiently certain to permit us to venture on it. Let the work of scientific progress go on, and wait patiently." On the threshold of a tempting generalisation of an idea which he has just expressed, he remarks, "The psycho-analytical explanation of neurosis has, however, nothing to do with considerations of so vast a range."

He always examines very carefully the objections that are made to his theories. You will find, for instance, in the last chapter of the *Introduction to Psycho-analysis* a remarkable discussion of the idea that all the discoveries of psycho-analysis might very well be a product of suggestion exercised on the patients. When you think of the weight of this objection, and then see the masterly way in which Freud replies to it, you cannot but have a feeling of confidence both in the uprightness and the power of his mind.

Yet it must be confessed: something still remains of Jules Romains's criticism, and there are certain defects of method in Freud, of which we must be fully conscious and for which we must allow, before we follow in his footsteps.

It is evident that we are dealing with a lively and bustling imagination, and one that sometimes reacts a little too quickly to the first results of an experiment. Reading Freud, you are struck with the rapidity of certain of his conclusions. Often from a single fact he will deduce an immediately general

Source: Jacques Riviere, "Notes on a Possible Generalization of the Theories of Freud," *The Criterion* 1 (1923), 329-334. Translated by F. S. Flint.

affirmation; often, too, it is quite sufficient, if he is able to interpret a fact in accordance with his theory, for him to regard any other interpretation as excluded.

Moreover, the undeniable victory over the enigmas of nature which his leading idea represents, gives him a kind of intoxication, and leads him into a sort of imperialism. I mean to say that he seeks to annex too many phenomena to his explanation. In especial, his interpretation of dreams and day-dreams, which is full of profound observations, nevertheless seems to me, taken as a whole, much more factitious and much less convincing than his theory of neurosis. And when I learn that, historically, he began by an explanation of neurotic symptoms, I wonder whether the whole of his theory of dreams and day-dreams may not be a somewhat arbitrary, or at least too systematic, extension of a just idea into a domain unfitted for it, at the very least in its textual form.

In other words, I am wondering whether the order in which Freud chose to expound his doctrine in his *Introduction to Psycho-analysis*, and which is, as is known, the following: frustrated acts, dreams, neuroses,—whether this order is not extremely specious, and calculated to mislead in regard to the real working of his mind during the course of his discoveries, and in regard to the actual value of those discoveries. Even if it seems logical to show first of all the unconscious at work in the most elementary acts of our normal daily life, this becomes an error of method, if it cannot be revealed with as much evidence in those acts as in pathological acts, if its intervention therein is more disputable, and if, in fact, it was not first of all disclosed in those acts.

I cannot help it: the theory of day-dreams and the theory of dreams appear to me as a sort of double gate constructed by Freud, as an afterthought, before the monument he had raised. He thought that this would make a more agreeable and more convincing approach to that monument; but, to my mind, he was mistaken, because, in this preliminary part, you do not receive strongly enough the impression that you are in contact with an invincible, irrefutable observation, with that observation which gave rise to the theory. You feel the subtlety of the author, but you do not feel sufficiently his justification.

For this reason, I think it necessary to keep in mind continually and principally his theory of neuroses, if you wish to seize his thought at its point of maximum intensity, and to realise all the consequences it implies, all the generalisations it is capable of bearing, its farthest reach, or, if you prefer it, its greatest explosive force.

In what follows, I desire, not to analyse in detail the Freudian doctrine, but, on the contrary, supposing it to be known to all my readers, to bring out, if I may say so, its potentialities. I desire to present the three great psychological discoveries which, it seems to me, we owe to Freud, and to reveal the wonderful light which they can project into the study of internal things, and, particularly, of the feelings. I desire, especially, to show how great is their extensibility, and how they may be made to take on a more supple, and, if I may say so, a still more generous, form than that given to them by Freud.

In the account of the facts which suggested to him the first idea of his theory, and which, as is known, are the body of manifestations of hysteria, Freud insists with especial force on the complete ignorance in which his patients were of the cause and purpose of the acts they were performing. "While she was carrying out the obsessive act," he writes, "its meaning was unknown to the patient, both as regards the origin of the act and its object. Psychic processes were therefore acting in her, of which the obsessive act was the product. She certainly perceived this product with her normal psychic organisation, but none of her psychic conditions reached her conscious mind. . . . It is situations of this kind which we have in mind when we speak of *unconscious psychic processes*." And Freud concludes: "In these symptoms of obsessional neurosis, in these ideas and impulses which spring from nowhere, as it seems, which are so refractory to any influence of normal life, and which appear to the patient himself like all-powerful guests from a foreign world, like immortals who have come to mingle in the tumult of this mortal life, how is it possible not to recognise an indication of a particular psychic region, isolated from all the rest, from all the other activities and manifestations of the inner life? These symptoms, ideas, and impulses lead infallibly to the conviction of the existence of a psychic unconsciousness."

It does not appear, at first sight, that there is in these passages any very extraordinary novelty, and it might be thought paradoxical that we should see in them one of the sublimities of the Freudian theory. The unconscious is not a discovery of Freud's. You may quote immediately names that appear to reduce to the slightest proportions his originality on this point: Leibnitz, for instance, Schopenhauer, Hartmann, Bergson, and many others.

Nevertheless, I reply:

(1) That there is a considerable difference between a metaphysical conception and a psychological conception of the Unconscious, that to admit the Unconscious as a principle, as a force, as an entity, is a far different thing from admitting it as a body of facts, as a group of phenomena.

(2) That, in reality, many contemporary psychologists, particularly Pierre Janet and his school, still refuse to admit a psychological unconsciousness.

(3) Finally, that, if we admit that a psychological unconsciousness has been recognised by everybody, as a kingdom, as a domain, Freud is the first to conceive of it:

(*a*) As a well-defined domain or kingdom, which has its own geography, or, dropping the metaphor, which contains extremely precise tendencies and inclinations directed towards particular aims;

(*b*) As a domain or kingdom which may be explored, starting from the consciousness, and, even, which must be explored, if the consciousness is to be understood.

Here, I recover confidence to affirm that the novelty of the theory seems to me entire, and formidably important. Remember that hitherto consciousness has been conceived as a closed chamber, wherein the objects, of a definite number, were, so to speak, entered on an inventory and had affinities only with each other, and that if it was desired to explain any incident of our psychic life, you could only go to some fact which you had previously perceived. Remember that the whole of psychology was limited to a logical explanation of our determinants. Remember the scanty causal stock it had at its disposal, and imagine its richness immediately Freud opened up to it the immense reservoir of submerged causes.

He is himself, moreover, conscious of the revolution which this mere proclamation of the definite reality of the unconscious may produce in the history of ideas, and he permits himself a touch of pride. "By attributing so much importance to the unconscious in our psychic life," he cries, "we have raised up against psycho-analysis the most ill-natured critics. . . ." And yet "the lie will be given to human megalomania by that psychological research which proposes to prove to the *ego* that he is not only not master in his own house, but that he is so little master there that he has to be content with rare and fragmentary information on what is going on, outside his consciousness, in his psychic life. The psycho-analysts are neither the first nor the only people who have launched this appeal for modesty and composure, but it seems to have fallen to their lot to defend this point of view with the greatest earnestness, and to produce in its support materials borrowed from observation and accessible to all."

Let us reflect a moment. Let us turn, if I may say so, against us this principle of the unconscious as the seat of definite tendencies that combine to modify the conscious, and let us confront it with our own observation. In other words: *let us think for a moment of all we do not know that we want.*

Is not our life a constant seeking after possessions, pleasures, and satisfactions that not only would we not dare to confess we desire, but that we do not know we desire, we are seeking? Is it not nearly always *a posteriori* and only when we are performing it that we are aware of the long psychic labour and of all the chain of latent feelings that led us to an act?

And again: at what moment does direct inspection of our consciousness inform us exactly on all we are experiencing and all we are capable of? Are we not in constant ignorance of the degree and even of the existence of our feelings? Are there not, even in passion, moments when we discover absolutely nothing left of that passion, when it appears to us a pure construction of our mind? And yet does not that passion exist at that very moment, in what I may be allowed to call an infinitely precise fashion, since the slightest accident that may happen to place an obstacle in its way or to postpone its object may instantly provoke a complete upheaval of our whole being, which will find expression even in our physical condition and will even influence the circulation of our blood?

In love, for example, is not a sincere lover often reduced to making experiments and almost to tricks in order to auscultate his feelings and to ascertain whether they still exist? And that too at the very moment when, if he were told that he must give up hope or that he has been deceived, he would perhaps find himself on the verge of crime.

Therefore, a first great discovery (which may perhaps be presented as a negative one, but negative discoveries are no less important than the others) must be placed to Freud's credit: it is that a considerable part of our psychic life takes place, if I may say so, outside us, and can only be disclosed and known by a patient and complicated labour of inference. In other words: we are never quite wholly available to our own minds, quite wholly objects of consciousness.

This first analysis should make clear the spirit in which I have tackled the study of Freud and how I intend to follow it up. I do not by any means profess to follow his thought step by step in all its developments. I simply seek and seize on, one by one, without troubling to point out their relations one with another, the points of his doctrine that seem to me capable of being enlarged into psychological truths of general interest. I am an outsider who egotistically pillages a treasure and carries it far away from the temple. I may be judged severely from the moral point of view; but in any case it is not, I imagine, incumbent upon me to adopt the slow and processional gait which is obligatory on the priests of Psycho-analysis.

Let us therefore proceed at once to the examination of another of Freud's ideas which seems to me of considerable importance; I mean the idea of repression, with which must be connected that of the censorship of dreams.

Its essential points are well known: basing himself on his observations as a practitioner, Freud asserts that there is in every subject who is analysed, or even questioned, an instinctive resistance to any question and any effort to penetrate to the background of his thought. This resistance is moreover subject to variations of intensity. The patient is more or less hostile, more or less critical, according as the thing which the doctor is endeavouring to bring to light is more or less disagreeable.

The resistance therefore seems to be the effect of a force, of a strictly affective nature, which opposes itself to the appearance in the open consciousness, to the illumination, of certain psychic elements which it considers incongruous, as impossible to be faced.

This force which is met when you set to work to cure the patient is the very same force that has produced the malady by repressing a psychic process which tended from the unconscious towards the conscious. The tendency thus baulked has in fact transformed, disguised, itself—in order to go at any rate a little farther—as a mechanical act, with no apparent meaning, but which the subject is helpless to avoid: it is the symptom: "The symptom has been substituted for that which has not been accomplished."

Freud therefore brings to light the presence in the consciousness of an activity that reduces or deforms our obscure spontaneity. He also shows it at work in our dreams, and then calls it the *censorship*. Just as the censorship, during the war, either mutilated newspaper articles, or else forced their authors to present their thought only in an approximate or veiled form, in the same way a secret power modifies and disguises our unconscious thoughts, and permits them to reach our mind only in the enigmatic forms of the dream.

"The tendencies exercising the censorship are those which the dreamer, with his waking judgment, recognises as being his own, with which he feels himself in agreement. . . . The tendencies against which the censorship of dreams is directed . . . are the reprehensible tendencies, indecent from the ethical, aesthetic, and social point of view . . . are things of which one dares not think of or of which one thinks only with horror."

The neurotic symptoms are "the effects of compromises, resulting from the interplay of two opposing tendencies, and they express both what has been repressed as well as the cause of the repression, which thus too contributed to their production. The substitution may be made for the greater benefit of one or other of these tendencies; it is seldom made for the exclusive benefit of one alone."

In the same way, the dream is a sort of composite of, or rather compromise between, the repressed tendencies, to which sleep gives strength, and the tendencies really representing the self, which continue their work by means of the distorting censorship.

In other words, neurotic symptoms and dreams correspond to an effort of our diverse sincerities to display themselves at the same moment.

The whole of this conception seems to me to be extraordinarily novel and important. It may be that Freud himself has not perceived it in all its general bearings.

The discovery in us of a deceptive principle, of a falsifying activity, may nevertheless furnish an absolutely new view of the whole of consciousness.

I shall at once exaggerate my idea: all our feelings are dreams, all our opinions are the strict equivalent of neurotic symptoms.

There is in us a constant, obstinate, inexhaustively inventive tendency that impels us to camouflage ourselves. At any cost, in every circumstance, we will ourselves to be, we construct ourselves, other than we are. Of course, the direction in which this deformation is exercised and its degree vary extraordinarily with different natures. But in all the same principle of deceit and embellishment is at work.

To start out on the study of the human heart without being informed of its existence and its activity, without being armed against its subterfuges, is like trying to discover the nature of the sea-depths without sounding apparatus and by merely inspecting the surface of the waters. Or better still, as Jules Romains says, it is like the traditional method of analysis, which "even when it searches the depths is guided by the showy indications of the surface. It suspects the presence of iron only when the rocks above are red with rust, of coal only when black dust is underfoot."

Who does not know this demon which Freud calls the censorship, and which so subtly and ceaselessly makes our moral toilet? Each instant, the whole of what we are, I mean the confused and swarming mass of our appetites, is taken in hand and tricked out by it. It slips into our lowest instincts enough of nobility to enable us to recognise them no longer. It furnishes us in abundance with the pretexts, the colours, that we need to cover up the petty turpitudes we must commit in order to live. It provides us with what we call our *bonnes raisons*. It maintains us in that state of friendship and alliance with ourselves without which we cannot live, and which is yet so completely devoid of justification that we do not understand how it can possibly take its rise.

But I feel that I am leaving Freud's idea far behind. The principle governing repression and the censorship, far from working for the triumph of our appetites, is, in his opinion, what combats them, stops them. It is the representative of the moral ideas, or, at least, of convention, so far from helping to circumvent it.

Nevertheless, there are cases in which it is beaten, partially at least: the neurotic symptom, the dream, the day-dream, correspond to relative successes over it by the lower part of ourselves. And if it is not directly an agent of hypocrisy, it becomes so in so far as it does not gain the victory.

When I maintain that all our feelings, all our opinions, are dreams or obsessive acts, I mean that they are impure, masked, hypocritical states; I mean, in fine, something that must be looked straight in the face: that hypocrisy is inherent in consciousness.

Taking Freud's idea to its logical end, I will say that to possess consciousness is to be a hypocrite. A feeling, a desire, enter the consciousness only by forcing a resistance of which they retain the deforming imprint. A feeling, a desire, enter the consciousness only on the condition that they do not appear to be what they are.

From this point of view, the chapter that Freud devotes to the processes employed by the censorship to distort the latent content of the dream and to render it unrecognisable would be worth developing to a considerable extent. Several of these processes are certainly used by us in the waking state, to enable us to conceive our feelings under an acceptable form. I will cite one only as an example: the displacement, the carrying over, of the accent to an aspect of what we feel, or need to feel, peacefully, which is not the *essential aspect*. In other words, the rupture by the imagination of the centre of gravity of our sentimental complexes.

Let it be said, in passing, that if I was a little severe at the beginning regarding the Freudian theory of dreams, it was very much because I regretted to see Freud apply too minutely to a particular phenomenon an idea that seemed to me to have an infinite bearing. His analysis of the symbolism of dreams goes much too far; it reintroduces into the consciousness, the suppleness and extreme convertibility of which he has shown, something fixed, which, it seems to me, has no place there. Freud's thought must be allowed to retain, if not a certain vagueness, at least a certain generality, if its value is to be fully understood.

Before leaving this idea of the censorship, one other aspect of it, of considerable importance, must be dealt with.

When I say that hypocrisy is inherent in consciousness, I either say too much or too little. The censorship, the force that controls repression, is partly made up of external contributions; they are created chiefly by education; they represent the influence of society upon the individual. Nevertheless, they are not altogether adventitious or artificial; they finally become one with the self. Freud even represents them as the tendencies constituting the ego.

And, in fact, it would be simplifying things very much to represent our lower instincts alone as constituting our personality. That which represses them is also part of us.

But then this conclusion is inevitable: that in so far as we are moral persons, and even in so far as we are persons merely, we are condemned to hypocrisy. We will no longer say hypocrisy, if you like; but we cannot avoid another word—impurity. To live, to act, if it is to be in one sole direction and with method and in such a way as to trace an image of ourselves on the retina of others, is to be composite and impure, is to be a compromise.

Sincere comes from a Latin word which means *pure*, speaking of wine. It may be said that there is no sincerity for man in his integrity. He becomes sincere again only in decomposition. Sincerity is, therefore, the exact contrary of life. You must choose between the two.

The third point in Freud's doctrine which we can, it seems to me, though to a less extent perhaps, *enlarge*, is the theory of sexuality.

The general lines of this theory may be recalled.

Inquiring into the nature of the tendencies which are stopped by repression and which are expressed by *substitution* in the symptoms and in dreams, Freud, it will be remembered, thinks that they can all be said to be of a sexual nature.

Several *nuances* should be noted here. Freud does not say, and even denies having said, that everything appearing in our dreams is of sexual origin. Only that which appears camouflaged is of sexual origin.

Moreover, Freud does not say, and denies having said (for example, in the letter published by Professor Claparède as an appendix to the brochure on psycho-analysis), that our whole being may be reduced to sexual tendencies, or even that "the sexual instinct is the fundamental impulse of all the manifestations of psychic activity." On the contrary: "In psycho-analysis it has never been forgotten that non-sexual tendencies exist; its whole edifice has been erected on the principle of a clear and definite separation between sexual tendencies and tendencies relating to the self, and it was affirmed, before any objections had been made, that the neuroses are the products, not of sexuality, but of the conflict between the *self* and sexuality."

Nevertheless, it remains true that the body of spontaneous and unconscious tendencies of the being is considered by him as fundamentally identical with the sexual instinct.

He is careful, however, to define that instinct very broadly, distinguishing it from the procreative instinct and even from strictly genital activity. In order clearly to mark its general character, he calls it *libido.*

The *libido* conception is evidently not absolutely clear. At times it has an almost metaphysical value, and the next instant it is used simply to designate the sexual appetite, desire properly so called.

But I am wondering whether, instead of reproaching Freud with this ambiguity, instead of trying to force him to hang on to this word *libido* an absolutely distinct and limited tendency, it would not be more fitting if we were grateful to him for the vagueness in which he leaves it and for the wider play it permits him. I am wondering whether his principal discovery, in the domain we are discussing, is not indeed precisely that of one sole transformable tendency, which perhaps forms the basis of our spontaneous psychic life.

In other words, the idea that desire is the motive-power of all our activity, at least of all our expansive activity, seems to me of an admirable novelty and truth. Or, better still, the idea that we are creators, producers, only in so far as we move in the direction of desire.

But we must be careful not to betray by too great precipitation Freud's idea itself, his conception of sublimation. I resume, therefore.

Freud, by a long analysis, strongly supported by experimental observations, which fills the whole of a small brochure, entitled, *Three Dissertations on the Sexual Theory*, establishes that the sexual instinct has at first neither the object nor the aim we know of it. He shows it first of all immanent, so to speak, in the body of the child, and neither seeking nor even suspecting any external satisfaction. This is the period of what he calls auto-eroticism.

He shows it at the same time irradiating confusedly and impartially into all the organs, and receiving satisfactions almost indifferently from all.

Then experience, which may, moreover, be preceded by foreign interventions, teaches the *libido* to externalise itself. But even after this leap forward it remains hesitating between several possible satisfactions, and places itself exclusively at the service of the genital act only at the moment of puberty and by a kind of very complex synthetic action, which is liable to be influenced by a crowd of accidental interferences.

This desire, which is beneath its object, and which at the same time exceeds it or even transcends it, is a conception marked with boldness and of magnificent depth.

How much it permits Freud to explain may be easily understood. If the *libido* is repressed, one of two things happens: it will either turn to a means of satisfaction which he calls pregenital, and you will have a perversion, by *fixation*; or it will produce an uneasiness which will generate neuroses.

But, on the other hand, the fact that it is not really bound up in any constitutional manner with the genital act will permit it to go beyond that act and to place itself at the service of intellectual activity, to irrigate, so to speak, our spiritual faculties. Its sublimation will consist, therefore, in this deviation of the *libido* to the advantage of the intelligence or even of morality.

The reflections inspired by this part of the Freudian theory might be presented in the following manner:

(1) It is of considerable importance, from the point of view of the psychology of creation, to have established that the source of all spiritual creation is carnal, if the word may be used. This is important, not because it degrades creation, but because it brings to light the unity of our psychic life, and because it makes clear that, all in all, we have at our disposal only one kind of energy, and all our liberty is confined to directing the use of it.

It is important because it explains aesthetic emotion before a great work, and because, whatever may be the object represented, it explains the sensual element the work always possesses when it is sincere.

It is important, even from the point of view of aesthetic criticism, because it teaches us to seek in the work, not the little smothered story which may be at its origin in the author—as has been done with too much precision, in my opinion, by those who have hitherto applied psycho-analysis to art—but the current of desire, the impulse in which it was born. And a sort of vague aesthetic criterion might be established, which would enable us to distinguish works born of an inclination from those manufactured by will, the aesthetic quality being, of course, reserved to the former.

(2) By analysing, on the one hand, all that the *libido* builds up in the subconsciousness under the shelter of repression, and, on the other hand, all that the repression of the *libido* may produce in the conscious life, Freud opens up to psychology a prodigious domain.

I do not think that the analysis of dreams, as practised by Freudian orthodoxy, can lead to much of any great interest—owing, especially, to the strange preliminary telegraphic code that imprisons interpretation.

But think of what might be discovered by a psychologist without prejudice (either Freudian or anti-Freudian), who is simply resolved not to ignore what I should like to call the sexual situation of the subjects he is studying. Think of that abyss, as yet so ill-explored, of sexual attractions, and perhaps especially of sexual hatreds. Think what an access to individual character, what a key to the whole conduct of a given subject, might be given by a knowledge of his or her sexual experiences, and especially of the consequences of and the reactions from these experiences.

The novelist hitherto, even if he did not note them down, has been careful to keep in mind, for his own guidance, the social situation, the material conditions, the business, and the parentage of each of his characters. It seems impossible to me, after Freud, that he can neglect to imagine, likewise beforehand, even if he is not to say a word about it during the course of his story (his story may even have for its object merely to suggest it), the sexual situation of each of his characters and its relation—you will understand that I am using the word in its most general sense—from the sexual point of view, with the rest.

(3) By detaching the *libido* from its object, Freud implicitly adopts a subjectivist conception of love. It is evident that this mobile, shiftable desire which he describes will need to receive nothing from the object it chooses, will even be unable to receive anything from it, and that it is from its own resources entirely that the image of the beloved object will be formed in the mind of the lover.

He speaks somewhere of the "over-estimation of the sexual object," and doubtless he intends this first of all in the physical sense, but he certainly has it in mind as well that all the moral beauties with which the lover embellishes the beloved object are the reflection of the projection on it of the *libido.* He admits, therefore, that all love is hallucinatory, and seeks in foreign beings only a pretext to fix itself. He does not admit the appeal, the attraction, of one being for another, or that love can ever be born of real, objective affinities.

We must now endeavour to embrace in one view the whole of Freud's doctrine and to appraise it.

Freud brings us two things: a new world of facts, a "new family of facts" (and here I am of an entirely different opinion from Jules Romains, who denies him this kind of discovery), and, if not a new "law" of these facts, at least a new method of exploring them, or, more vaguely, a new attitude to take regarding them.

The new world is the world of the unconscious, conceived and shown for the first time as a system of definite facts, of the same nature and the same stuff as those appearing in the consciousness, and in constant relation, in constant *exchange*, with the conscious facts.

Among these unconscious facts, Freud reveals the wonderful flora of the sexual tendencies and complexes. Even if he describes them with too much precision (this is always somewhat of a defect of his), and if he typifies them too much, it is an admirable novelty merely to have unveiled them.

Others may follow his footsteps, with more lightness and a more acute sense of the individual, into this strange garden. But he has already indicated to those others—and it is his second contribution, which is equally priceless—the attitude of approach to take up in order to make good observations. He warns us of the force that is at work in us to deceive us about ourselves; he teaches us its ruses and the means of circumventing them.

More generally, he sketches a new introspective attitude, which may be the point of origin of an entirely new direction to psychological studies. This attitude consists in endeavouring to know oneself, if I may be permitted the expression, only by the signs. Instead of attending to the feeling or sensation itself, Freud seeks for it in its effects only, in its symptoms.

Of course, long before him, attempts had been made to observe psychic phenomena, for greater safety, indirectly, particularly in their conditions. The whole of psycho-physiology was an endeavour to obtain information about the consciousness by starting from the exterior, from something that was not of it, but which had this advantage that it could be touched, measured, and made to vary. But the error of psycho-physiology, as Bergson has so well observed, was to ignore the differences of quality in the phenomena.

Bergson's own error perhaps (I indicate it here only in the most prudent and hypothetical manner) was to plunge with too much confidence into the pure psychological flow, and too naively to expect knowledge from mere embracing contact. Can you mark the course of a river by swimming in it?

Freud escapes the error of the psycho-physiologists by accepting as information concerning the psychic life only psychic facts. He builds up an independent, autonomous psychology; and that is one of the reasons for the opposition he has met with.

But, on the other hand, he does not believe in these psychic facts; I mean, he does not accept them at face value. He regards them, *a priori*, both as deceitful and as explicable. He uses them as signs enabling him to trace back inductively to a deeper and more masked psychic reality. He strives in an opposite direction to the vital current.

And thus he gives back to the intelligence that active role, that role of mistrust and of penetration, which in all the orders of intelligence has always been the only one that permitted and favoured knowledge. There would be much to say on his complete faith in psychological determinism. But as a method, to be used as long as it is possible, determinism is unassailable. It is only by this method that you can hope to make headway, with any distinction and advantage for thought, in the chaos that our soul sends out to meet us.

Freud, the Revisionists, and Social Reality

Will Herberg

Although psychoanalysis emerged first as a therapy and then as psychological research into the anomalous and pathological in human behavior, its impact on social thought was immense from the very beginning. I have reference not so much to the early ventures of Freud and the Freudians into anthropology and the cultural disciplines; these were usually too rash and ungrounded to be more than suggestive. What I have in mind rather are the profound changes that Freud and the psychoanalytic movement effected in the very outlook and methods of the social sciences, to the point indeed where these sciences may almost be dated as pre- or post-Freud. It is not for nothing that two eminent American sociologists, acknowledging their debt to the "work of the great founders of modern social science," name Freud along with Durkheim and Max Weber. If Talcott Parsons and Edward A. Shils can hail Freud as one of the Founding Fathers of modern social science, it is surely appropriate in this colloquium marking the centenary of his birth to try to assess, if not the full scope of his influence—that would obviously be impossible on this occasion—at least some of its major aspects. In a rough sort of way, I think, we can describe the varied impact of Freud and his thinking on the social sciences along the following lines:

1. It has transformed the social sciences—as it has psychiatry—into disciplines concerned with dynamics rather than with statics. It would not be entirely true to say that before Freud the social sciences were exclusively devoted to classifying, formulating, and methodologizing, but neither would it be entirely false. Official sociology was largely preoccupied with conceptualized description of social phenomena and with interminable debates as to the nature of sociology and its methods. The academic psychology of the time was almost useless for an understanding of any but the very simplest forms of human behavior and was, therefore, of little help to the sociologists, even if the sociologists had been inclined to avail themselves of it. With Freud, the picture changed drastically; human behavior could now be studied in its dynamic complexity, and a new depth-psychology of man emerged. Directly and indirectly, this new psychology made a deep impress on the social sciences, above all by showing the real possibility of a dynamic approach to an understanding of man in society.

Source: Will Herberg, "Freud, The Revisionists, and Social Reality," in Benjamin Nelson, ed., *Freud and the Twentieth Century* (Cleveland: The World Publishing Company, 1957), pp. 143-159.

2. Psychoanalysis has given substance to this dynamic approach by discovering, isolating, and defining certain basic mechanisms of human behavior. So familiar have these become to us that we rarely bethink ourselves of their source and origin. We speak of frustration and aggression, of guilt, insecurity, and anxiety, of projection and displacement, of repression, reaction formation and transference, of unconscious impulses, wish fulfillments, and defense mechanisms, without always realizing that most of these, at least in the sense in which we use them today, are the coinage of the Freudian mint, usually the work of the master himself.

3. Very much the same is true of the motivational factors with which we work in the social sciences and cultural disciplines today. They largely stem from the psychoanalytic researches that began with Freud. The central place given nowadays to such motivational forces as sex and aggression, with all their derivatives, is evidence of the new outlook, which has become virtually universal despite differences as to evaluation and interpretation.

4. The Freudian influence has tended to add a new dimension to our social thinking—a time-dimension, it might almost be said, though not in the historical sense. Freud's own early ventures into anthropology—or rather into pseudo-anthropology—helped open the way. The psychoanalytic penchant for finding the same basic mechanisms and motivations operating in primitive cultures as in contemporary societies has been subjected to all sorts of criticism, much of it not altogether unjust. It has had this consequence, however: it has encouraged us to look upon our own culture in the same anthropological spirit as we have become accustomed to employing in dealing with so-called primitive societies. We do not always realize how essentially new such an approach is in the social sciences. If today we have anthropologists and sociologists studying a New England community, the medical profession, or the Hollywood movie industry in much the same way as they would study the cultural system of a primitive tribe, this, too, we owe in large part to Freud, who was always striving to discern the contemporary in the archaic and the archaic in the contemporary.

5. Least enduring, perhaps, of all Freud's contributions have been his own particular ventures into sociological and anthropological speculation. There is no need in this day to repeat the severe strictures that have been levelled at the bold and arbitrary constructions to which Freud was so prone. No doubt these strictures have been well deserved; I should not care to defend Freud's visions of the primal horde, the primordial murder of the father by the band of brothers, the emergence of totem and taboo, religion and conscience, out of this ancient crime, and its endless repercussions through the ages. We may well admit that these visions have little or no relation to sober anthropological fact, and never can be taken seriously as science. And yet even the most unfounded of Freud's anthropological speculations have not been void of a certain significance, even in some cases of a certain truth. The seventeenth- and eighteenth-century notions of the

"state of nature" and the "social compact" initiating society were every bit as unfounded and scientifically untenable as Freud's primal horde and band of brothers; yet these anthropological myths not only influenced social philosophy very deeply, but also revealed aspects of contemporary social reality that had not hitherto been seen or appreciated. To some extent, at least, the same may be said of Freud's anthropological speculations. When we are dealing with genius, even the errors, even the extravagant errors, may prove illuminating.

In all of these ways, and no doubt many others, Freud's impact on the social and cultural sciences has been immense. In recent years, however, there have arisen in psychoanalytic and related circles various movements and tendencies challenging not merely this or that theory or teaching of Freud's, but his very presuppositions and postulates. These revisionist tendencies have generally claimed support and substantiation from the social sciences, or at least have insisted that their approach was more in accord with the spirit of modern social science than Freud's own outlook. In view of this challenge, it would appear to me that a fruitful way of assessing the significance—the strength and the weakness—of Freud's thinking in the field with which we are now concerned might be to examine Freud's views on the basic issues of social life, and compare what he has to say with revisionist thinking on the same issues. For this purpose, I have chosen the problem of culture, which, as Freud treats it, means the problem of the foundations of society and the individual's place in it.

All of Freud's thinking in the field of human relations hinges on a twofold dualism—the dualism of ego and id within the psyche, and the dualism of individual and society within the culture. The two dualisms are not unrelated.

In his mature account of the structure and "topography" of the psyche, Freud repeatedly emphasizes that it is the id, the repository of the deep instinctual drives, which is the dynamic part of the self. It is the "oldest of mental provinces or agencies"; it is wholly unconscious, and supplies the psychic energy for the functioning of the entire organism. Its law is to press for immediate gratification: "instinctual cathexes seeking discharge—that is all the id contains." The id knows only the pleasure principle, which it obeys "inexorably." Such gratification of instinct is "happiness"; indeed, Freud often speaks as though it were the very purpose of life. At any rate, it is the primodial law of the organism.

But the immediate satisfaction for which these instincts in the id press might well imperil the entire organism by bringing it into conflict with its environment, natural and social. Of this, the id knows and cares nothing, since it is governed entirely by the pleasure principle. The survival of the organism requires the functioning of another agency that will restrain and control the unregulated activity of the id. This agency is the ego, described by Freud as "a special organization which henceforward acts as an intermediary between the id and the external world." The ego is concerned with "keeping down the instinctual claims of the id," but also with "discovering the most favorable and least perilous method of obtaining satisfaction, taking the external world into account." It is the ego's task, in short, "to mediate between the pretensions of the id and the preventions of the outer world." To achieve this purpose, the mature ego operates with the reality principle; it is the rational element of the self. In a word, "the ego stands for reason and circumspection, while the id stands for the untamed passions."

The deep tension, and frequently open conflict, that prevails between these two agencies of the psyche is thus grounded in their diverse natures and principles of operation. The picture becomes even more complicated as we recognize a third agency in the psychic life, the superego, the internalized "successor and representative of the parents and educators," which stands over the ego, out of which it has emerged, as an inward monitor, exercising the powers of "observation, criticism, and prohibition." Assailed by the id, which demands instinctual gratification, and harassed by the superego, which keeps watch and threatens lest its rigid standards be violated, the poor ego must try to promote "reason and sanity" as best it can in order to preserve the organism and see that instinctual gratification takes place in a manner and form that is not self-destructive.

The conflict between the ego and the id takes place within the depths of the psyche, but it is a conflict writ large in the life of society. The ego, let us remember, is described by Freud as an "intermediary between the id and the external world," while the superego, in whose shadow the ego operates, is characterized as the outcome of a process by which "part of the inhibiting forces of the outer world become internalized." The inner tension between the ego and the id is thus reflected in the tension between the self and the outer world.

The immanent purpose of life, as Freud sees it, is happiness—that is, the gratification of instinctual desires. But, as we have noted, the unregulated gratification of instinctual drives would not only disrupt the psyche; it would make the coexistence of man with man in society impossible. Something of the same functions of the ego and the superego must now be exercised in the larger context of social life.

For Freud, the very possibility of social coherence constitutes a grave problem. He sees community emerging under the influence of the two great forces Eros and Ananke, love and necessity—the former by giving the male a "motive for keeping the female, or rather his sexual object, with him"; the latter by bringing about some sort of cooperation for work. But these forces, Freud feels, would hardly of themselves avail to sustain society and "domesticate" man, who is basically so anti-social. Something more is necessary if men are to survive. That something is supplied with the emergence of culture or civilization.

Freud has many wise and penetrating things to say about culture scattered through his writings, to which I cannot possibly do justice in this essay. But essentially what Freud means by culture in this connection is the vast and intricate complex of psychological and institutional devices by which society is held together and man converted into a social being. The superstructure of culture is vast and far-reaching, and no one admires the fine flowers of cultural creativity more than Freud; but no one can be more pragmatic, more realistic—more crude, some would say—than he in interpreting its social function.

Culture demands increasing amounts of instinctual renunciation, Freud insists. This is so because, as we have seen, unrestricted and unregulated instinctual expression would make social life impossible; it is so for the even more important reason that cultural creativity requires a considerable amount of psychic energy, which can only be obtained by diverting it from its primary instinctual aims and using it for cultural purposes. This applies to both sex and aggression, the two great drives with which Freud is concerned.

Culture requires extensive "restrictions on the sexual life," for sexual activity must be carefully regulated if any society at all is to be possible. Moreover, "culture obeys the laws of psychological economic necessity in making the restrictions, for it obtains a great part of the mental energy it needs by subtracting it from sexuality." A good deal of the sexual energy is sublimated, and a good deal is transformed into what Freud calls "aim inhibited libido," which helps "strengthen communities by bonds of friendship." Freud is fascinated by the "law of love," the injunction to love one's neighbor as oneself. This "law" is absurd on all "rational" grounds, he feels, yet is necessary for civilization. "Culture has to call up every possible reinforcement," he says. "Hence its system of methods by which mankind is to be driven to identifications and aim-inhibited love relationships; hence the restrictions on sexual life; and hence too, its ideal command to love one's neighbor as oneself, which is really justified by the fact that nothing is so completely at variance with original human nature as this." Unfortunately, however, Freud acknowledges, such extremities of "aim-inhibited friendliness" as are required by the "law of love" are possible only for a very few. St. Francis is Freud's favorite example.

For, after all, it is man's aggressiveness that is the "most powerful obstacle to culture" and social cooperation. What means does civilization employ to control this self-destructive aggressiveness of man? As with sex, aggression has to be sharply restricted by repression and by other devices. As with sex, too, there are carefully regulated and socially sanctioned forms of expression. As Zilboorg points out, "direct violence in a social setting becomes temporarily an ally of both the ego and the superego; the individual participating in an act of violence of a social nature—whether he is a striker, a revolutionist, a soldier, or a policeman—rarely if ever feels guilty about it. . . . The superego lends its full support to the ego in that it justifies the violence on grounds of ethical social principles." In this way, a certain amount of aggression can be gratified not only without endangering the position of the ego and social control, but even in a manner that strengthens it.

But the most subtle device by which society protects itself against the disruptive forces of human aggression is the introjection of the latter into the superego. The aggressiveness is turned inward and supplies the stern superego with its power to keep the ego, and through the ego the id, in line. "Civilization," Freud points out with telling effect, "thus obtains mastery over the dangerous love of aggression in individuals by enfeebling and disarming it, and setting up an institution within their minds to keep watch over it, like a garrison in a conquered city." The supreme irony of the situation is, of course, that the power by which aggressiveness is controlled is derived from the very aggressiveness itself.

Freud goes even further, and in his studies of group psychology suggests that an essential factor in the cohesion of the group, and in group relations generally, is to be found in the projection by all members of the group of their superego on to a single figure, the leader. Where no leader figure emerges as a sort of corporate superego, the culture is enfeebled and confused, Freud feels. This was his criticism of American civilization in the 1920's, where (according to him) "leading personalities failed to acquire the significance that should fall to them in the process of group formation."

Such are the complex, hidden ways in which the culture of a society, according to Freud, operates to tame, divert, and make use of man's instinctual drives to preserve the social life that these very instinctual drives threaten to destroy. Culture accomplishes an amazing work, but can it do more than establish a precarious balance always threatened from below? Freud points to "the difficulties inherent in the very nature of culture, which," according to him, "will not yield to any efforts at reform." "Every culture," he repeats, "is based on coercion and instinctual renunciation." He therefore warns against the panacea-mongers with their patented schemes of achieving a perfect adjustment of man and society. "A great part of the struggles of mankind," he says, "centers around the single task of finding some expedient . . . solution between . . . individual claims and those of the civilized community; it is one of the problems of man's fate whether this solution can be arrived at in some particular form of culture, or whether the conflict will prove irreconcilable. . . . The fateful question of the human species seems to me to be whether and to what extent the cultural process developed in it will succeed in mastering the derangements of communal life caused by the human instinct of aggression and self-destruction." On this ominous note, Freud ends his remarkable tract, *Civilization and Its Discontents*, upon which I have mainly drawn for this exposition.

When we turn to Erich Fromm, the most influential of the neo-Freudian revisionists, we enter, as it were, a new world. Where Freud is dualistic,

Fromm is harmonistic; where Freud is somber, even pessimistic, Fromm exhibits an amazing confidence in the possibilities of human progress; where Freud assumes the posture of a disillusioned observer, Fromm is always the reformer.

To Fromm, man is not divided against himself in the very structure of the psyche, but is essentially unified, intact, perfect. The imperfections and distortions of human nature Fromm traces to the corrupting effects of the culture.

It is not easy to get a clear picture of man from Fromm's writings because he still wavers between his earlier conception that it is "the social process which creates man" and his more recent view that there is a normative human nature which "is the same for man in all ages and all cultures." But, however man is understood, he is seen as essentially good and rational; there is practically no vestige in Fromm of the dark Freudian picture of the primordial struggle between ego, superego, and id in the depths of the self. All man really wants, according to Fromm, is to "relate to the world lovingly, . . . [to] use his reason to grasp reality objectively, [to] experience himself as a unique individual entity, and at the same time one with his fellow man, [to be] one who is not subject to irrational authority, [but who] accepts willingly the authority of conscience and reason. . . ." That is all man wants to be, but society will not let him, at least society hitherto has not let him. The locus of evil and irrationality is thus not in man, but in society. The real conflict, in other words, is between the good, healthy human nature, on the one side, and a "sick" society, on the other. It is the evil society that corrupts the perfection of "normative" man.

Viewing man and society from this agle, Fromm sees the real problem not as the taming of the innate destructive drives in man through the devices and institutions of civilization—which is Freud's view—but as the reconstruction of society to fit normative human nature. Once such reconstruction is achieved—and to Fromm it is well within the realm of human achievement—once society is restored to "sanity," the essential goodness and rationality of human nature will have the opportunity of expressing itself in social life. Sane men in a sane society might be described as Fromm's vision, and men become sane when they have a sane society to live in, a society that will not distort or corrupt their personalities. "Fromm promises the advent of loving, creative, and reasonable man"—Paul Kecskemeti thus acutely summarizes the argument of *The Sane Society*—"on condition that society recognizes the sovereignty of the individual. Society can become perfect because human nature already is: the only thing needed is that society no longer dim the light of natural perfection." "It is our first task," Fromm himself writes, "to ascertain what is the nature of man, and what are the needs which stem from this nature. We then must proceed to examine the role of society in the evolution of man and to study its furthering role for the development of man as well as the recurrent conflicts between human nature and society, and the consequences of these conflicts, particularly as far as modern society is concerned."

It is possible to pay tribute to the brilliance, even profundity, of much of Fromm's criticism of contemporary culture without overlooking the incredible naïveté of the panacea he offers for our ills, for it is a panacea that he offers. "The only alternative to the danger of robotism," he says, "is humanistic communitarianism"—by which he means the restructuring of society into small and decentralized communal entities, in which the functions of worker and manager will somehow be combined. How, in the face of all experience, Fromm can believe such combination to be really possible, or why he should assume that were it achieved there would be nothing in life but love, rationality, and creativeness is very far from clear. But sufficiently clear is Fromm's basic understanding of man and society that emerges from it. And it is so different from Freud's that one might not know that it is the same man and same society that they are talking about.

Both Freud and Fromm see a conflict between man and society. But Freud sees the conflict as one between man's "biological self," with its destructive instinctual drives, and society, with its apparatus of coercion and enforced renunciation. Freud's view, in short, is Hobbesian both in its conception of man and in its notion of the function of society. "There are present in all men," Freud says, "destructive and therefore anti-social and anti-cultural tendencies." "Civilized man has exchanged some part of his chances of [instinctual] happiness for a measure of security." Fromm, on the other hand, is a most manifest Rousseauean, for to him natural man is born free and good, only to be enslaved and corrupted by an evil society. Freud finds the evil drive in man's biologic nature ("The tendency to aggression is an innate independent and instinctual disposition in man"), whereas Fromm sees man's aggressive anti-social tendencies to be the result of social pressures, particularly social frustration and insecurity. As a natural consequence, Freud refuses to reassure us by holding out the possibility of a cure for the "discontents" of civilization; Fromm has his program all ready.

Freud and Fromm are both essentially rationalists, but of very different kinds. It may seem strange to call Freud a rationalist, but David Riesman not only calls him that, but goes on to say that "it would be difficult to find anyone in the Enlightenment who was more so." And Riesman is right, though perhaps his statement is somewhat extreme. For reason is Freud's god, and truth—which he identifies with scientific truth—the only epiphany he recognizes. He is even a little Platonic in his rationalism. "One might," he says with a gesture at Plato's celebrated figure of the rider and the chariot, "one might compare the relation of the ego to the id with that between a rider and his horse," the ego standing for reason . . . [and] the id standing for the untamed passions." His whole conception of the psychoanalytic cure is rationalistic, for (as Fromm correctly points out) Freud's "psychoanalysis is the attempt to uncover the truth about oneself. . . . The aim of the cure is the restoring of health, and the remedies are truth and reason."

Freud's rationalism is reflected in the way he envisages the conflict within the self. To him it is essentially a division between the ego and the id, between the reason and the passions, with the superego hovering above as a stern internalized monitor. He does not see that the conflict in the self is not merely the assault of the ego by the id and its harrassment by the superego, but is actually the self divided against itself at all levels—the cleft runs *through* the ego, superego, and id alike, and not merely between them. Because he does not see this, his picture of the ego strikes one as altogether too simple. The ego can hardly be as rational as Freud makes it out to be.

Even in his view of society, Freud permits a strong element of rationalism to creep in, particularly when he deals with religion. To Freud, religion is, of course, an "illusion," but it is an illusion that somehow does not meet with the same kind of tolerant understanding that he extends to most other forms of human self-deception. Now suddenly he discovers that "the time has probably come to replace the consequences of repression by rational mental effort," and so he makes a plea, in just so many words, for "a purely rational basis for cultural laws." This from Freud, who more than anyone else has convinced us that society and culture cannot have any "purely rational basis"! So Freud's rationalism penetrates even his social thinking. Yet, on the whole, it is a secondary and muted influence in this area, at least so long as he keeps away from religion.

Fromm is somewhat less rationalistic in his view of the self, for his picture of human normativeness is complex and many-sided. But he makes up for that by his extreme social and political rationalism, which expresses itself in an embarrassingly simple-minded utopianism with its optimistic faith in schemes of social reconstruction and their power to achieve perfection. Freud is, in a sense, a Voltairean: reason should rule society through a rational, scientific-minded elite, taking account of, perhaps even manipulating, the illusions and self-deceptions of mankind. Fromm, I emphasize again, is essentially a Rousseauean, who envisages man in his perfection and looks to the perfect—that is to say, the rational—society to release this perfection in man.

Both Freud and Fromm are naturalists—Freud a biologistic naturalist and Fromm largely of the culturistic variety. Freud's biologism may be interpreted as a reflection of the late nineteenth-century scientific outlook which pervades all his thinking, for this outlook was biological, evolutionary, and materialistic to the very core. Fromm's culturism may, in much the same way, be understood as a reflection of the current scientific mind, with its sociological bias and its uneasy hesitation between a sociological relativism and the effort to find some fixed point in a "pan-human" normative type. But while all this is no doubt true, it is important not to overlook that Freud's biologism and Fromm's culturism both have deeper significance. Insistence on the "biological self" is so necessary for Freud because, as Lionel Trilling has pointed out, for Freud this "biological self" represents the "hard, irreducible, stubborn core from which the culture may be judged." Freud "needed to believe that there was some point at which it was possible to stand beyond the reach of culture," and that point he found in the inviolable "biological self." Fromm, on the other hand, no doubt finds culturism so appealing because he is so passionately concerned with believing that the perfect society (which he imagines to be achievable in history) will provide the necessary and sufficient condition for the emergence of the perfect man. The type of naturalism each espouses is thus seen to be closely related to his underlying interest and concern.

Both—each from his own standpoint—very largely misunderstand the inner relation of the self and society. To Freud, the self is essentially individualistic, a self-contained entity prior to society and requiring community for external reasons. Society is, at bottom, no more than an aggregation of individual selves held together by derivative psychological bonds. Freud does not see what the Jewish-Christian tradition has always stressed, and what modern existentialists such as Buber, Jaspers, and Marcel are reemphasizing, that the human self emerges only in community and has no real existence apart from it, in isolation. The self is not prior to society, but in fact coeval with it. On the other hand, Fromm sometimes tends to lose the self by identifying the community which the self needs for its being with the "right" social order, according to his blueprint, never realizing that even the best social order necessarily institutionalizes—and therefore corrupts as well as implements—the I-Thou community which is the self's true context of being. In neither system does the self really come into its own.

Nor can either system do full justice to the self in its "dramas of history." Freud's interest in history is primarily to find invariant patterns biologically grounded amidst the multifarious diversity of the historical process. Fromm's interest in history, like Rousseau's, is to lay bare the many ways in which evil institutions have corrupted man in the past, and to draw appropriate lessons from the story. Neither understands the full historicity of man, or what this implies for human existence, because neither has a proper sense of human transcendence, of the capacity of the self to transcend all the coherences, social as well as biological, through which it is provisionally defined and in which it is provisionally enclosed. Freud tries to guarantee the integrity of the human self by making it biological, Fromm by grounding it in a normative human nature "the same for man in all ages and all cultures." Neither sees it where Christian thinkers like Reinhold Niebuhr have placed it, in the dimension of being where the "self is enabled to stand as it were above the structures and coherences of the world," facing the eternal.

In such an ultimate perspective, Freud's biologism must be judged inadequate and misleading. Yet it does contain a profound insight which idealistic and spiritually-minded people often tend to forget: man's inescapable biological grounding. If it is true, as Paul Tillich insists, that "in man nothing is 'merely biological,'" it is also true, as Tillich goes on to say, that "in man ... nothing is 'merely spiritual.'" "Every cell of [man's] body participates in his freedom," but "every act of his spiritual creativity is nourished by his vital dynamics." We have Freud to thank for emphasizing one half, and an important half, of this truth. In a way, Fromm stresses the other half, for he does constantly emphasize that man's nature is more than biological. Yet, because he remains a prisoner of his naturalistic presuppositions, he cannot really say what this "beyond-the-biological" is and how it manifests itself in man.

Freud's biologism serves him well in fostering his stubborn insistence that the trouble lies deep in man, and is not simply the result of adverse social conditions. Freud's view is, as Robert Merton has suggested, a "variety of the 'original sin' doctrine," a biologistic variety. Despite its obvious short comings, it preserves him from the pitfalls of social utopianism and perfectionism. It brings him to the Niebuhrian insight, although reached on vastly different grounds, that history remains ambiguous to the very end.

Fromm's Rousseauism very largely blinds him to this hard wisdom. For him, "sin" is socially derived, and history is ultimately redeemable through human effort. Yet Fromm is surely right in feeling that human aggression and destructiveness are not merely biological, but are somehow emergent out of the human situation, which, however, Fromm wrongly takes to be identical with the social situation. What both Freud and Fromm need, it seems to me, is some of the insight that has come down to us in the biblical tradition and that has recently been restated by some of the existentialists. It is not the biological constitution of man that can be held responsible for the human evil we know in history, nor is it the social order, although both the biological constitution and the social order are involved. It is out of man's existential situation—out of the tension between his self-transcending freedom and the inherent limitations of his creatureliness—that is engendered the basic insecurity and anxiety that prompt him to "pride" (in the theological sense) and self-aggrandizement. This "pride" and self-aggrandizement enlists all suitable biological impulses in its service and exploits all serviceable social institutions, but it is not itself the product of either. It emerges out of the depths of the self in its existential situation. Freud and Fromm each see an aspect of this truth, but Freud, I feel, somehow comes closer.

If we were to permit ourselves a theological vocabulary in this discussion, we might, I think, without going too far afield, say that Freud tends to a Manicheanism of a biologistic kind, while Fromm tends to an extreme Pelagianism, which glorifies human autonomy and holds man to be sufficient unto himself for salvation. Freud does not see that the tragic dualism pervading human life, however real and beyond the reach of idealistic conjurations, can be neither the first nor the last word of human destiny. Fromm, on the other hand, does not seem able to understand that man's desperate plight today, which he describes with such feeling, is, in great part at least, the direct consequence of the unbridled pretentions to autonomy that have characterized modern culture from the Renaissance to our time. Freud sees man desperately caught between the upper and nether millstones of his instincts and his culture; Fromm exalts the "all-powerful 'I.'" Neither really understands man's plight or man's hope.

Zilboorg is right in pointing out that what Freud devoted his life to studying was the "psychological reactions of man *in the state of sin*"; but Freud was not fully aware of what he was doing. He took the "fallen," the sinful condition of man, the condition of man with which he was empirically acquainted—man at war with himself and his fellow men—as man's original, and so to speak, normative condition. His attitude was therefore deeply pessimistic, though his pessimism was inconsistently relieved here and there by a dim hope engendered out of an irrational faith in reason. Fromm, on the other hand, like all idealists, simply ignores the "fall," brushes aside the insistent lessons of experience, and deals with man as though he were still in his "original rightness." No wonder the glorious creature he described—glorious when uncorrupted by an evil society—bears little resemblance to Freud's aggressive, libidinal being whom civilization has to beguile and restrain. Fromm's man is essentially untouched by the dreadful disorder of human sinfulness; Freud's man is an imperfectly tamed beast desperately trying to domesticate himself. Neither Freud nor Fromm sees man in his full complexity and his many dimensions, in both his "grandeur" and his "misery," *at once* in the perfection of his original creation and in the radical imperfection of his actual existence. It was to them both that Pascal was speaking, when he described his purpose and vocation:

> If anyone exalts man, I humble him; if he humbles man, I exalt him; and I always contradict him until he understands what an incomprehensible monster man is.

8 Positivism and Behaviorism: Methodological Metaphors

In connection with both seventeenth-century science and Newtonian science, we saw that some thinkers were less interested in applying specific concepts from the natural sciences to other disciplines than in using investigative techniques and methods borrowed from the sciences. The importation of techniques from the physical sciences into history, moral philosophy, social philosophy, and political theory increased in frequency throughout the eighteenth and nineteenth centuries and was often urged by men who explicitly rejected the gross use of analogical or metaphorical reasoning exemplified in the writings of the social Darwinists or men like Henry Adams.

Thomas Reid, for example, one of the eighteenth-century Scottish philosophers whose principal aim was to create a science of the mind—what we now call psychology—was fully aware of man's tendency to think metaphorically but unalterably opposed to its use in studying mental phenomena. He knew that men, whose common knowledge was largely concerned with the physical world, would be greatly tempted to derive their notions of mental phenomena from analogies with physics and physiology and objected to the fact that such analogies would "be apt to impose upon philosophers, as well as upon the vulgar, and lead them to materialize the mind."

Reid had many reasons, both religious and philosophical, for believing that the mind was immaterial and not subject to the same laws as physical objects; consequently he denied that concepts from physics were directly applicable to the mind; and he and his followers led British and American psychology into the "scientific" but introspective rather than materialistic pattern which it assumed throughout the nineteenth and early twentieth centuries.

The basic contention of Reid and of increasing numbers of scientists, social thinkers and historians during the nineteenth century was that the only reliable source of human knowledge was empirical and experimental science. The aim of such science—of all science—was simply to discover the connections between phenomena and to express them as general laws of nature. Experimental science was not to consider the underlying efficient causes of phenomena; i.e., it was not to ask the *why* of things, but merely to organize and systematize our knowledge of *how* phenomena took place. It was to determine the necessary observable conditions for the manifestation of the phenomenon under consideration in order to enable men to predict and control the course of nature.

With their emphasis on description and prediction rather than explanation, the proponents of this type of science were more than willing to admit that different sciences were characterized by their own special problems and objects of investigation, and that all they really needed to share was a basic method which did not allow the intrusion of unobservable entities. Metaphors or analogies might suggest hypotheses for testing—i.e., they might be valuable heuristic devices—but they had no explanatory value in themselves. In spite of their general suspicion of analogy, however, the new proponents of empirical science—who began to be known under the vaguely defined label of positivists—clearly based their methodological precepts on a belief that all phenomena bore a sufficient resemblance to physical phenomena, that the same methods used in studying material objects were appropriate to the study of mental and social phenomena. Similarly, they agreed that the ends of all knowledge—the discovery of general laws—were identical with the ends of the natural scientists' quest. Thus, in a very important sense, positivism rests at bottom on a tacit acceptance of the broadest and vaguest analogies of all: (1) that all human knowledge is essentially analogous to the kind of knowledge which natural scientists have of the material world, and (2) that all phenomena are subject to laws analogous to the kind of deterministic relations that have been discovered in the material world.

Positivism clearly had deep historical roots in the scientific revolution, but it did not reach fruition until the mid- to late-nineteenth century in the writings of Claude Bernard (whose *Introduction to the Study of Experimental Medicine* (1865) became the model for French discussions of scientific method well into the twentieth century) and of Ernst Mach (whose *The Science of Mechanics: A Critical and Historical Account of Its Development* (1883) established the anti-metaphysical, anti-causal emphasis of positivism as the dominant theme not only among scientists but among almost all philosophers interested in theories of knowledge).

The great impact of positivism on Western thought has been carried into the twentieth century in part through the philosophical writings of the Vienna Circle, whose members, Maurice Schlick, Otto Neurath, Phillip Frank, Ludwig Wittgenstein, and Rudolph Carnap, developed Mach's rejection of unobservable concepts into the powerful and sophisticated form of logical-positivism during the 1920's. But logical-positivism, while important as a guide to certain forms of social theory, has been of primary concern to a relatively small intellectual elite. A much more pervasive, and thus in a sense more important, role has been played by positivism. It has provided the philosophical system which underlies the widely heralded "behavioral" or "behavioristic" approach of modern psychology and "political science."

The first selection of this section, Emile Zola's essay on "The Experimental Novel" (1880) demonstrates the wide-ranging impact of Claude Bernard's *Introduction to the Study of Experimental Medicine*. Zola takes Bernard's work as a model for explaining the needs and possibilities of adopting the experimental approach even in the traditionally humanistic literary realm. In the second selection, taken from B. F. Skinner's *Science and Human Behavior* (1953), a foremost behavioral psychologist clearly discusses the positivists' reinterpretation of the meaning of causation and points up both the positivists' demands for directly observable concepts and their demands for prediction and control.

The final article from Sheldon Wolin's "Political Theory as Vocation" discusses the desire of many contemporary students of politics to emphasize scientific methodology and criticizes this tendency in the belief that there may be legitimate—even important—ends of human understanding which cannot be met within the limitations of a scientific approach.

The Experimental Novel

Emile Zola

In my writing on literary subjects I have often discussed the experimental method as applied to the novel and the drama. The return to nature, the naturalistic evolution, which is the main current of our age, is gradually drawing all manifestations of human intelligence into a single scientific course. However, the idea of literature determined by science is likely to be surprising unless clearly defined and understood. It therefore seems useful to be explicit about what the experimental novel means, as I see it.

My remarks on this subject will be only an adaptation, for the experimental method has been set forth with force and marvelous clarity by Claude Bernard in his *Introduction to the Study of Experimental Medicine*. This book by a scientist of decisive authority will provide me with a solid base. In it the whole question is treated, and I shall confine myself to giving such quotations from it as are necessary as irrefutable arguments. This dis-

Source: Emile Zola, "The Experimental Novel", translated by George J. Becker in George J. Becker, *Documents of Modern Literary Realism* (Princeton, N. J.: Princeton University Press, 1963), pp. 162-175.

cussion therefore will be no more than a compilation of texts; for on all points I intend to fall back on Claude Bernard. Usually it will be sufficient for me to replace the word "doctor" by the word "novelist" in order to make my thought clear and to bring to it the rigor of scientific truth.

What has led me to choose the *Introduction* is the fact that to many people medicine, like the novel, is still an art. All his life Claude Bernard worked and fought to set medicine on the road of science. With him we see the first stammerings of a science disengaging itself slowly from empiricism and coming to rest on truth, thanks to the experimental method. Claude Bernard demonstrates that this method as applied in the study of inanimate bodies, in chemistry and physics, ought equally to be used in the study of living bodies, in physiology and medicine. I shall attempt to prove in my turn that if the experimental method leads to knowledge of physical life, it may also lead to knowledge of passional and intellectual life. It is only a question of gradation on the same scale from chemistry to physiology, and then from physiology to anthropology and sociology. The experimental novel comes at the end.

For greater clearness I shall briefly summarize the *Introduction* here. The reader will better comprehend the application I make of my quotations by knowing the plan of the work and the material which it treats.

Claude Bernard, after asserting that medicine will in the future follow the road of science by relying on physiology, using the scientific method, first sets forth the differences that exist between the observational sciences and the experimental sciences. He comes to the conclusion that fundamentally an experiment is nothing but a forced observation. All experimental reasoning is based on doubt, for the experimenter should have no preconceived idea about nature and should always maintain an open mind. He merely accepts phenomena which occur once they are verified.

Then in the second section he begins his real subject by demonstrating that the spontaneity of living bodies does not preclude the use of experimentation. The difference comes only from the fact that an inanimate body is to be found in the common external environment whereas the elements of higher organisms exist in an interior and highly developed environment, though one which is also endowed with the same constant physiochemical properties as the external environment. Thus there is absolute determinism in the conditions of existence of natural phenomena both for living beings and for inert matter. By "determinism" he means the cause which determines the appearance of phenomena. This proximate cause, as he calls it, is nothing but the physical and material condition of the existence or manifestation of phenomena. The goal of the experimental method, the end of all scientific research, is therefore identical for living beings and for inert bodies: it consists in finding the relations which link any phenomenon whatsoever to its proximate cause, in other words, in determining the conditions necessary for the occurrence of this phenomenon. Experimental science should not concern itself with the *why* of things; it explains the *how*, nothing more.

After setting forth the experimental conditions common to living bodies and inert bodies, Claude Bernard goes on to the experimental considerations special to living bodies. The great and unique difference is that in the organism of living beings a harmonious ensemble of phenomena is to be considered. He then treats of the technique of experimentation to be used on living beings, of vivisection, the preparatory anatomical conditions, the choice of animals, the employment of mathematics in the study of phenomena, and the physiological laboratory.

Finally, in the last part of the *Introduction*, Claude Bernard gives examples of experimental physiological investigation in support of the ideas he has formulated. He then gives examples of a critique for experimental physiology and ends by showing the philosophical obstacles that experimental medicine encounters. In the first rank he places the false application of physiology to medicine and scientific ignorance, as well as certain illusions of the medical mind. He concludes by saying that empirical medicine and experimental medicine, far from being incompatible, should on the contrary be inseparable from each other. The last assertion of the book is that experimental medicine is not dependent on any medical doctrine or philosophical system.

Such, very simply, is the skeleton of the *Introduction* divested of its flesh. I hope this rapid summary will suffice to fill in the gaps that my method of discussion will inevitably leave; for naturally I shall take from the book only those quotations necessary to define and comment on the experimental novel. I repeat that this is only the foundation on which I base my case, a foundation most rich in arguments and proofs of all sorts. Even though stammering, experimental medicine alone can give us an exact idea of experimental literature, which, still in embryo, has not even begun to stammer.

The first question above all to consider is this: Is experimentation possible in literature, where heretofore observation alone seems to have been used?

Claude Bernard has a long discussion of observation and experiment. There is a very sharp line of demarcation between them. Here it is: "We give the name of *observer* to him who applies simple or complex procedures of investigation to the study of phenomena which he does not cause to vary and which he merely collects as they are provided by nature; we give the name of *experimenter* to him who uses simple or complex procedures of investigation to vary or modify natural phenomena for whatever purpose, and to make them appear in circumstances and under conditions not found in nature." For example, astronomy is a science of observation, because we cannot imagine an astronomer acting on the stars; whereas chemistry is an experimental science since the chemist acts on nature and modifies it. According to Claude Bernard, this is the only really important distinction between the observer and the experimenter.

I cannot follow him into his discussion of the various definitions given in the past. As I have said, he concludes that an experiment is basically only a provoked observation. I quote: "In the experimental method, the search for facts—that is, investigation—is always accompanied by reasoning, with the result that usually the experimenter sets up an experiment in order to control or verify the value of an experimental idea. Thus we may say that in this case the experiment is a forced observation to provide a control."

In addition, to be able to find out what there may be of observation and experimentation in the naturalist novel, I need only the following passages:

"The observer sets down purely and simply the phenomena he has before his eyes.... He ought to be the photographer of phenomena; his observation ought to represent nature exactly.... He listens to nature and writes at her dictation. But once the fact is set down and the phenomenon is carefully observed, ideas come into play, reasoning intervenes, and the experimenter appears in order to interpret the phenomenon. The experimenter is he who, by virtue of a more or less probable but anticipatory interpretation of observed phenomena, institutes an experiment in such manner that in the logical framework of prevision, it furnishes a result which serves as a control for the hypothesis or preconceived idea.... From the moment that the result of the experiment is manifest, the experimenter confronts a true observation which he has induced, one which he must set down without preconceived idea like any other observation. The experimenter ought thus to disappear or rather transform himself instantly into observer, and it is only after he has set down the results of the experiment in absolutely the same fashion as he would an ordinary observation that his mind will once more proceed to reason, compare, and judge whether the experimental hypothesis has been verified or invalidated by these same results."

There you have the whole procedure. It is somewhat complicated, and Claude Bernard is led to say: "When all this takes place at once in the head of a scientist who is an investigator in a science as confused as medicine still is, there is such a tangle between what results from observation and what belongs to experiment that it would be impossible and indeed useless to try to analyze each of these terms in their inextricable mingling." In short, we can say that observation "shows," and experiment "informs."

Now, coming back to the novel, we can see equally well that the novelist is both observer and experimenter. The observer in him presents data as he has observed them, determines the point of departure, establishes the solid ground on which his characters will stand and his phenomena take place. Then the experimenter appears and institutes the experiment, that is, sets the characters of a particular story in motion, in order to show that the series of events therein will be those demanded by the determinism of the phenomena under study. It is almost always an experiment "in order to see," as Claude Bernard puts it. The novelist starts out in search of a truth. I shall take as an example the figure of Baron Hulot in Balzac's *La Cousine Bette*. The general fact observed by Balzac is the ravages that a man's amorous nature produces in himself, his family, and society. Once he has chosen his subject, he has departed from observed facts, for he has initiated his experiment by subjecting Hulot to a series of tests, having him pass through various situations in order to show how the mechanism of his passion works. It is therefore evident that we here have not only observation but also experimentation, since Balzac does not restrict himself to being a photographer of facts gathered by himself, but he intervenes directly to place his character in conditions over which he maintains control. The problem is to learn what such and such a passion, acting in such and such a milieu under such and such circumstances, will bring about in terms of the individual and of society; and an experimental novel, such as *La Cousine Bette*, for example, is simply the record of the experiment which the novelist repeats before the eyes of the public. In short, the whole operation consists of taking facts from nature, then studying the mechanism of the data by acting on them through a modification of circumstances and environment without ever departing from the laws of nature. At the end there is knowledge, scientific knowledge, of man in his individual and social action.

To be sure, we are here far from the certainties of chemistry, or even physiology. We do not yet know the reactive agents which will break up the passions and permit us to analyze them. Often in this study I shall remind the reader that the experimental novel is even younger than experimental medicine, which, however, has scarcely been born. But I do not intend to set forth achieved results; I desire simply to give a clear exposition of a method. If the experimental novelist still gropes his way in the most obscure and complex of sciences, that does not prevent that science from existing. It is undeniable that the naturalistic novel, such as we know it at this time, is a true experiment which the novelist makes on men, with the support of observation.

Moreover, this opinion is not mine alone, but is equally that of Claude Bernard. He says somewhere: "In the conduct of their lives men do nothing but make experiments on one another." And what is more conclusive, here is the whole theory of the experimental novel: "When we reason about our own acts, we have a certain guide, for we have consciousness of what we think and what we feel. But if we wish to judge the acts of another man and know the motives of his action, it is quite a different thing. No doubt we have before our eyes the movements of such a man and his behavior, which are, we are sure, modes of expression of his sensibility and will. In addition, we admit that there is a necessary rapport between his acts and their cause; but what is that cause? We do not feel it in ourselves; we are not conscious of it as in ourselves; we are thus obliged to interpret it, to suppose it on the basis of the movements we see and the words we hear. Thus we must use the acts of this man as controls one upon the other; we consider how he acts under a given condition, and in short we have recourse to the experimental method." All that I have advanced above is summed up in that last sentence, which is by a scientist.

I shall also quote an image of Claude Bernard's which has greatly struck me: "The experimenter is the examining magistrate of nature." We novelists are the examining magistrates of men and their passions.

But see how things begin to clear up when you take the position of the experimental method in the novel, with all the scientific rigor of the physical sciences. A stupid reproach made against us naturalist writers is that we wish to be merely photographers. In vain have we asserted that we accept temperament and personal expression; people go right on answering us with imbecile arguments about the impossibility of the strictly true, about the necessity of arrangement of facts to make any work of art whatever. Well, with the application of the experimental method to the novel all argument comes to an end. The idea of experiment carries with it the idea of modification. We begin certainly with true facts which are our indestructible base; but to show the mechanism of the facts, we have to produce and direct the phenomena; that is our part of invention and genius in the work. Thus without having recourse to questions of form and style, which I shall examine later, I state right here that when we use the experimental method we must modify nature without departing from nature. If the reader will recall the definition: "Observation shows, experiment instructs," we can henceforth claim for our works the high instruction of experiment.

Far from being diminished, the writer is hereby singularly increased. An experiment, even the most simple one, is always based on an idea, which in turn comes from observation. As Claude Bernard says, "The experimental idea is in no way arbitrary or purely imaginary; it must always have a basis in observed reality, that is, in nature." It is on this idea and on scepticism that he rests his whole method. "The appearance of the experimental idea," he says further on, "is quite spontaneous and completely individual in nature; it is an individual sentiment, a ***quid proprium***, which constitutes the originality, the inventiveness, the genius of each man." Next, scepticism is necessary as the great scientific lever, "The sceptic is the true scientist; he doubts only himself and his interpretations but he believes in science; he even admits, in the experimental sciences, an absolute criterion or principle, the determinism of phenomena, absolute in the phenomena of living beings as in those of inert bodies." Thus instead of binding the novelist tightly, the experimental method leaves to him all his intelligence as thinker and all his genius as creator. He must see, understand, invent. An observed fact will bring forth the idea of the experiment to try, of the novel to write, in order to arrive at complete knowledge of a truth. Then, when he has discussed and laid out the plan of his experiment, he will at all times judge the results with the freedom of intelligence of a man who accepts only facts in conformity with the determinism of phenomena. He has begun in doubt in order to arrive at absolute knowledge; he does not cease to doubt until the mechanism of passion, taken to pieces and put back together again by him, functions according to laws fixed by nature. There is no broader or freer task for human intelligence. We shall see later the miserable state of the scholastics, the systematizers and theoreticians of the ideal, in comparison with the triumph of the experimenters.

I sum up this first section by repeating that the naturalistic novelists observe and experiment, and that their whole task begins in the doubt which they hold concerning obscure truths, inexplicable phenomena, until an experimental idea suddenly arouses their genius and impels them to make an experiment, in order to analyze the facts and become master of them.

Such then is the experimental method. But it has long been denied that this method can be applied to living bodies. This is the important point which I am going to examine with the help of Claude Bernard. The chain of reasoning will be very simple: if the experimental method has been capable of extension from chemistry and physics to physiology and medicine, then it can be carried from physiology to the naturalist novel.

To cite only one scientist, Cuvier maintained that experimentation, while applicable to inert bodies, could not be used with living bodies; according to him, physiology had to be purely a science of anatomical observation and deduction. The vitalists still posit a vital force which in living bodies is in constant struggle with physio-chemical forces and neutralizes their action. Claude Bernard, on the contrary, denies the existence of any mysterious force and asserts that experimentation is applicable everywhere: "I do not propose," he says, "to establish that the science of life phenomena can have other bases than the science of phenomena of inert bodies and that there is in this regard any difference between the principles of the biological sciences and the physio-chemical sciences. In fact the goal which the experimental method sets for itself is the same everywhere; it consists of linking by experiment natural phenomena to their conditions of existence or to their proximate causes."

It seems to me unnecessary to enter into the complex explanation and reasoning of Claude Bernard. I have said that he insisted on the existence of an internal milieu in the living being. "In experimentation on inert bodies," he says, "one has to take only one milieu into account: that is the external cosmic milieu; whereas in the higher living bodies there are at least two milieux to consider: the external or extra-organic and the internal or intra-organic. The complexity arising from the existence of an intra-organic milieu is the sole reason for the great difficulties we encounter in experimental determination of phenomena of life and in the application of measures capable of modifying it." He goes on from there to prove that there are fixed laws for the interior physiological elements as there are fixed laws for the chemical elements in the exterior milieu. Thus one may experiment on the living being just as on the inert body; it is a question merely of achieving the desired conditions.

I insist on this because, I repeat, the important point of the matter is there. Claude Bernard, writing about the vitalists, says this: "They consider life to be a mysterious and supernatural influence which acts arbitrarily and freed from all determinism, and they berate as materialists all those who make an effort to bring vital phenomena under determined organic physio-chemical conditions. Those are false ideas which it is not easy to root out once they have lodged themselves in the mind; only the progress of science will make them disappear." And he sets down this axiom: "In living bodies as well as in inert bodies the conditions of existence of every phenomenon are determined in an absolute fashion."

I limit my remarks so as not to complicate the reasoning unduly. This then is the progress of science. In the last century a more exact application of the experimental method has created chemistry and physics, which disengage themselves from the irrational and the supernatural. Thanks to analysis, it has been discovered that there are fixed laws; we become masters of phenomena. Then we push ahead another step. Living bodies, in which the vitalists still admitted a mysterious influence, are in their turn brought and reduced to the general mechanism of matter. Science proves that the conditions of existence of all phenomena are the same for living bodies as for inert bodies; and thenceforth physiology takes on little by little the certainty of chemistry and physics. But are we to stop there? Evidently not. When we shall have proved that man's body is a machine, when we shall someday be able to take it to pieces and put it together again at the will of the experimenter, we shall then have to go on the passional and intellectual acts of man. Then we shall enter into a domain which up to now has belonged to philosophy and literature; that will be the decisive conquest by science of the hypotheses of the philosophers and writers. We have experimental chemistry and physics; we shall have experimental physiology; later still we shall have the experimental novel. That is a progress which imposes itself, the last term of which it is easy to foresee even now. Everything holds together; it was necessary to start with the determinism of inert bodies in order to arrive at the determinism of living bodies; and since scientists like Claude Bernard now show that fixed laws govern the human body we can assert without fear of mistake that the day will come when the laws of thought and the passions will be formulated in their turn. One and the same determinism must govern the stone in the road and the brain of man.

This opinion is stated in the *Introduction*. I cannot repeat too often that I am taking all my arguments from Claude Bernard. After explaining that the most specialized phenomena can be the result of the more and more complex union or association of organized elements, he writes this: "I am convinced that the obstacles which surround experimental study of psychological phenomena are in large part due to difficulties of this kind; for, in spite of their marvelous nature and the delicacy of their manifestations, it is impossible, as I see it, not to bring cerebral phenomena, like all the phenomena of living bodies, under the laws of a scientific determinism." That is clear. Later, no doubt, science will discover the determinism of all the cerebral and sensory manifestations of man.

From that time on science will therefore enter into the domain of us novelists, who are now the analysts of man in his individual and social action. By our observations and our experiments we continue the task of the physiologist, who has continued that of the physicist and the chemist. To a certain extent we are doing scientific psychology, as a complement to scientific physiology; and to complete the evolution we have only to bring to our studies of the nature of man the decisive tool of the experimental method. In short, we must operate with characters, passions, human and social data as the chemist and the physicist work on inert bodies, as the physiologist works on living bodies. Determinism governs everything. It is scientific investigation; it is experimental reasoning that combats one by one the hypotheses of the idealists and will replace novels of pure imagination by novels of observation and experiment.

Certainly I have no intention of formulating laws at this point. In the present state of the science of man confusion and obscurity are still too great for us to risk the least synthesis. All that we can say is that there is an absolute determinism for all human phenomena. From that point on investigation is a duty. We have the method; we must go forward even if a whole lifetime of effort produces the conquest of only a tiny bit of truth. Consider physiology: Claude Bernard made great discoveries, and he died admitting that he knew nothing, or nearly nothing. On each page he confesses the difficulties of his task. "In phenomenal relations," he says, "as nature offers them to us there always reigns a more or less considerable complexity. In this respect the complexity of mineral phenomena is much less great than that of vital phenomena; that is why sciences studying inert bodies have come into being so much more quickly. In living bodies the phenomena are of enormous complexity, and in addition the mobility of the vital properties makes them much more difficult to get hold of and determine." What should we say then of the difficulties which the experimental novel may expect to meet, since it takes from physiology its studies on the most complex and delicate organs and treats of the most elevated manifestations of man as individual and as member of society? Evidently analysis becomes even more complicated here. Thus if physiology is just being constituted today, it is natural that the experimental novel is only at its very beginning. It may be foreseen as a fated consequence of the scientific evolution of the age; but it is impossible to base it on certain laws. When Claude Bernard speaks "of the limited and precarious truths of biological science," we may well confess that the truths of the science of man with respect to the mechanism of the mind and passions are even more precarious and limited. We are still babbling; we are the latest comers; but that should merely be an additional spur to push us to exact studies as soon as we have the tool, the experimental method, and our goal is very clear, to know the determinism of phenomena and to make ourselves masters of those phenomena.

Without risking the formulation of laws, I believe that the question of heredity has a great influence in the intellectual and passional behavior of man. I also accord a considerable importance to environment. Here it would be necessary first to consider Darwin's theories; but this is only a general study of the experimental method as applied to the novel, and I would be lost if I tried to go into detail. I shall simply say a word about environments. We have just seen the decisive importance given by Claude Bernard to the study of intra-organic environment, which must be taken into account if we are to find out the determinism of phenomena among living beings. Very well! In the study of a family, of a group of living beings, I believe the social environment also has capital importance. One day physiology will no doubt explain the mechanism of thought and the passions; we shall know how the individual machine of a man works, how he thinks, how he loves, how he goes from reason to passion to madness; but these phenomena, these data of the mechanism of the organs acting under the influence of the internal environment do not occur outside in isolation and in a vacuum. Man is not alone; he lives in a society, in a social milieu, and hence for us novelists the social milieu endlessly modifies phenomena. Indeed our great study is there, on the reciprocal influence of society on the individual and of the individual on society. For the physiologist the external environment and the internal environment are purely chemical and physical, which permits him to discover their laws easily. We have not reached the point of being able to prove that the social milieu also is nothing but chemical and physical. It certainly is, or rather it is the variable product of a group of living beings who themselves are absolutely subject to physical and chemical laws which govern living bodies just as much as inert bodies. Thus we shall see that we can act on the social environment by acting on the phenomena over which we shall have control in man. And that is what makes the experimental novel: to have the mechanism of phenomena in men, to show the working of the intellectual and sensory manifestations as physiology will explain them to us under the influences of heredity and the surrounding circumstances, then to show man living in the social milieu which he himself has produced, which he modifies every day, and in the midst of which he in his turn undergoes continuous modification. Therefore we lean heavily on physiology; we take man in isolation from the hands of the physiologist in order to carry forward the solution of the problem and resolve scientifically the question of knowing how men behave themselves once they are in society.

These general ideas are enough to guide us today. Later on when science will have gone forward, when the experimental novel will have given decisive results, some critic will make a definitive statement of what I merely sketch out today.

Moreover, Claude Bernard confesses how difficult the application of the scientific method is to living beings. "The living being," he says, "above all among the higher animals, never falls into physio-chemical indifference toward the external environment: he undergoes incessant movement, an organic evolution in appearance spontaneous and constant, and although that evolution has need of external circumstances to manifest itself, it is nonetheless independent in its progress and in its modality." And he concludes as I have said: "In résumé, it is only in the physio-chemical conditions of the internal milieu that we will find the determinism of the external phenomena of life." But whatever the complexities which present themselves, even when special phenomena occur, the application of the experimental method remains rigorous. "If vital phenomena are more complex than and apparently different from those of inert bodies, they only present this difference by virtue of conditions, determined or determinable, which are proper to them. Thus if the vital sciences should differ from the others by their applications and their special laws, this makes no difference to the scientific method."

I must still say a word about the limits which Claude Bernard sets for science. As he sees it, we shall always be ignorant of the *why* of things; we can know only the *how*. He expresses this in these terms: "The nature of our minds leads us to seek out the essence or the *why* of things. In that respect we aim further than the goal which it is given us to attain; for experience soon teaches us that we may not go further than the *how*, that is, beyond the proximate cause or the conditions of existence of phenomena." Further on he gives this example: "If we cannot know *why* opium and its derivatives induce sleep, we shall be able to know the mechanism of that sleep and know *how* opium or its principles produce sleep; for sleep takes place only because the active substance comes in contact with certain organic elements which it modifies." And the practical conclusion from this: "Science has precisely the privilege of teaching us what we do not know, by substituting reason and experiment for feeling, and by showing us clearly the limits of our present knowledge. But by a marvelous compensation, in the measure that science thus reduces our pride she increases our power." All these considerations are strictly applicable to the experimental novel. So as not to stray into philosophical speculations, so as to replace idealist hypotheses by the slow conquest of the unknown, it must refrain from searching for the *why* of things. That is its exact role; from that it draws, as we shall see, its raison d'être and its morality.

Why Organisms Behave

B. F. Skinner

The terms "cause" and "effect" are no longer widely used in science. They have been associated with so many theories of the structure and operation of the universe that they mean more than scientists want to say. The terms which replace them, however, refer to the same factual core. A "cause" becomes a "change in an independent variable" and an "effect" a "change in a dependent variable." The old "cause-and-effect connection" becomes a "functional relation." The new terms do not suggest *how* a cause causes its effect; they merely assert that different events tend to occur together in a certain order. This is important, but it is not crucial. There is no particular danger in using "cause" and "effect" in an informal discussion if we are always ready to substitute their more exact counterparts.

We are concerned, then, with the causes of human behavior. We want to know why men behave as they do. Any condition or event which can be shown to have an effect upon behavior must be taken into account. By discovering and analyzing these causes we can predict behavior; to the extent that we can manipulate them, we can control behavior.

There is a curious inconsistency in the zeal with which the doctrine of personal freedom has been defended, because men have always been fascinated by the search for causes. The spontaneity of human behavior is apparently no more challenging than its "why and wherefore." So strong is the urge to explain behavior that men have been led to anticipate legitimate scientific inquiry and to construct highly implausible theories of causation. This practice is not unusual in the history of science. The study of any subject begins in the realm of superstition. The fanciful explanation precedes the valid. Astronomy began as astrology; chemistry as alchemy. The field of behavior has had, and still has, its astrologers and alchemists. A long history of prescientific explanation furnishes us with a fantastic array of causes which have no function other than to supply spurious answers to questions which must otherwise go unanswered in the early stages of science. . . .

Inner "Causes"

Every science has at some time or other looked for causes of action inside the things it has studied. Sometimes the practice has proved useful,

Source: B. F. Skinner, *Science and Human Behavior* (New York, The Free Press, 1965), pp. 23-24, 27-39.

sometimes it has not. There is nothing wrong with an inner explanation as such, but events which are located inside a system are likely to be difficult to observe. For this reason we are encouraged to assign properties to them without justification. Worse still, we can invent causes of this sort without fear of contradiction. The motion of a rolling stone was once attributed to its *vis viva*. The chemical properties of bodies were thought to be derived from the *principles* or *essences* of which they were composed. Combustion was explained by the *phlogiston* inside the combustible object. Wounds healed and bodies grew well because of a *vis medicatrix*. It has been especially tempting to attribute the behavior of a living organism to the behavior of an inner agent, as the following examples may suggest.

Neural Causes

The layman uses the nervous system as a ready explanation of behavior. The English language contains hundreds of expressions which imply such a causal relationship. At the end of a long trial we read that the jury shows signs of *brain fag*, that the *nerves* of the accused are *on edge*, that the wife of the accused is on the verge of a *nervous breakdown*, and that his lawyer is generally thought to have lacked the *brains* needed to stand up to the prosecution. Obviously, no direct observations have been made of the nervous systems of any of these people. Their "brains" and "nerves" have been invented on the spur of the moment to lend substance to what might otherwise seem a superficial account of their behavior.

The sciences of neurology and physiology have not divested themselves entirely of a similar practice. Since techniques for observing the electrical and chemical processes in nervous tissue had not yet been developed, early information about the nervous system was limited to its gross anatomy. Neural processes could only be inferred from the behavior which was said to result from them. Such inferences were legitimate enough as scientific theories, but they could not justifiably be used to explain the very behavior upon which they were based. The hypotheses of the early physiologist may have been sounder than those of the layman, but until independent evidence could be obtained, they were no more satisfactory as explanations of behavior. Direct information about many of the chemical and electrical processes in the nervous system is now available. Statements about the nervous system are no longer necessarily inferential or fictional. But there is still a measure of circularity in much physiological explanation, even in the writings of specialists. In World War I a familiar disorder was called "shell shock." Disturbances in behavior were explained by arguing that violent explosions had damaged the structure of the nervous system, though no direct evidence of such damage was available. In World War II the same disorder was classified as "neuropsychiatric." The prefix seems to show a continuing unwillingness to abandon explanations in terms of hypothetical neural damage.

Eventually a science of the nervous system based upon direct observation rather than inference will describe the neural states and events which immediately precede instances of behavior. We shall know the precise neurological conditions which immediately precede, say, the response, "No, thank you." These events in turn will be found to be preceded by other neurological events, and these in turn by others. This series will lead us back to events outside the nervous system and, eventually, outside the organism. In the chapters which follow we shall consider external events of this sort in some detail. We shall then be better able to evaluate the place of neurological explanations of behavior. However, we may note here that we do not have and may never have this sort of neurological information at the moment it is needed in order to predict a specific instance of behavior. It is even more unlikely that we shall be able to alter the nervous system directly in order to set up the antecedent conditions of a particular instance. The causes to be sought in the nervous system are, therefore, of limited usefulness in the prediction and control of specific behavior.

Psychic Inner Causes

An even more common practice is to explain behavior in terms of an inner agent which lacks physical dimensions and is called "mental" or "psychic." The purest form of the psychic explanation is seen in the animism of primitive peoples. From the immobility of the body after death it is inferred that a spirit responsible for movement has departed. The *enthusiastic* person is, as the etymology of the word implies, energized by a "god within." It is only a modest refinement to attribute every feature of the behavior of the physical organism to a corresponding feature of the "mind" or of some inner "personality." The inner man is regarded as driving the body very much as the man at the steering wheel drives a car. The inner man wills an action, the outer executes it. The inner loses his appetite, the outer stops eating. The inner man wants and the outer gets. The inner has the impulse which the outer obeys.

It is not the layman alone who resorts to these practices, for many reputable psychologists use a similar dualistic system of explanation. The inner man is sometimes personified clearly, as when delinquent behavior is attributed to a "disordered personality," or he may be dealt with in fragments, as when behavior is attributed to mental processes, faculties, and traits. Since the inner man does not occupy space, he may be multiplied at will. It has been argued that a single physical organism is controlled by several psychic agents and that its behavior is the resultant of their several wills. The Freudian concepts of the ego, superego, and id are often used in this way. They are frequently regarded as nonsubstantial creatures, often in violent conflict, whose defeats or victories lead to the adjusted or maladjusted behavior of the physical organism in which they reside.

Direct observation of the mind comparable with the observation of the nervous system has not proved feasiblc. It is true that many people believe that they observe their "mental states" just as the physiologist observes neural events, but another interpretation of what they observe is possible. Introspective psychology no longer pretends to supply direct information about events which are the causal antecedents, rather than the mere accompaniments, of behavior. It defines its "subjective" events in ways which strip them of any usefulness in a causal analysis. The events appealed to in early mentalistic explanations of behavior have remained beyond the reach of observation. Freud insisted upon this by emphasizing the role of the unconscious—a frank recognition that important mental processes are not directly observable. The Freudian literature supplies many examples of behavior from which unconscious wishes, impulses, instincts, and emotions are inferred. Unconscious thought-processes have also been used to explain intellectual achievements. Though the mathematician may feel that he knows "how he thinks," he is often unable to give a coherent account of the mental processes leading to the solution of a specific problem. But any mental event which is unconscious is necessarily inferential, and the explanation is therefore not based upon independent observations of a valid cause.

The fictional nature of this form of inner cause is shown by the ease with which the mental process is discovered to have just the properties needed to account for the behavior. When a professor turns up in the wrong classroom or gives the wrong lecture, it is because his *mind* is, at least for the moment, *absent*. If he forgets to give a reading assignment, it is because it has slipped his *mind* (a hint from the class may *remind* him of it). He begins to tell an old joke but pauses for a moment, and it is evident to everyone that he is trying to make up his *mind* whether or not he has already used the joke that term. His lectures grow more tedious with the years, and questions from the class confuse him more and more, because his *mind* is failing. What he says is often disorganized because his *ideas* are confused. He is occasionally unnecessarily emphatic because of the force of his *ideas*. When he repeats himself, it is because he has an *idée fixe*; and when he repeats what others have said, it is because he borrows his *ideas*. Upon occasion there is nothing in what he says because he lacks *ideas*. In all this it is obvious that the mind and the ideas, together with their special characteristics, are being invented on the spot to provide spurious explanations. A science of behavior can hope to gain very little from so cavalier a practice. Since mental or psychic events are asserted to lack the dimensions of physical science, we have an additional reason for rejecting them.

Conceptual Inner Causes

The commonest inner causes have no specific dimensions at all, either neurological or psychic. When we say that a man eats *because* he is hungry, smokes a great deal *because* he has the tobacco habit, fights *because* of the instinct of pugnacity, behaves brilliantly *because* of his intelligence, or plays the piano well *because* of his musical ability, we seem to be referring to causes. But on analysis these phrases prove to be merely redundant descriptions. A single set of facts is described by the two statements: "He eats" and "He is hungry." A single set of facts is described by the two statements: "He smokes a great deal" and "He has the smoking habit." A single set of facts is described by the two statements: "He plays well" and "He has musical ability." The practice of explaining one statement in terms of the other is dangerous because it suggests that we have found the cause and therefore need search no further. Moreover, such terms as "hunger," "habit," and "intelligence" convert what are essentially the properties of a process or relation into what appear to be things. Thus we are unprepared for the properties eventually to be discovered in the behavior itself and continue to look for something which may not exist.

The Variables of Which Behavior is a Function

The practice of looking inside the organism for an explanation of behavior has tended to obscure the variables which are immediately available for a scientific analysis. These variables lie outside the organism, in its immediate environment and in its environmental history. They have a physical status to which the usual techniques of science are adapted, and they make it possible to explain behavior as other subjects are explained in science. These independent variables are of many sorts and their relations to behavior are often subtle and complex, but we cannot hope to give an adequate account of behavior without analyzing them.

Consider the act of drinking a glass of water. This is not likely to be an important bit of behavior in anyone's life, but it supplies a convenient example. We may describe the topography of the behavior in such a way that a given instance may be identified quite accurately by any qualified observer. Suppose now we bring someone into a room and place a glass of water before him. Will he drink? There appear to be only two possibilities: either he will or he will not. But we speak of the *chances* that he will drink, and this notion may be refined for scientific use. What we want to evaluate is the *probability* that he will drink. This may range from virtual certainty that drinking will occur to virtual certainty that it will not. The very considerable problem of how to measure such a probability will be discussed later. For the moment, we are interested in how the probability may be increased or decreased.

Everyday experience suggests several possibilities, and laboratory and clinical observations have added others. It is decidedly not true that a horse may be led to water but cannot be made to drink. By arranging a history of severe deprivation we could be "absolutely sure" that drinking would occur. In the same way we may be sure that the glass of water in our experiment will be drunk. Although we are not likely to arrange them experimentally, deprivations of the necessary magnitude sometimes occur outside the laboratory. We may obtain an effect similar to that of deprivation by speeding up the excretion of water. For example, we may induce sweating by raising the temperature of the room or by forcing heavy exercise, or we may increase the excretion of urine by mixing salt or urea in food taken prior to the experiment. It is also well known that loss of blood, as on a battlefield, sharply increases the probability of drinking. On the other hand, we may set the probability at virtually zero by inducing or forcing our subject to drink a large quantity of water before the experiment.

If we are to predict whether or not our subject will drink, we must know as much as possible about these variables. If we are to induce him to drink, we must be able to manipulate them. In both cases, moreover, either for accurate prediction or control, we must investigate the effect of each variable quantitatively with the methods and techniques of a laboratory science.

Other variables may, of course, affect the result. Our subject may be "afraid" that something has been added to the water as a practical joke or for experimental purposes. He may even "suspect" that the water has been poisoned. He may have grown up in a culture in which water is drunk only when no one is watching. He may refuse to drink simply to prove that we cannot predict or control his behavior. These possibilities do not disprove the relations between drinking and the variables listed in the preceding paragraphs; they simply remind us that other variables may have to be taken into account. We must know the history of our subject with respect to the behavior of drinking water, and if we cannot eliminate social factors from the situation, then we must know the history of his personal relations to people resembling the experimenter. Adequate prediction in any science requires information about all relevant variables, and the control of a subject matter for practical purposes makes the same demands.

Other types of "explanation" do not permit us to dispense with these requirements or to fulfill them in any easier way. It is of no help to be told that our subject will drink provided he was born under a particular sign of the zodiac which shows a preoccupation with water or provided he is the lean and thirsty type or was, in short, "born thirsty." Explanations in terms of inner states or agents, however, may require some further comment. To what extent is it helpful to be told, "He drinks because he is thirsty"? If to be thirsty means nothing more than to have a tendency to drink, this is mere redundancy. If it means that he drinks because of a state of thirst, an inner causal event is invoked. If this state is purely inferential—if no dimensions are assigned to it which would make direct observation possible—it cannot serve as an explanation. But if it has physiological or psychic properties, what role can it play in a science of behavior?

The physiologist may point out that several ways of raising the probability of drinking have a common effect: they increase the concentration of solutions in the body. Through some mechanism not yet well understood, this may bring about a corresponding change in the nervous system which in turn makes drinking more probable. In the same way, it may be argued that all these operations make the organism "feel thirsty" or "want a drink" and that such a psychic state also acts upon the nervous system in some unexplained way to induce drinking. In each case we have a causal chain consisting of three links: (1) an operation performed upon the organism from without—for example, water deprivation; (2) an inner condition—for example, physiological or psychic thirst; and (3) a kind of behavior—for example, drinking. Independent information about the second link would obviously permit us to predict the third without recourse to the first. It would be a preferred type of variable because it would be nonhistoric; the first link may lie in the past history of the organism, but the second is a current condition. Direct information about the second link is, however, seldom, if ever, available. Sometimes we infer the second link from the third: an animal is judged to be thirsty if it drinks. In that case, the explanation is spurious. Sometimes we infer the second link from the first: an animal is said to be thirsty if it has not drunk for a long time. In that case, we obviously cannot dispense with the prior history.

The second link is useless in the *control* of behavior unless we can manipulate it. At the moment, we have no way of directly altering neural processes at appropriate moments in the life of a behaving organism, nor has any way been discovered to alter a psychic process. We usually set up the second link through the first: we make an animal thirsty, in either the physiological or the psychic sense, by depriving it of water, feeding it salt, and so on. In that case, the second link obviously does not permit us to dispense with the first. Even if some new technical discovery were to enable us to set up or change the second link directly, we should still have to deal with those enormous areas in which human behavior is controlled through manipulation of the first link. A technique of operating upon the second link would increase our control of behavior, but the techniques which have already been developed would still remain to be analyzed.

The most objectionable practice is to follow the causal sequence back only as far as a hypothetical second link. This is a serious handicap both in a theoretical science and in the practical control of behavior. It is no help to be told that to get an organism to drink we are simply to "make it thirsty" unless we are also told how this is to be done. When we have obtained the necessary prescription for thirst, the whole proposal is more complex than it need be. Similarly, when an example of maladjusted behavior is explained by saying that the individual is "suffering from anxiety," we have still to be told the cause of the anxiety. But the external conditions which are then invoked could have been directly related to the maladjusted behavior. Again, when we are told that a man stole a loaf of bread because "he was hungry," we have still to learn of the external conditions responsible for the "hunger." These conditions would have sufficed to explain the theft.

The objection to inner states is not that they do not exist, but that they are not relevant in a functional analysis. We cannot account for the behavior of any system while staying wholly inside it; eventually we must turn to forces operating upon the organism from without. Unless there is a weak spot in our causal chain so that the second link is not lawfully determined by the first, or the third by the second, then the first and third links must be lawfully related. If we must always go back beyond the second link for prediction and control, we may avoid many tiresome and exhausting digressions by examining the third link as a function of the first. Valid information about the second link may throw light upon this relationship but can in no way alter it.

A Functional Analysis

The external variables of which behavior is a function provide for what may be called a causal or functional analysis. We undertake to predict and control the behavior of the individual organism. This is our "dependent variable"—the effect for which we are to find the cause. Our "independent variables"—the causes of behavior—are the external conditions of which behavior is a function. Relations between the two—the "cause-and-effect relationships" in behavior—are the laws of a science. A synthesis of these laws expressed in quantitative terms yields a comprehensive picture of the organism as a behaving system.

This must be done within the bounds of a natural science. We cannot assume that behavior has any peculiar properties which require unique methods or special kinds of knowledge. It is often argued that an act is not so important as the "intent" which lies behind it, or that it can be described only in terms of what it "means" to the behaving individual or to others whom it may affect. If statements of this sort are useful for scientific purposes, they must be based upon observable events, and we may confine ourselves to such events exclusively in a functional analysis. We shall see later that although such terms as "meaning" and "intent" appear to refer to properties of behavior, they usually conceal references to independent variables. This is also true of "aggressive," "friendly," "disorganized," "intelligent," and other terms which appear to describe properties of behavior but in reality refer to its controlling relations.

The independent variables must also be described in physical terms. An effort is often made to avoid the labor of analyzing a physical situation by guessing what it "means" to an organism or by distinguishing between the physical world and a psychological world of "experience." This practice also reflects a confusion between dependent and independent variables. The events affecting an organism must be capable of description in the language of physical science. It is sometimes argued that certain "social forces" or the "influences" of culture or tradition are exceptions. But we cannot appeal to entities of this sort without explaining how they can affect both the scientist and the individual under observation. The physical events which must then be appealed to in such an explanation will supply us with alternative material suitable for a physical analysis.

By confining ourselves to these observable events, we gain a considerable advantage, not only in theory, but in practice. A "social force" is no more useful in manipulating behavior than an inner state of hunger, anxiety, or skepticism. Just as we must trace these inner events to the manipulable variables of which they are said to be functions before we may put them to practical use, so we must identify the physical events through which a "social force" is said to affect the organism before we can manipulate it for purposes of control. In dealing with the directly observable data we need not refer to either the inner state or the outer force.

The material to be analyzed in a science of behavior comes from many sources:

(1) Our *casual observations* are not to be dismissed entirely. They are especially important in the early stages of investigation. Generalizations based upon them, even without explicit analysis, supply useful hunches for further study.

(2) In *controlled field observation*, as exemplified by some of the methods of anthropology, the data are sampled more carefully and conclusions stated more explicitly than in casual observation. Standard instruments and practices increase the accuracy and uniformity of field observation.

(3) *Clinical observation* has supplied extensive material. Standard practices in interviewing and testing bring out behavior which may be easily measured, summarized, and compared with the behavior of others. Although it usually emphasizes the disorders which bring people to clinics, the clinical sample is often unusually interesting and of special value when the exceptional condition points up an important feature of behavior.

(4) Extensive observations of behavior have been made under more rigidly controlled conditions in *industrial, military, and other institutional research*. This work often differs from field or clinical observation in its greater use of the experimental method.

(5) *Laboratory studies of human behavior* provide especially useful material. The experimental method includes the use of instruments which improve our contact with behavior and with the variables of which it is a function. Recording devices enable us to observe behavior over long periods of time, and accurate recording and measurement make effective quantitative analysis possible. The most important feature of the laboratory method is the deliberate manipulation of variables: the importance of a given condition is determined by changing it in a controlled fashion and observing the result.

Current experimental research on human behavior is sometimes not so comprehensive as one might wish. Not all behavioral processes are easy to set up in the laboratory, and precision of measurement is sometimes obtained only at the price of unreality in conditions. Those who are primarily concerned with the everyday life of the individual are often impatient with these artificialities, but insofar as relevant relationships can be brought under experimental control, the laboratory offers the best chance of obtaining the quantitative results needed in a scientific analysis.

(6) The extensive results of *laboratory studies of the behavior of animals below the human level* are also available. The use of this material often meets with the objection that there is an essential gap between man and the other animals, and that the results of one cannot be extrapolated to the other. To insist upon this discontinuity at the beginning of a scientific investigation is to beg the question. Human behavior is distinguished by its complexity, its variety, and its greater accomplishments, but the basic processes are not therefore necessarily different. Science advances from the simple to the complex; it is constantly concerned with whether the processes and laws discovered at one stage are adequate for the next. It would be rash to assert at this point that there is no essential difference between human behavior and the behavior of lower species; but until an attempt has been made to deal with both in the same terms, it would be equally rash to assert that there is. A discussion of human embryology makes considerable use of research on the embryos of chicks, pigs, and other animals. Treatises on digestion, respiration, circulation, endocrine secretion, and other physiological processes deal with rats, hamsters, rabbits, and so on, even though the interest is primarily in human beings. The study of behavior has much to gain from the same practice.

We study the behavior of animals because it is simpler. Basic processes are revealed more easily and can be recorded over longer periods of time. Our observations are not complicated by the social relation between subject and experimenter. Conditions may be better controlled. We may arrange genetic histories to control certain variables and special life histories to control others—for example, if we are interested in how an organism learns to see, we can raise an animal in darkness until the experiment is begun. We are also able to control current circumstances to an extent not easily realized in human behavior—for example, we can vary states of deprivation over wide ranges. These are advantages which should not be dismissed on the a priori contention that human behavior is inevitably set apart as a separate field. . . .

A Distinction between Political Science and Political Wisdom

Sheldon Wolin

The purpose of this paper is to sketch some of the implications, prospective and retrospective, of the primacy of method in the present study of politics and to do it by way of a contrast, which is deliberately heightened, but hopefully not caricatured, between the vocation of the "methodist" and the vocation of the theorist. My discussion will be centered around the kinds of activity involved in the two vocations. During the course of the discussion various questions will be raised, primarily the following: What is the idea which underlies method and how does it compare with the older understanding of theory? What is involved in choosing one rather than the other as the way to political knowledge? What are the human or educational consequences of the choice, that is, what is demanded of the person who commits himself to one or the other? What is the typical stance towards the political world of the methodist and how does it compare to the theorist's? . . .

Although one might be troubled by the kind of human concern which would provoke a confrontation between "political wisdom" and "political *science*," the antithesis has the merit of opening the question, What is political wisdom? Put in this vague form, the question is unanswerable, but it may be reformulated so as to be fruitful. The antithesis between political wisdom and political science basically concerns two different forms of knowledge. The scientific form represents the search for rigorous formulations which are logically consistent and empirically testable. As a form, it has the qualities of compactness, manipulability, and relative independence of context. Political wisdom is an unfortunate phrase, for the question is not *what* it is but *in what* does it inhere. History, knowledge of institutions, and legal analysis were mentioned. Without violating the spirit of the quotation, knowledge of past political theories might also be added. Taken as a whole, this composite type of knowledge presents a contrast with the scientific type. Its mode of activity is not so much the style of the search as of reflection. It is mindful of logic, but more so of the incoherence and contradictoriness of experience. And for the same reason, it is distrustful of rigor. Political life does not yield its significance to terse hypotheses, but is elusive and hence meaningful statements about it often have to be allusive and intimative. Context becomes supremely important, for actions and events occur in no other set-

Source: Sheldon Wolin, "Political Theory as Vocation," *The American Political Science Review* 63 (1969), 1062, 1070-1078.

ting. Knowledge of this type tends, therefore, to be suggestive and illuminative rather than explicit and determinate. Borrowing from Polanyi, we shall call it "tacit political knowledge."

The acquisition of tacit political knowledge is preeminently a matter of education of a particular kind and it is on this ground that the issue needs to be joined with the political methodist. The mentality which is impatient of the past and of traditional political theory is equally curt with the requirements of tacit political knowledge which is rooted in knowledge of the past and of the tradition of theory. The knowledge which the methodist seeks is fairly characterized in his own language as composing a "kit of tools" or a "bag of tricks." To acquire knowledge of techniques is no small matter, for they are often difficult and require considerable "retooling," which is to say that they imply a particular kind of program of instruction in specific methods.

Tacit political knowledge, on the other hand, accrues over time and never by means of a specified program in which particular subjects are chosen in order to produce specific results. Whatever may be the truth of the adage that he who travels lightest travels farthest, diverse, even ill-assorted baggage is needed because the life of inquiry preeminently demands reflectiveness, that is, an indwelling or rumination in which the mind draws on the complex framework of sensibilities built up unpremeditatedly and calls upon the diverse resources of civilized knowledge. But if the life of inquiry is narrowly conceived as the methodical "pursuit" of knowledge, it is likely to become not a pursuit but an escape from the spare and shabby dwelling which Descartes literally and symbolically occupied when he composed his *Meditations*. Even those who would wish to address their minds to "data" are aware that data are constituted by abstractions, and that usually what has been culled from the phenomena are the subtle traces of past practices and meanings which form the connotative context of actions and events.

To recognize the connotative context of a subject matter is to know its supporting lore; and to know the supporting lore is to know how to make one's way about the subject-field. Such knowledge is not propositional, much less formulary. It stands for the knowledge which tells us what is appropriate to a subject and when a subject-matter is being violated or respected by a particular theory or hypothesis. Although appropriateness takes many forms, and we shall return to some of them, it is impossible to reduce its contents to a check-list of items. For example, can we say with exactness what is the precise knowledge which makes us uneasy with statements like the following?

> The interesting issues in normative political theory are in the end generally empirical ones. . . . There does exist, however, *one interesting problem in political theory* which is strictly normative. That is the problem of evaluating

> mixes of desiderata. . . . It may be called the "utility problem" or in still more modern terminology, the "dynamic-programming problem." . . . On this strictly normative problem of program packages more progress has been made in the past half century than in all the previous 2,000 years of political theory put together.

Although these assertions may appear absurd, it is not easy to say why, except that some important political and theoretical questions are being rendered unrecognizable. Behind the assertions, however, lie some revealing attitudes towards knowledge. These bear upon the contrast between methodistic knowledge and the forms of theory congenial to it, and, on the other hand, the kind of knowledge characteristic of tacit political knowledge and the forms of theory built upon it. The methodistic assumption holds that the truth of statements yielded by scientific methods has certain features, such as rigor, precision, and quantifiability. The connection between the statements and their features is intimate so that one is encouraged to believe that when he is offered statements rigorous, precise, and quantifiable, he is in the presence of truth. On the other hand, an approach to the "facts" consisting of statements which palpably lack precision, quantifiability, or operational value is said to be false, vague, unreliable, or even "mystical." In actuality, the contrast is not between the true and the false, the reliable and the unreliable, but between truth which is economical, replicable, and easily packaged, and truth which is not. Methodistic truth can be all these things because it is relatively indifferent to context; theoretical truth cannot, because its foundation in tacit political knowledge shapes it towards what is politically appropriate rather than towards what is scientifically operational.

Questions concerning appropriateness, context, and respect for a subject do not concern effete matters, but very practical ones. They involve the resources, or the lack thereof, which we draw upon when the decision concerns matters for which there can be no certitude. What "belongs" to a given inquiry is one such matter, and how to decide between one theory and another, or between rival methods, are others. Yet the kind of knowledge necessary to these decisions, tacit political knowledge, is being jeopardized by the education increasingly being instituted among political scientists. To illustrate the problem, we might consider the implications for tacit political knowledge of a typical proposal for increasing the student's mastery of methods. Our example is a recent volume on survey research methods for undergraduates and graduates in political science. In the spirit of Descartes' *regulae*, the authors describe it as a "handbook" or "manual," "a check-list" or inventory of "do's and don'ts," whose aim is to encourage the "empirical emphasis" in political science. Not content with offering a manual of technical instruction, the authors claim advantages of an educational and vocational kind will be promoted if survey research is made part of the curriculum. Thus the instructor, impaled by the twin demands of teaching and research, is reassured that the two can be reconciled if students are put to work learning survey methods while conducting his research. Further, the method is extolled as a way of overcoming the shortcomings of "the lone scholar" whose skills are inadequate for dealing with the size and range of problems confronting empirical political science. The imperative, "resources must be increased," decrees that the lone scholar be replaced by "group activity and teamwork." In the same vein, it is claimed that "the educational advantages for students are impressive" and among the putative advantages are the acquisition of an *ingenium* with traits congenial to the new emphasis:

> . . . students gain the opportunity to learn more about themselves . . . . Too few students get the experience of fighting to remain neutral while carefully probing attitudes hostile to their own. Such instruction in self-control is valuable for the headstrong and overprotected.

Despite the tenor and direction of this conception of education, it is insisted that the new generation of students will be able to do "what was not expected of the previous generation of college students—i.e., to discover new knowledge as well as to acquire old."

But will they? As for acquiring "the old," the authors bemoan the fact that political science departments have been hampered by the "lack of knowledge of research skills" and that the conventional academic calendar does not afford sufficient time for students to learn "sampling, interviewing, coding, analysis, etc." Exactly how the student will "acquire the old" when the demands of the "new" are so great is not discussed. . . .

Although the invention of methods, like the invention of theories, demands a high order of creativity and is entitled to the highest praise, something important, perhaps ironical, occurs when that discovery is institutionalized in a training program. The requirements for those who are to use the theory or method are very different from the talent which discovered them, although, paradoxically, the technical skills may be the same. Descartes noted that a child might become as proficient as the genius in following the rules of arithmetic, but he never argued that the child could discover the rules. This is so, not simply because of the chance element in discovery, but because of the more baffling questions of the personal and intellectual qualities of the discoverer and of the cultural conditions of discovery.

In this context the contemporary methodist's notion of training becomes significant. The idea of training presupposes several premeditated decisions: about the specific techniques needed and how they will be used; about what is peripheral or irrelevant to a particular form of training; and about the desired behavior of the trainee after he has been released from his apprenticeship. The idea of theorizing, on the other hand, while it presupposes skills, cannot specify briefly and simply the skills needed, their degree, or combinations. Kepler's followers could be contemptuous of their master's Platonism and astrology, as Newton's admirers were of his religious fascinations; but it would be risky to discount the influence of these extra-scientific considerations upon the formation of the respective theories.

The impoverishment of education by the demands of methodism poses a threat not only to so-called normative or traditional political theory, but to the scientific imagination as well. It threatens the meditative culture which nourishes all creativity. That culture is the source of the qualities crucial to theorizing: playfulness, concern, the juxtaposition of contraries, and astonishment at the variety and subtle interconnection of things. These same qualities are not confined to the creation of theories, but are at work when the mind is playing over the factual world as well. An impoverished mind, no matter how resolutely empirical in spirit, sees an impoverished world. Such a mind is not disabled from theorizing, but it is tempted into remote abstractions which, when applied to the factual world, end by torturing it. Think of what must be ignored in, or done to, the factual world before an assertion like the following can be made: "Theoretical models should be tested primarily by the accuracy of their prediction rather than the reality of their assumptions." No doubt one might object by pointing out that all theorizing does some violence to the empirical world. To which one might reply, that while amputations are necessary, it is still better to have surgeons rather than butchers.

It is not enough, therefore, to repeat commonplaces, viz., that facts are senseless without theoretical concepts, or that the meaning which facts acquire from a theory is purchased at the price of shaping the facts by the theoretical perspective employed. It is not enough because so much depends upon the kind of theory being used and the personal and cultural resources of the user. Perhaps it is some debilitating legacy of Puritanism that causes us to admire "parsimony" in our theories when we should be concerned that the constitution of the factual world depends upon the richness of our theories which, in turn, depends upon the richness of the inquiring mind. This concern may well be what fundamentally unites the scientific theorist and the so-called traditional theorist.

When a scientist observes a fact, he "sees" it through concepts which are usually derived from a theory. Facts are, as one philosopher has neatly put it, "theory-laden." Kepler, for example, observed many of the same facts as his predecessors, but because he viewed them differently a new era of science was ushered in. The same might be said of Machiavelli, as well as of every major theorist from Plato to Marx. Some theorists, as Tocqueville suggested, see differently, others see farther. All would probably have agreed with Tocqueville that, for the theorist, nothing is more difficult to appreciate than a fact, and nothing, it might be added, is more necessary as a condition for theorizing than that facts not be univocal. If they were, creativity and imagination would play a small role and it would be appropriate to speak of theorizing as a banal activity, as "theory-construction." If facts were simply "there" to be collected, classified, and then matched with a theory (or with the observation-statements derived from it), the political scientist might well declare, "Whether [a] proposition is true or false depends on the degree to which the proposition and the real world correspond." But although everyone is ready to acknowledge that facts depend upon some criteria of selection or of significance, what is less frequently acknowledged is that such criteria usually turn out to be fragments of some almost-forgotten "normative" or "traditional" theory.

Because facts are more multi-faceted than a rigid conception of empirical theory would allow, they are more likely to yield to the observer whose mental capacities enable him to appreciate a known fact in an unconventional way. As one philosopher has said, "Given the *same* world it might have been construed differently. We might have spoken of it, thought of it, perceived it differently. Perhaps facts are somehow moulded by the logical forms of the fact-stating language. Perhaps these provide a 'mould' in terms of which the world coagulates for us in definite ways." Once again we are confronted by the warning that the richness of the factual world depends upon the richness of our theories: "The paradigm observer is not the man who sees and reports what all normal observers see and report, but the man who sees in familiar objects what no one else has seen before." Thus the world must be supplemented before it can be understood and reflected upon.

Vision, as I have tried to emphasize, depends for its richness on the resources from which it can draw. These extra-scientific considerations may be identified more explicitly as the stock of ideas which an intellectually curious and broadly educated person accumulates and which come to govern his intuitions, feelings, and perceptions. They constitute the sources of his creativity, yet rarely find explicit expression in formal theory. Lying beyond the boundaries circumscribed by method, technique, and the official definition of a discipline, they can be summarized as cultural resources and itemized as metaphysics, faith, historical sensibility, or, more broadly, as tacit knowledge. Because these matters bear a family resemblance to "bias," they become sacrificial victims to the quest for objectivity in the social sciences. If scientists have freely acknowledged the importance of many of these items, how much more significant are these human creations for the form of knowledge, political science, which centers on the perplexities of collective life, on objects which are all too animate in expressing their needs, hopes, and fears.

Doubtless the objection will be raised that if a discipline is to be empirical its practitioners must be equipped to "handle" data in ways approximative of the sciences which have been more successful, and that to suggest otherwise is to consort with the heresy of saying that philosophical and moral knowledge may lead to a better empiricism. Yet we might consider the following.

Throughout the history of political theory a student will find a preoccupation with the phenomenon of "corruption." Today, however, we scarcely know how to talk about it except when it flourishes in non-Western societies. Yet it is a common and documented fact that "organized crime" exerts significant power and influence, controls enormous wealth, and exhibits many of the same features which ordinarily arouse the interest of political scientists, e.g., organization, authority, power, kinship ties, rules, and strong consensus. Despite the promising research possibilities, no textbook on American government provides a place for organized crime in "the system," no study of "polyarchy" or community-power has taken cognizance of it. It is not far-fetched to suggest that this empirical oversight is connected with the belief that moral knowledge is empirically irrelevant.

Or, to take another example, one can think of many fine empirical studies which have never been conducted because contemporary political science has substituted the bland, status quo-oriented concept of "political socialization" for the ancient idea of "political education." If, instead of blinkering the inquiring eye with a postulate that "conduct is *politicized* in the degree that it is determined by considerations of power indulgence or deprivation of the self by others," we took seriously an old-fashioned hypothesis, such as that advanced by J. S. Mill, that "the first element of good government . . . being the virtue and intelligence of the human beings composing the community, the most important point of excellence which any form of government can possess is to promote the virtue and intelligence of the people themselves," we might be better sensitized to the importance of genuinely empirical studies of truly fundamental political concern. For example, think of the empirical richness of an inquiry into the current structure of income taxes, especially in terms of the moral and political implications it holds for civic education. The structure of income taxes is a registry of the power and powerlessness of our social, economic, and ethnic groups; of the official way we rate the value of various social activities by the one standard generally accepted. It is also a system of incentives for behavior that define what is virtuous, unvirtuous, and morally indifferent; and, by tacitly encouraging behavior otherwise deemed blameworthy, encourages the gradual legitimation of that behavior, and thereby shapes what used to be called "the virtue of the citizenry." It would be difficult to imagine a richer field for behavioral inquiry, or one more likely to yield important knowledge about the quality of life in this republic. Yet it remains unharvested because our impoverished understanding of civic virtue and education has caused us to neglect the field.

Finally, one cannot help wondering whether political science, having jettisoned "metaphysical" and "normative" preoccupations about justice in favor of research into "judicial behavior" and the "judicial process," are not reaping the results: an inability to address a major phenomenon like the dangerous rash of *political* trials in America today and to reflect upon what these trials signify for the future of the authority and legitimacy of the state.

If the presence or absence of the moral and philosophical element affects the process by which theories constitute the empirical world, the choice among theories would seem to be a serious matter. But again, the contemporary mood trivializes what is involved in a theory's formulation and thereby obfuscates the importance of the choice among rival ways of constituting the world. The following quotation may be extreme but it does disclose the fantasies of the behavioral scientist about theories:

> In a report entitled *Communication Systems and Resources in the Behavioral Sciences*, the Committee on Information in the Behavioral Sciences outlines an ideal system that would in effect provide researchers with a computer analogue of the intelligent, all-informed colleague. Such a colleague would read widely, have total recall, synthesize new ideas, always be accessible, and be sensitive to each researcher's needs.... The computer based system could respond to an individual's direct request for facts, data, and documentation; it could take the initiative and stimulate the researcher by suggesting new ideas, facts, or literature of interest; it could react intelligently to a scientist's work (analyze its logic, trace implications, suggest tests); and it could help disseminate ideas and provide feedback from the scientific community.

If we can safely assume that choosing a theory or a method is not quite the same as choosing a helpful friend who, as Nietzsche taught, must be worthy of being your enemy, we might want to press the question further. When we choose a theory or a method, are we choosing something momentous, like a self, or something innocuous, like an "intellectual construct" or "conceptual scheme"? or something depersonalized, like "a series of logically consistent, interconnected, and empirically verifiable propositions," or like "a generalized statement of the interrelationships of a set of variables"?

Undoubtedly these characterizations tell us something about the formal features of a theory, but they are deceptive in their parsimony. If the question is slightly reformulated to read, what is the human significance of choosing a theory?, then it becomes evident that much more is involved. Choosing a theory is significant for two conflicting reasons: it initiates new ways of thinking, evaluating, intuiting, and feeling; and it demands a substantial sacrifice in the existing forms of these same human processes. The first point is obvious, the second less so. This is because, like the law of treason, history books tend to be written by the victors and hence the sacrifices which accompany the triumph of a new theory are apt to be overlooked or bathed in a kind of Jacobite nostalgia.

The history of political theory is instructive on this score, for many of the great innovative theorists were highly self-conscious about choosing among theoretical alternatives. They knew that the true drama of theorizing involved offering a theory which could not be accommodated within prevailing values and perceptions of the world. When Hobbes allowed that his readers would be "staggered" by his theory, he was not merely stating the obvious fact that his views concerning religion, authority, rights, and human nature were incompatible with traditional religious and political notions, but the more profound point that unless his readers were prepared to revise or discard those notions, they would not be able to grasp the full meaning of the theory and the theory itself could not become an effective force in the world. The same general assumptions had been made by Plato in his challenge to traditional Greek values and to the democratic ethos of Athens, and by Augustine in his effort to demolish classical notions of history, politics, virtue, and religion. Among more recent writers, none has been as sensitive as Max Weber to the emotional and cultural losses attendant upon the commitment to scientific rationalism.

Where our contemporary way of talking has not obscured the drama and demands of theorizing, it has trivialized them. Theories are likened to appliances which are "plugged into" political life and, since it is the nature of appliances to be under sentence of built-in obsolescence, "theories are for burning," leaving only a brief funereal glow which lights the way to "more scientific theories and more efficient research procedures." If adopting a theory were equivalent to "trying out an idea," testing an hypothesis, or selecting a technique, there would be little reason to object to treating it casually.

At the very least, a theory makes demands upon our time, attention, energy, and skills. More fundamentally, the adoption of a theory signifies a form of submission with serious consequences both for the adopter and for those who imitate him, as well as for the corner of the world which the theory seeks to change our mind about. A certain sensibility is needed, qualities of thinking and feeling which are not readily formulable but pertain to a capacity for discriminative judgment. Why is this so? To compress the answer severely, in political and social matters we tend to think in one of two ways: in trying to explain, understand, or appraise we may ask, what is it like?; or we may ask, what is appropriate? The first way invites us to think metaphorically, e.g., Hobbes' argument that a representative is like an agent, or the contemporary notion of a political society as a system of communications. Ever since Plato, theorists have recognized the fruitfulness of metaphorical thinking, but they have also come to realize that at certain crucial points a metaphor may become misleading, primarily because the metaphor has a thrust of its own which leads to grotesque implications for the object or events which it is supposed to illuminate. A recent example of this pitfall is provided by Professor Deutsch's *Nerves of Government*, which argues for the concept of a communications system as a useful and proper model for political theory. The argument rests on a combination of metaphors and the success of the argument depends upon a confusion of the two. The first metaphor consists in likening the nature of human thinking and purposive action to the operation of a communications system, e.g., the "problem of value" is like a "switchboard problem," or "consciousness" is "analogous" to the process of feedback. The second metaphor involves the reverse procedure: a communications system may be treated like a person. Human qualities, such as "spontaneity," "freedom of the will," and "creativity," can be "built into" a machine, and then it becomes possible to propose empirical propositions about society derived from the operations of the machine. But the whole argument depends upon, first, mechanizing human behavior and, second, humanizing mechanical processes. Once this is accomplished, grotesque results follow, e.g., internal rearrangements in a system, or in a person, which reduce goal-seeking effectiveness are described as "pathological" and resemble "what some moralists call 'sin.' "

A second way of judging asks, what is appropriate? Appropriateness of judgment cannot be encapsulated into a formula. This is because it depends upon varied forms of knowledge for which there is no natural limit. This dependence is rooted in the basic quest of political and social theory for theoretical knowledge about "wholes" made up of interrelated and interpenetrating provinces of human activity. Whether the primary theoretical task be one of explanation or critical appraisal, the theorist will want to locate "divisions" in the human world and embody them in theoretical form. For example, what aspects of that division which we call "religion" have a significant bearing on the activity called "economic"? Perforce, a political theory is, among many other things, a sum of judgments, shaped by the theorist's notion of what matters, and embodying a series of discriminations about where one province begins and another leaves off. The discriminations may have to do with what is private and what is public, or they may be about what will be endangered or encouraged if affairs move one way rather than another, or about what practices, occurrences, and conditions are likely to produce what states of affairs. The difficulty is the same regardless of whether the theoretical intention is to provide a descriptive explanation, a critical appraisal, or a prescriptive solution. By virtue of their location in a whole, one province shades off from and merges into others: where, for example, does the cure of souls end and the authority of the political order over religion begin? where do the effects of technical education merge into questions about ethics and character? where does the autonomy of administrative and judicial practices start and the "mysteries of state" stop? how much of the impetus for the Crusades is to be assigned to religious motives and how much to political or economic considerations?

If, as Plato suggested long ago, the task of theory is to locate "the real cleavages" in things and to "avoid chopping reality up into small parts" or drawing false boundaries, then the sense of what is appropriate is critical. Given the theorist's preoccupation with wholes, the interconnectedness of human provinces, the values and expectations with which men have invested each of their provinces, and the ultimate bewildering fact that man is single but his provinces are multiple, a theoretical judgment which, by definition, must discriminate can only be restrained from rendering inappropriate determinations if it is civilized by a meditative culture. To be civilized is not only the quality of being sensitive to the claims and characters of many provinces, but, according to an older definition, rendering what is proper to a civil community.

If the preceding analysis has any merit it will have suggested that the triumph of methodism constitutes a crisis in political education and that the main victim is the tacit political knowledge which is so vital to making judgments, not only judgments about the adequacy and value of theories and methods, but about the nature and perplexities of politics as well. Here lies the vocation of those who preserve our understanding of past theories, who sharpen our sense of the subtle, complex interplay between political experience and thought, and who preserve our memory of the agonizing efforts of intellect to restate the possibilities and threats posed by political dilemmas of the past. In teaching about past theories, the historically-minded theorist is engaged in the task of political initiation; that is, of introducing new generations of students to the complexities of politics and to the efforts of theorists to confront its predicaments; of developing the capacity for discriminating judgments discussed earlier; and of cultivating that sense of "significance" which, as Weber understood so well, is vital to scientific inquiry but cannot be furnished by scientific methods; and of exploring the ways in which new theoretical vistas are opened.

For those who are concerned with the history of political theories, the vocation has become a demanding one at the present time. . . . In the formative period of their education students are required to master textbooks rather than to familiarize themselves with the creative writings of the great scientists of the past. The characteristic teaching of scientific textbooks, according to Kuhn, is to show how the great achievements of the past have prepared the way for the present stage of knowledge and theory. As a result, discontinuities are smoothed over, discarded, or unsuccessful theories are assumed to have been inferior, and the idea of methodical progress dominates the entire account.

How easy it is to impoverish the past by making it appear like the present is suggested by the way in which social scientists have lapsed into the same idiom as Kuhn's scientific textbooks. "As Aristotle, the first great behavioral scientist, pointed out a long time ago . . ." or, again, "the behavioral persuasion in politics represents an attempt, by modern modes of analysis, to fulfill the quest for political knowledge begun by the classical political theorists," although it is admitted that classical theory is "predominantly prescriptive rather than descriptive." What seems to have been forgotten is that one reads past theories, not because they are familiar and therefore confirmative, but because they are strange and therefore provocative. If Aristotle is read as the first behavioralist, what he has to say is only of antiquarian interest and it would be far more profitable to read our contemporaries.

What we should expect from a reading of Aristotle is an increase in political understanding. What we should expect from the study of the history of political theories is an appreciation of the historical dimension of politics. The cultivation of political understanding means that one becomes sensitized to the enormous complexities and drama of saying that the political order is the most comprehensive association and ultimately responsible as no other grouping is for sustaining the physical, material, cultural and moral life of its members. Political understanding also teaches that the political order is articulated through its history; the past weighs on the present, shaping alternatives and pressing with a force of its own. At the present time the historical mode is largely ignored in favor of modes of understanding which are inherently incapable of building upon historical knowledge. One of the most striking features of game theory, communications models, and mechanical systems is that in each case the organizing notion is essentially history-less.

The threat to political understanding is not to be denied by arguing that we can substitute more precise functional equivalents for older language or that we can translate older notions into more empirical terms. From time immemorial writers have talked of the "burdens" of ruling, the "anguish" of choosing, and the "guilt" of actors who must employ coercion. To assimilate these actions to the calculations of gamesters or to describe them as "decision-making" or "outputs" is to distort both sides of the analogy. If in game-playing, for example, anguish, burdens, and guilt were recurrent features, the whole connotative context surrounding the idea of a game would be lost and nobody would "play." The ancient writer Philostratus once remarked of painting that no one could understand the imitative techniques of the painter without prior knowledge of the objects being represented. But when the attempt is being made to convey knowledge, not by imitative techniques, but by abstract signs and symbols which stand for objects commonly understood, everything depends on whether one truly understands what the symbol means. Does he understand, for example, the kinds of discriminative judgments which have been suspended when the symbol of an "input" is made to stand equally for a civil rights protest, a deputation from the National Rifle Association, and a strike by the U.A.W.? Does he understand that what allows him to discriminate between these "inputs" is a tacit knowledge derived from sources other than systems theory? Again, will he be able to compensate for the fact that systems theory makes it possible to talk about an entire political society without ever mentioning the idea of justice, except in the distorting form of its contribution to "system maintenance"? Is he aware that if one can focus on the American political order as a *system*, he does not have to confront the unpleasant possibility of it as an *imperium* of unsurpassed power? If, in rebuttal, the political scientist claims that the sort of studies referred to above really do presuppose the knowledge which would make political sense out of formal methods, then it is necessary to reply that the contemporary political scientist threatens to chalk around himself a vicious circle: his methods of study presuppose a depth of political culture which his methods of education destroy.

9 The New Physics of the Twentieth Century: Yet Another Reassessment of the Nature of Reality and the Extent of Human Understanding

Throughout the second half of the nineteenth century, a growing sense of unease permeated the European and American community of scientists. Bitter and profound philosophical debates split the community as some scientists developed theories—like the kinetic theory of gases—which assumed the existence of unobservable entities and which thus violated the accepted positivistic canons of method. Moreover, there were a series of problems which stubbornly resisted solution in terms of traditional physical theories. After nearly a century of work, physicists were still unable to discover the structure of the luminiferous ether—the medium in which light was assumed to travel. And new problems arose in connection with the study of cathode-rays, which stubbornly resisted attempts to classify them as either waves or collections of moving particles for nearly a quarter of a century.

Beginning in 1881, moreover, when young American naval officer Albert A. Michelson made his first attempt to measure the velocity of the earth through the ether, a series of experimental discoveries began to undermine the foundations of the secure structure of classical physics. The very existence of the stationary ether, that "continuous substance filling all space, which can vibrate light, which can be shared into positive and negative electricity, which in whirls constitutes matter, and which transmits by continuity and not by impact every action and reaction of which matter is capable," was called into question by Michelson's experiment. And since much nineteenth-century physical theory depended on the ether's assumed existence, this raised fundamental problems.

Again in 1887, Heinrich Hertz made an odd and inexplicable discovery in the course of confirming some of the most impressive predictions of classical electromagnetic theory. He found that when ultra-violet light shone on the antenna of his apparatus it became a more sensitive detector of electromagnetic waves; but there was nothing in electromagnetic theory to explain why. Even more puzzling and disturbing were the discoveries that began in 1895 when Wilhelm Konrad Röntgen of Munich detected a mysterious radiation—X rays—emanating from electric discharge-tubes. During the next year, Henri Becquerel's investigations showed that uranium gave off another kind of unknown radiation without any kind of external stimulation; and within the next few years three different forms of natural radioactivity, α-rays, β-rays, and γ-rays, had been isolated if not understood. In 1900 the experimental work of Rubens and Kurlbaum seemed to show that even the heat radiation from ordinary materials had properties which were inconsistent with the most fundamental and general of theories, classical thermodynamics.

Throughout the later nineteenth century, one further set of experimental results began to bother the physicists as well. As early as 1822 Joseph Fraunhoffer had discovered that different elemental substances in the gaseous phase emit unique and characteristic sharp lines of light—spectra—when they are burned or heated. By 1860 these characteristic spectra were being used to detect the presence of elements even in minute traces and to investigate the composition of the sun and other stellar bodies. It was natural that physicists should try to discover what it was in the structure of the elements that gave rise to these characteristic spectral admissions. Unfortunately there were almost no clues to work from. In 1885 Johan J. Balmer, one of many searchers, discovered an empirical mathematical relation among some of the frequencies of light emitted in the spectrum of one element—hydrogen—but the quantitative relations discovered by Balmer and others seemed to complicate rather than simplify the attempt to discover the atomic structure from within classical physical theory.

These discoveries and many more made during the first decades of the twentieth century demanded dramatic changes in physical theories; changes which were not surprisingly paralleled by and to some degree mirrored in changes in social thought, literature, and art.

The two physical theories which seemed to call forth the greatest response among nonscientists were the special theory of relativity, first advanced by Albert Einstein in 1905 in response to problems closely related to the Michelson-Morely experiment, and quantum mechanics, a body of theories developed in response to problems associated with the interpretation of Hertz's work, radioactivity, atomic spectra, and with the pattern of radiation emitted by hot bodies. Because of space limitations, I will confine my selections and comments to some of the semi-popular responses to the theory of relativity and to the "indeterminacy" or "uncertainty" principle from quantum mechanics.

There can be no doubt that relativity and quantum theory have had profound effects on formal aspects of contemporary philosophy nor that modern science has provided dramatic analogs outside the limits of these two theories; but to most moderately well-educated men in Western society, "relativity" and "indeterminacy" are the most common and important terms which provide a link to the abstruse theories of men like Einstein, Max Planck, Niels Bohr, and Werner Heisenberg. And these terms or concepts have played a very important role in giving the modern intellectual a sense of his own limitations and a sense—whether it be real or illusory—of his own freedom to act and to control his destiny.

Einstein's theory clearly overthrew a variety of traditional beliefs. One physical implication, for example, was that the law of conservation of energy was not strictly valid and that mass and energy were somehow interconvertible. Similarly, the theory called into question our basic notions of space and of time. As Herbert Minkowski wrote in 1908, for the relativistic physics, "space by itself and time by itself, are doomed to fade away into mere shadows, and only a kind of union of the two will preserve an independent reality." This interrelation between space and time undermined the common-sense notion of simultaneity; for in relativity theory, two events which are simultaneous in one frame of reference, may occur at different times in another. In a like manner, the space-time relation overthrew the common sense notion of length; for a body might have different spatial "lengths" in different frames of reference. Perhaps most fundamentally, Einstein's assumption of the invariable speed of light in a vacuum made it impossible to expect that any mechanistic structure of the ether might be developed to account for the propagation of electromagnetic waves. In fact, it precluded all possibility of traditional mechanistic explanations and forced its adherents to accept the idea that a theory need be no more than a set of mathematical functions in a four-dimensional space along with rules for interpreting the values of those functions.

In spite of the radical revisions which relativity theory called for in traditional physical ideas, however, it does not in any way challenge the fundamental lawfulness of physical phenomena, nor does it justify a belief in the subjectivity of physical experience. The theory of relativity is premised on the assumption that no preferred frame of reference exists from which absolute motion or rest can be determined, but this does not mean that physical phenomena as seen from various vantage points in space-time are in any sense incompatible with one another or dependent on unspecifiable, personal aspects of the observer. In fact, the theory explicitly demands that all physical laws have an identical form for all frames of reference and thus provides a way to discover precisely what the description of a physical event will be in any arbitrary frame of reference once it is described in one frame. Vast confusion on this issue reigned even among competent physicists during the early years of the twentieth century; and many popularizers and scientific critics provided the license which lay readers used to turn Einstein's theory on its head as a justification for a belief in the subjectivity of all knowledge.

A good example of the kind of distortion of the theory of relativity which reached the nonscientific public was L. T. More's statement that,

> Both Professor Einstein's theory of relativity and Professor Planck's theory of Quanta are proclaimed somewhat noisily to be the greatest revolutions in

scientific method since the time of Newton. That they are revolutionary there can be no doubt, insofar as they substitute mathematical symbols as the basis of science and deny that any concrete experience underlies these symbols, thus replacing an objective by a subjective universe."[1]

More was correct insofar as he saw relativity as a challenge to traditional materialism, but the notion expressed here that Einstein's theory is less "objective" than mechanistic theories is absurd if one uses this term in its ordinary signification. The term objective usually means: based on experienced fact and free from personal biases, and this is precisely what relativity theory is—it is more in conformity with experienced fact than classical physics, and it by no means implies that personal attitudes should play a role in physical theory. Yet the implication that relativity theory was somehow subjective formed the basis for the lay reception. Thomas Jewell Cravens' "Art and Relativity" provides an excellent example of the use of relativistic physics to justify the rejection of all absolute aesthetic standards and to applaud purely personal expression.

One of the most important cultural uses of relativity theory was made by historians to reinforce a trend away from the positivistic and scientistic attempts to write history that had been popular during the nineteenth century. In "Charles A. Beard, the 'New Physics,' and Historical Relativity" Hugh I. Rodgers explains the background to this movement and provides a case study which shows how relativistic notions filtered down through scientific popularizations into important aspects of social thought.

The principle of uncertainty or indeterminancy enunciated by Werner Heisenberg in 1927 as a fundamental principle of quantum mechanics has been no less important than Einstein's theory of relativity in calling into question important traditional beliefs and in giving new hope to those who see a deterministic and materialistic universe as unbearable and valueless. According to the most widely held interpretation of quantum mechanics, the uncertainty principle challenges the very notion of cause and effect which underlies almost all classical science and much traditional social and theological thought. In addition, by guaranteeing our inability to perfectly stipulate the state of any isolated part of the universe, the uncertainty principle weakens the argument of determinists in all fields and seems to give a new support for speculations about free will.

The basic contention of the uncertainty principle is almost incredibly simple. To any physical quantity Q which we might investigate, there corresponds another quantity P, such that we can never know the magnitudes of both P and Q exactly. In fact, the product of the uncertainties in the measurements of P and Q can never be less than the so-called quantum of action, a constant discovered by Max Planck in 1901 in connection with his studies of radiant heat. In particular, the principle implies that if we know

---

[1] "The Theory of Relativity," *The Nation* XLIV (January-June, 1912), 370.

the position of a particle with perfect accuracy, we can predict nothing about its momentum, or if we know the energy of an entity exactly, we cannot predict at what time the energy was being carried.

The uncertainty principle—and with it, all of quantum theory—can be interpreted in several different ways. According to one interpretation, statements from quantum theory can only be interpreted statistically. The statistical laws given by quantum theory may be perfectly determined, although the results of any single event cannot be known. Our uncertainty may only be the result of ignorance which is unavoidable because of the nature of all physical techniques of measurement. In order to measure or observe an event, the observer must disturb the situation he seeks to study; for example, in measuring the position of a particle he inevitably changes its velocity. This interpretation of quantum theory places an emphasis on the inseparability of observer and observed and reinforces the "subjectivist" notions of relativity. It makes the scientist play a more crucial role within scientific theory and is used to justify the interposition of personal values and biases within social theory. At the same time it leaves intact the basic belief in a deterministic universe.

This statistical interpretation of quantum mechanics, however, faces serious challenges on empirical grounds, and many scientists as well as laymen see in quantum theory the justification for a belief in a fundamentally noncausal, indeterministic element in the structure of reality—an element which reinforces humanistic yearnings for a new legitimizing of the notion of free will. This second, more radical, interpretation of quantum theory has had the greatest impact on humanistic thought and is represented in both of my selections.

In his "Quantum Mechanics, Free Will, and Determinism," Henry Margenau, a physicist who is deeply concerned with the philosophical foundations and implications of modern physics, probes the impact of quantum theory on modern interpretations of our experience of freedom. The article is complex, but no less involved argument can do justice to the fundamental problems considered. The selection "Heisenberg's Recognitions: the End of the Scientific World View" from John Lukac's *The Historical Consciousness* is much more general than Margenau's argument and provides a succinct summary of the major implications drawn from quantum mechanics by humanistic scholars—implications which seem to signal a possible reversal of the three-hundred year dominance of scientific, objective methods in many fields of human endeavor.

Art and Relativity

Thomas Jewell Craven

Professor Einstein's revolutionary theory is the latest example of the eternal kinship between art and science. His principle of relativity, essentially valid in the unbounded realm of mechanics, leads portentously to an aesthetic analogue which has hitherto received no critical attention. It has long been recognized in the plastic arts that the potentialities of linear alteration are governed by the design, a fact as familiar to the psychologist as to the painter; the relativity of colour values is equally well known, but this interdependence, because of its endless range, has never been fully catalogued. While the celebrated physicist has been evolving his shocking theories of the courses of natural phenomena, the world of art has suffered an equivalent heterodoxy with respect to its expressive media. This revolt has sprung from the conviction that the old art is not necessarily infallible, and that equally significant achievements may be reached by new processes and by fresh sources of inspiration.

The term organization, applied universally by the modernists—and sometimes, it must be confessed, quite absurdly—to the disposition of the forms selected for pictorial treatment, is one instance of the radical change in the artist's attitude toward his work. Originally the term connoted the idea of the relationship of the constituent elements; with the men of the recent schools this idea has developed into an end in itself, as distinguished from a means, as it was employed in the art of the past. With this concept in mind it is readily seen why so much of the diligence and experimental activity of the younger painters results in merely technical combinations, and it also helps to explain why so much of their work is fragmentary and often apparently superficial. When organization is regarded as an end, the business of relating the material chosen for presentation becomes of supreme importance, and points the way to a new set of co-ordinates without which no new forms can be created.

In his special theory of relativity Professor Einstein has demonstrated with brilliant finality that Newton's laws of inertia are true only for a Newtonian system of co-ordinates; that is, when the gravitational field is disregarded, and when the description of motions is definitely referable to a point on a rigid body of specification; he has shown that these laws are adequate for practical measurements but incompatible with the law of the propagation of light unless the Lorentz transformation be substituted. In his

Source: Thomas Jewell Craven, "Art and Relativity," *The Dial* 70 (1921), 535-539.

general theory he has defined the limited validity of the special principle, and has made clear that the laws of natural phenomena cannot be formulated with absolute accuracy unless the old co-ordinates are abolished and a new system devised wherein the reference-bodies are no longer fixed in relative motion. In connecting the equations of an abstract science like mathematics with philosophy the symbolical method must be followed; in the case of art the same plan is retained, and with even more striking results. When one considers the reflective aspect of art and its close affinity with the general thought of its time, this connection will not seem strange. The plastic world is, of course, compounded of manifold details gathered from the forms of perceptional experience, but the processes involved in harmonizing these details are purely psychic and inseparably bound to all other psychic factors of the age. It is hardly necessary to add that neither scientific nor mathematical formulae are directly concerned with this reaction to life, and that the quest for new relations in art-forms is guided almost entirely by feeling after the first intellectual step has been taken.

The fixed co-ordinates upon which the Newtonian measurements were erected have their parallel in more than one aesthetic manifestation. It is of no consequence that these manifestations have differed in tendency—there has always existed a common bond of interest, a rigid system of judgements corresponding to an immovable reference-body, and it is this abstract quality which establishes the analogy between the old art and classical mechanics. Professor Einstein's general theory of relativity has shaken the whole physical structure; similarly has the modern painter broken the classical traditions.

Although the artists of the past, in striving for enduring beauty, never regarded organization as an end, nevertheless they were conscious of its importance; and in every period the creative will has received its impetus from specific and rigid tenets. Most of these principles since the days of Giotto have been founded upon verisimilitude, architectural proportion, and the like; they have been born of the belief that truth could not be attained except by strict adherence to the dictates of experience. Co-ordinates from which further relations were constructed have varied from time to time, but in every movement to the present they have had inception in inflexible ideas, such as the logic of light and shade, correct anatomical structure, and perspective. Even in rhythm the balancing actions and counter-actions have become standardized, and composition has deteriorated into mechanical pattern-making. The artist of to-day is not seeking the impossible, the overthrow of the past; he asks that the relativity of individual truths be acknowledged; he is convinced that the real meaning of art lies beyond precise lines of definition, and is searching for a new point of departure, a system of co-ordinates which allows him to achieve coherence without falling back on the laws of visual experience, knowing that these laws invariably become static and conventionalized when severed from the field of personal action where they originate. It is undeniable that the great man of former periods has broken the laws of his age, has revolted against the aesthetic dogma handed down to him; but what has signalized his genius has not been the construction of a new and moving reference-body, but a change in the direction from a fixed basis.

It is at last recognized that the truth of art from a constructive point of view is a matter of coherence, of inevitable relationships, and that to intensify its value as a reflection of life, art can no longer proceed from the traditional loci. Instead of clinging to the rigid laws of photographic vision for a logic of creative activity, the modernist is ever mindful of his psychic responses to experience. For example: a painter has chosen for a theme a specific landscape consisting, say, of two houses, a prominent tree, a brook, and a bridge, items which may be delineated in several ways, and which may be held together pictorially by following a precise scheme of light and shade, by obedience to correct perspective, or by certain recurrent accents of lines. Each of these methods is compatible with the old doctrine of art, and each is adequate for graphic rendition; but it is not to be inferred that the primary inspiration of the painter was the simple idea of representation. What made it his own theme was the fact that the landscape aroused his perceptive powers and stirred his emotions—it had characteristics peculiar to him alone. It is here that a factor enters the old system of co-ordinates which is quite as disturbing as Professor Einstein's introduction of the time element into the Euclidean laws of spatial calculation. The personal feeling of the artist must be injected to arrive at greater truth, a truth far beyond that of mere vision, for the latter quality, while it serves all purposes of illustration, reveals nothing psychologically.

In the landscape mentioned above, the painter feels the predominance of certain forms; some objects attract him and stimulate his imagination profoundly—others are instinctively allotted a secondary position; the forms which excite him are contemplated, one might say, out of perspective, out of the pure logic of vision—they assume a magnitude that transcends all reality. Obviously the artist's conception of the real and living truth cannot be portrayed by conformity to any laws of actual appearance—it is compassed in a different fashion. Nor can the goal be reached by the simple device of accentuations, for here he is confronted with the fundamental requisite of coherence which insists on the relativity of the constituent parts in spite of all emphasis. He must, therefore, discover some point of reference that will provide for the desired accentuations and at the same time preserve unity and sequence without which art is inconceivable.

It is here that organization becomes a decidedly conscious process, and proclaims the necessity for a new and mobile basis identified with the personal element. We must not conclude that such an element has been absent in the old art; but not until modern times has painting been regarded as a vehicle for psychological truth, has it been made the reflection of the artist's mental states in the presence of simple objects of experience. The message of the former periods, notably in the great ages of productivity, has been spiritual in the collective sense—pervaded with religious thought; to-day it testifies to individual psychology and mirrors scientific experiment. Seizing the old system of visual co-ordinates, the modern painter has infused into it the personal element with a high degree of premeditation, and in place of the static pivot, correct in architectural symmetry, sound in aerial perspective, and logical in light and shade, he has given us a moving body of specification, independent of naturalism of any sort, and by which the integral forms are bound together by flowing sequences of line and colour.

Recognizable objects find their way as often as not into the new works of art, but they are never servile to realistic appearance, and it is unlikely that the painter will ever again attempt the ancient efforts to reproduce nature literally. Endowed with the system of co-ordinates gradually evolved since the death of Cézanne he has at his command the most plastic medium of expression that the world has ever known.

Charles A. Beard, the "New Physics," and Historical Relativity

Hugh I. Rodgers

Charles Austin Beard was such a provocative and seminal thinker that the passage of time has increased rather than diminished the interest of social scientists and historians in his work. Several studies have appeared in recent years dealing with the origins and development of his political, economic, social and historical thought. Beard's controversial doctrine of historical relativity received special attention in the works of Morton G. White and Cushing Strout. Both viewed Beard's relativism as an aspect of the "pragmatic revolt" in American thought.

Certainly the pragmatic and evolutionary traditions are important for the development of relativism in Beard's thought. Carl Becker, whose ideas often anticipated and paralleled many of Beard's, noted in 1910 that men influenced by evolutionary theory had come to ask if truth, reality, and the "very facts themselves" were not subject to change. Becker indicated that the truth of historical synthesis lay in its usefulness; it was "true relatively to the needs of the age which fashioned it." To the relativism of the pragmatists

Source: Hugh I. Rodgers, "Charles A. Beard, the 'New Physics,' and Historical Relativity," *The Historian* 30 (1968), 545-560.

a new emphasis was added during the 1920s and 1930s. This new emphasis was on the involvement of the observer with the "facts" observed.

The stress on subjectivity was an important contribution of the new physics to the attack on the empirical method in historiography and the social sciences. Writing in the middle of the 1930s, Tobias Dantzig summarized the profound transformation wrought by Einstein in basic scientific assumptions by noting that modern science differed from its classical predecessor in recognizing "the anthropomorphic origin and nature of human knowledge." It was "inherently impossible" to establish "whether any universe of human discourse possessed objective reality" at all. One would never be able to assert that he had "exhaustively determined all (his) limitations and discounted all (his) foibles." The new science of Einstein was skeptical of objectivity and frankly admitted the subjective nature of observations and thus of knowledge itself.

Cushing Strout has suggested that Charles Beard was not sufficiently aware of these developments in modern physical science to draw support from them in his struggle against determinism and objectivism in historical and social science writing. If true, this unawareness seems strange in view of the widespread discussion of the new doctrines. To the disgust of some physicists and the delight of others the theory of relativity provoked "strange speculations" in philosophy, psychology, politics, ethics, and race theory. Among physicists themselves, Sir Arthur Stanley Eddington and Sir James Hopwood Jeans tried to draw philosophic conclusions from the new physics for the layman. A veritable flood of essays appeared in the Twenties and Thirties from the pens of venturesome writers offering explanations and applications of relativity.

The academic disciplines were affected by the new ideas. Some writers spoke of the relativity of human nature. Political scientists found metaphysical implications in the quantum theory and the doctrine of relativity. The immediate successor of Charles Beard as president of the American Political Science Association, William Bennett Munro of Harvard, entitled his presidential address in 1927, "Physics and Politics: An Old Analogy Revisited." Declaring outmoded the physics of Bagehot's day, Munro announced that the revolution produced by relativity in the physical world "must inevitably carry its echoes into other fields of human knowledge." He proceeded to apply the new physics to political science with rather startling results.

Historians were also affected by the new concepts although it required the better part of the decade for Clio's votaries to adapt relativity to historiography. In 1923, Edward P. Cheyney held out to the American Historical Association the hope of an "immutable, self-existent law" in history, but by 1933, Charles A. Beard urged the same body to opt for an "act of faith." In the intervening decade historians questioned the idea of law in history, became increasingly skeptical about history as a science, and above all, tackled the problem of objectivity in the writing and interpretation of history.

The optimistic trust in "law" collapsed rapidly. Samuel Flagg Bemis defended the old hope in 1926 when he maintained that some "future Newton" would discover a master principle by which to interpret history. This appeared a vain expectation as the findings of Einstein, Heisenberg, and Bohr put the Second Law of Thermodynamics and the Rule of Phase, which Henry Adams tried to apply to history, on quite shaky legal legs. Sir Charles Oman demonstrated the historian's loss of certainty by declaring flatly that history was "cataclysmic and not evolutionary."

Carl Becker, who had hinted at the lack of detachment in the writing of history before World War I, provoked a spirited debate among historians. Condemning Becker for holding that historical facts are figments of the historian's mind, Carl Stephenson trumpeted, "No fact, no history!" Despite such histrionics the goal of impartiality implied in Ranke's command to write history as it actually happened eluded historians.

The historian selected the facts to fit whatever aim he had in mind, James Harvey Robinson announced. Harry Elmer Barnes maintained in 1927 that whether one tried to evaluate an historical event in terms of its own age or in terms of the present, "serious distortion" resulted. He found that "tests of the significance of historical material are relative and pragmatic. Is there no absolute and transcendental test? Apparently not. . . ." Confessions of the sin of subjectivity multiplied. Generalization in history was admitted to depend on facts selected by a historian whose attitude, colored by the intellectual climate of his times, governed his interpretation of the facts.

The papers read at the annual meeting of the American Historical Association in 1927 reflected an active interest in problems of method, relation, outlook, and philosophy of history. In his presidential address that year Henry Osborn Taylor called the concept of relativity in physics, together with its dethronement of solid matter, a representation of the modern temper. Such concepts might not satisfy the spirit of another age, he stated, making the doctrine of relativity itself relative to its historical setting.

Taylor went on to remark that written history was also a product of the "intellectual conditions (which are the actualities) of the time of its composition." But Taylor did not stop after pointing out the relativity of conditions. He approached a subjectivist position by defining knowledge as experience; each man had a different experience, therefore, different knowledge. This was the disturbing "human equation" which made the study of the facts dependent upon the observer's ability and bias.

Carl Becker magnified the note of subjectivism sounded by Taylor when he informed the American Historical Association in 1931 that every man was his own historian. Becker's position was pushed to its logical conclusion the next year by Edward Maslin Hulme, who told the Pacific Coast Branch of the American Historical Association that detachment and objectivity in the writing of history was a myth. Not only time and place, but the personal characteristics of the historian colored his work. Noting that even physicists now took the personal equation into account, Hulme concluded: "Historic truth is relative and subjective, not absolute and objective." The cause of "scientific" history seemed doomed.

It remained for the economic historian Charles Woolsey Cole to demonstrate a direct connection between the new physics and historiography. He did so in an essay published in 1933 entitled "The Relativity of History." As in the past when Newton and Darwin had provided useful patterns of thought, Cole suggested the time had come once more to draw new points of view from the "austere domain of physics." Cole found in the new physics three principles: relativity, indeterminism, and discontinuity. Upon elaborating these principles, Cole showed that Einstein had made it clear the physical facts of time, distance, motion, and measurement had no meaning. "They are not absolute. They take on significance only when related to some frame of reference." The same thing was true of historical facts. They are made meaningful "when by order, selection, and interpretation they are related to a frame of reference." Facts, in other words, had the meaning which historians bestowed on them.

Historians thus moved from absolute to relative truth and from pretensions of detachment to admissions of subjectivity. The debate about the nature of historical knowledge was the context in which Charles A. Beard's doctrine of historical relativity was formed. Naturally, ascertaining just when and under what circumstances an individual develops a particular line of thought is difficult at best. Important clues to the progress of Beard's thought, however, are provided in his book reviews and shorter essays, especially if they are approached in chronological order.

No overt hint of Beard's later suspicion of objectivity is apparent in his works prior to the First World War. Beard's associate, James Harvey Robinson, emphasized in 1904 that historical matter could be truly and impartially stated. Robinson and Beard hoped to make the morning paper more understandable with their *Development of Modern Europe* which they pointedly subtitled, *An Introduction to the Study of Current History.* But granting that this aim was held, the authors asserted there would be no "distortion of the facts in order to bring them into relation to any particular conception of the present." The new history, pragmatic and utilitarian, did not aim at the rejection of objectivity.

Beard held that the economic interpretation of history was a most excellent aid in reaching the goal of objectivity. In his study of the Constitution, Beard criticized the scientific historians for being overly concerned with mere classifying and arranging of facts. He proposed to explain their "proximate or remote causes and relations." Beard accepted the causal nature of history while objecting to the narrowness of the

Rankean type of history writing. Beard was even confident that the historian might discover a law which would "reduce history to order" just as natural scientists had found such laws governing the physical world.

Beard was an evolutionist in 1920, although not necessarily an optimistic one. He wrote of a fast changing world in which the confusing elements of history constantly evolved "together in terrible fascination." The years from 1922 to 1925 found Beard becoming skeptical about the possibility of objectivity when writing about those elements. Instead of brashly talking about explaining proximate and remote causes, Beard now found that selection of material was a most "delicate" matter. "Each of us," he admitted, "must perforce see what is behind his own eyes."

Beard did not abandon wholesale or at once his attachment to the old standards. In July 1925, he could still praise a volume for meeting the "requirements of exact science." But in November of that same year he inquired into an author's "philosophy of values" which had determined both what he had included and what he had left out of his history. Beard began to deprecate the writer who made pretensions to a "Buddha-like detachment," seeing things simply as they were. He now preferred one who applied the standards of his own milieu when making value judgments.

In July 1926, Beard had occasion to make his first direct reference in print to the new physics when he advised historians that research into the origins of World War I would be a complicated job. One should not dash in lightly and start at once to pontificate, just as he should not rush into "the matter of 'relativity' or the 'quantum theory'" with a deficient knowledge of mathematics. For, as Beard confessed later that year to the American Political Science Association, the historian can never present all the facts. He must select a few from the number that had survived the destructive work of time and circumstance. "And any selection," Beard warned, "is an interpretation."

By the end of 1926 Beard had moved some distance from his pre-1920 position. The historian, he now held, never used all the facts. Those he selected were picked according to the historian's system of values. The philosophy of values in part depended upon the historian's position in time and place. Impartiality in the writing of history was a pretension. Beard's attack on objective history became sharper. It was the individual historian, Beard insisted, who selected, interpreted, and made the facts come alive "for those for whom he is writing in the age for which he is writing." The dictum of writing history as it actually happened he blasted as a "clever saying of Ranke's," labelling it the "fruit of vain imagination." He reiterated that it was the historian of a particular time and place who assigned value to the facts.

The years of the mid-1920s during which the gradual shift in Beard's thinking occurred coincided with the publication of several works of theoretical science and numerous popularizations of them. A new edition of

Sir James Hopwood Jeans' *The Mathematical Theory of Electricity and Magnetism* (1925), the appearance of Sir Arthur Stanley Eddington's *The Internal Constitution of the Stars* (1926), and two editions of Ernest William Hobson's *The Domain of Natural Science* (1923 and 1926) are examples of these publications. Beard was familiar with these works. He considered the Jeans and Eddington volumes as belonging to a group of books which had "had the most influence on thought and action during the last half century." Beard cited the Hobson book several times in his own study of the nature of the social sciences—a work marking a critical milestone in Beard's development as theorist.

The Jeans and Eddington books are specialized scientific treatises and are not directed to a popular audience. Both deal with Einstein's work and the relativity and quantum theories. Hobson attempted in his volume to explain developments in natural science and theoretical mathematics to the interested layman. Natural science, in Hobson's view, was not concerned with the nature of reality or with schemes of causation and determination. Science, like other branches of knowledge, was subjective. Its data were not absolute fact but consisted of conceptions of phenomena perceived by the senses. Hobson pointed out that whereas in the Newtonian system the scientist described the phenomena of motion by using an absolute frame of reference, in the Einstein system all frames of reference were "on a parity," thus making for complete relativity. Hobson cautioned against extending Einstein's theory beyond the scope of physics.

A number of these topical scientific themes were touched upon by Bertrand Russell in his essay contributed to *Whither Mankind*, a volume Charles Beard edited in 1928. Russell, after summarizing Einstein's work, remarked that the "old glad certainty" of science was gone. Natural law did not consist of immutable principles but of human conventions and statistical averages. Physicists had demonstrated the absence of causation in the natural world.

Beard continued to take note of developments in natural science, although he complained in 1931 that physicists resembled Supreme Court Justices in the diversity of their opinions. Beard, by that time deeply concerned about the depression crisis, was searching for some illumination "on the strange map of life." He told an assembly of history teachers that year that historians and social scientists were being outstripped in this search by the natural scientists. Jeans had ranged over complex areas of time and space to conclude that the universe consisted, not of familiar Newtonian matter, but of pure thought. Whitehead, Beard revealed, had collapsed before God as the "ultimate limitation," while Eddington had admitted the absence of causality in the material world. The three philosopher-scientists agreed there was a basic reality of some kind underlying the appearance of life, but that it was unknowable for the most part.

In the bewildering struggle of the Titans of science what could a mere historian do? Beard queried. "He knows nothing of relativity, electrons, and symbolic logic." But he went on to declare that the only discipline able to shed light on mankind's path was history. History revealed to Beard "contingency, the possibility of choices ever present in the stream of life." Beard stressed this point again in 1933 when he emphasized that the events of history, "the unfolding of ideas and interests in time-motion," could never be brought into a deterministic pattern. It was therefore possible for man to find a way out of his crises.

Beard categorically proclaimed the end of determinism in the social sciences. The bondage of social scientists was the result of transferring to their field assumptions "once deemed applicable to the world of physics." Social processes could not be analyzed in the manner of engineering problems. For too long, Beard remonstrated early in 1933, social thinkers had been dominated by the idea that once all the facts had been assembled, conclusions would inevitably flow from them. Boldly Beard challenged, "*All* the facts? Obviously, an impossibility. The obtrusive facts? Obtruded on whose vision? The important facts? Important to whom and to what ends?"

The limitations of the observer implied in these last statements came to receive more emphasis in Beard's thought. Early in 1933, Beard decried "the coming crisis in empirical method." He found that even in chemistry and astro-physics where the observer might be supposed to stand outside the discrete things observed, the scientist brought to his study a set of ideas and emotions which colored his perception of phenomena. Much less possible was objectivity in the social sciences where the "thinker floats in the streams of facts, so-called, which he observes. His thoughts are parts of the thing thought about." Such statements indicate Beard's deepening concern about the subjective nature of knowledge.

The position which historical thinking and social science theory had reached by 1933 makes Beard's famous address to the American Historical Association that year seem anticlimactic. Nearly everything Beard said then had already been stated. This is true of his attack on scientific method and Rankean objectivity and of his use of the well-known phrase, "frame of reference." But the address did serve as a fitting denouement to the long discussion about philosophy and method in historiography.

Beard called upon his listeners to throw off the shackles of analogies drawn from Newtonian physics and Darwinian biology. At one point he attempted, like Taylor six years earlier, to confound the doctrine of relativity itself. But Beard went on to formulate some of the classic statements of that doctrine. The scientific method was extremely limited and could never produce a science of history. The historian was not a detached, objective observer. In selecting and arranging his facts, the historian was guided "inexorably by the frame of reference" in his mind. This frame of reference included "things deemed necessary, things deemed possible, and things deemed desirable." Within this frame three conceptions of history were possible: history as meaningless chaos, a cyclic interpretation, or, the view that history moves in some direction. There was no test of validity, no inherent logic of the data, to guide the historian in making his choice. The selection made constituted the historian's "act of faith" and conditioned the kind of history he would write.

Beard's remarks threw into clear relief the gulf which separated historians from those halcyon days when Cheyney could speak of law. The prospect frightened some. The objectivists refused the new teaching and resolved to go down fighting with Rankean colors firmly nailed to the masthead. Beard clarified his position in the resulting conflict. He shrank from pushing relativism to its ultimate conclusion. He denied that history was chaos and asserted that there are not "as many schemes of reference as there are historians." Yet Beard would not retreat from his view that the idea of absolute truth in history was an illusion. Above all, Beard rejoiced that the pretension to objectivity had been "wrecked beyond repair."

Beard was encouraged to reject the theory of mechanistic causation in history because he knew the physicists were no longer sure about causation. Contemporary physics, Beard found, was "chary about the use of terms cause and effect." It merely described events in the order of their occurrence. What could not be done in the physical could not be done in the social sciences. The data of the social sciences, Beard insisted, could not be forced into a "closed circle of deterministic sequences."

The new science also aided Beard in his search for a solution to the impasse in empirical method. Much of the speculation in higher physics, Beard discovered, was a result of "subjective feeling rather than the outcome of objective observation." Modern scientists had unhesitatingly dismissed the old concept of complete objectivity. They no longer believed "'that the eye of any beholder is disinterested.'" Such admissions of subjectivism by natural scientists reinforced Beard's skepticism about objectivity in historiography and encouraged him in his bold view of written history as the historian's subjective act of faith in the future.

There can be little doubt that Beard was cognizant of trends in contemporary physics and of their implications. All the ingredients of Beard's relativism—the attack upon objectivity, the denial of determinist schemes and thereby of a science of history, the frequent use of phrases like "frame of reference"—all point to the impact of the new physics. The very terminology of Beard's statements fits easily into the discourse of modern physics and its popularizers. To suggest that Beard drew upon contemporary physics does not, of course, exclude other and perhaps ultimately more important influences on his thought. The sources of Charles A. Beard's inspiration and even phraseology, however, were varied, and in any consideration of his relativism the contribution of Einstein's physics must be taken into account.

Quantum Mechanics, Free Will, and Determinism

Henry Margenau

There are, I think, eternal questions, but no eternally valid answers. Philosophy shows itself to be alive when it raises, again and again, the deep concerns that plague man's reason; it dies when it presumes to have resolved them with finality. Determinism and freedom in their conjunction pose one of the eternal questions, and the partial answer I propose affords no stagnant resolution. It is intended to show compatibilities between disciplines that seemed discrepant, and by bringing recent scientific insights to bear upon the problem of freedom I hope to revive its fascination and perhaps to suggest ways in which progressively it can be solved, not in closed form to be sure, but along the vista of an infinite series of approximations.

A very limited aspect of the problem of freedom will enter our discussion. Modern usage has made this word so utterly meaningless that any useful treatment must be preceded by a statement of the sense we wish to convey and the nonsense we wish to avoid when the word's oracular sound appears. First a trivial point. There will never in the sequel be a reference to free will; we pass by the old psychological controversy based on a compartmentalization of human faculties. Man as an agent is free or bound; the question is whether his action is fully and under all circumstances conditioned by antecedent states. If not, the precise location of the elements of chance within the psychological complex will not form part of our inquiry: we shall not ask whether the causal hiatus occurs in the acting, or in the willing, or in the selection of alternatives, or in the reasoning that precedes the act. These, to be sure, are interesting questions; but I fear that the present state of psychology does not tolerate their formulation, let alone their solution. At any rate they lie beyond my competence.

A clear distinction must be made between at least two kinds of freedom. One of these I shall call, for want of a better word, *elemental*, the other *practical*. Practical freedom is overt freedom and resides in the absence of visible constraints or coercions. Usually these constraints are external or physical; they may result from application of force by others, from uncontrollable factors in a situation, or from personal inability to achieve a chosen goal. Or they may be internal, stemming from habituation, addiction, incompetence, and so on.

These constraints which inhibit practical freedom are objectively

Source: Henry Margenau, "Quantum Mechanics, Free Will, and Determinism," *Journal of Philosophy* 64 (1967), 714-725.

discernible, and their absence can likewise be certified. Elemental, or, in a certain sense, metaphysical freedom, may or may not be present even when practical freedom is at hand. Indeed if Kant is right, it is entirely indifferent to the presence or absence of coercion. This is what led him to claim that only the will is free and, therefore, the seat of moral responsibility and of moral qualities. Stripped of unnecessary impediments, elemental freedom involves the existence of genuine, causeless alternatives in any prevoluntary situation that is dominated by conscious deliberation relative to the performance of an act or the making of a decision. Only elemental freedom involves us in deep metaphysical problems; practical freedom (or freedoms!), although extremely important in ethics, jurisprudence, and politics, will not concern us here.

This places the largest part of the legal concept of freedom outside of our present purview. Although I do not share this view, it can perhaps be maintained that responsibility is based solely upon the presence of practical freedom. If that is true, this paper does not touch upon the problem of human responsibility.

Nor does it present a semantic analysis of the prevalent meanings of the words 'freedom' and 'determinism'. I doubt if problems can ever be solved, even progressively, by an inspection of the usages of words. They are often clarified, sometimes unmasked as pseudo-problems, nearly always proliferated, but never, by that method, carried up to the level of creative understanding. The reward of semantic analysis is a satisfied smile, not the excitement of discovery. Nothing is gained by seeing how the word 'freedom' is used in common parlance, except perhaps to note that it is not employed in a uniform sense at all. There are easy ways of making it, indeed of making all so-called eternal questions, seem silly; but it is the business of philosophy to phrase them in pregnant ways which are adequate to refined experience and by this token to preserve rather than destroy their meaning. Indeed, may not the philosopher be pardoned if he prides himself on having insights that transcend the market place, and therefore claims the right to cast them in a coin of language?

Persons unfamiliar with the literature pertaining to the problem of freedom may feel that reminders of the foregoing effusive kind are out of place in a disciplined discourse because they are an attack against windmills. This, unfortunately, is not the case. Nearly all the errors just criticized are made, for example, in a book by Stace—to cite a notable instance—who, probably because of a basic confusion between elemental and practical freedom and a belief in the peculiar redemptive qualities of language, proceeds to argue that freedom cannot be established because it has not been properly defined. Man, he says, could not be certified as existing if he were defined as a five-legged animal. Redefine him and he will be there! Suggestions of this sort, if significant, should greatly appeal to scientists who, after a weary search, fail to discover their quarry: they ought to console themselves because they merely used an erroneous definition. Or should they have fixed up their definition to begin with and not have searched at all?

Let us turn now to a positive statement of the problem to be attacked—not solved. Like many philosophic challenges, this one springs from an evident incompatibility between a certain ubiquitous kind of protocol experiences and the conceptual formalism available for their explanation. The uncoordinated *P*-fact in question is the clear and present awareness of freedom or feeling of choice that accompanies every conscious decision. It stands in contradiction to certain highly successful and widely accepted scientific theories which entail universal causation and, hence, a deterministic chain of events, even mental events—unless provision is made for an unpalatable dualism contrasting the mental and the material.

An acceptable resolution of the dilemma between the self-declarative protocol evidence of freedom and the constructed and powerfully documented thesis of determinism could involve one of the following tasks: one might show that customary theory can be extended without violating already approved and explained tests of it, and yet accommodate the subjective fact of freedom; or one might demonstrate that the customary rules of correspondence which link protocol facts with constructs are in need of revision, again without prejudice to what is already accepted; or a mixture of both procedures may be necessary. In the sequel we endeavor to show that a slightly mixed strategy of the latter kind is indicated, that in fact the extension of theory had already taken place without stimulus from the freedom quarter, and that a slight modification of the rules of correspondence between states of awareness and physical fact (verified constructs, verifacts) opens an avenue of attractive inquiry promising harmony between the inchoate components of the freedom paradox.

What is not permitted is to deny offhand the veridicality of the freedom awareness. That this is a cherished nostrum among many writers need here not be emphasized; it is the vogue among hard-boiled scientists, chiefly psychologists, and it feeds upon the claim that reality exhibits no namable counterpart to the intimate experience of choice. To this one can only answer that there is no preestablished concrete reality to which this particular protocol refers. For the scientist's reality, when philosophically analyzed, turns out to be the minimal set of posits (well-connected constructs) that render our experience, including the experience of freedom, coherent. Hence no appeal to it is possible until the problem of reality itself is solved. To deny the claim of felt freedom a hearing because it is an introspective phase of consciousness is to kill the spark of every incipient scientific venture. Subjective experiences of this very sort are both the first glimmer and the last instance of appeal in all acts of scientific validation.

Another attempt to eject this candidate from the halls of scientific respectability was once made by Professor Wood, who wrote:

> . . . The belief that there are genuine alternatives of action and that the choice between them is indeterminate is usually stronger in prospect and in retrospect than at the time of actual decision. The alternatives exist in prospect as imaginatively envisaged possibilities of action and in retrospect as the

> memory of the state of affairs before the agent had, so to speak, "made up his mind." Especially in retrospect does the agent recall his earlier decision with remorse and repentance, dwelling sorrowfully upon rejected possibilities of action which now loom up as opportunities missed (390).

Were I to comment on this passage I would have to say that it is out of accord with my own recollections. The agony, the accentuated awareness of risk that accompanies a truly important decision appear to me more memorable than any prospect or recall; indeed it is that intensified peak of consciousness to which regret and remorse later attach themselves as secondary feelings. Even more realistically, repentance in most instances deplores the circumstance that we have *failed* without ourselves to create the protocol consciousness of choice at the proper time, not that we have forgotten it as unimportant. But even if Wood's argument were correct, the fact that one tends to forget a feeling or a fact does not make it irrelevant.

After these preliminaries, let us turn to an examination of the theories in terms of which a solution of the problem now posed will be sought. Although the main thesis of this article presupposes a clear understanding of classical determinism and of quantal indeterminacy, limitations of space forbid their thoroughgoing discussion. Hence I merely state here the principal facts that enter as preambles into the later arguments concerning freedom.

(1) The ordinary doctrine of determinism (Laplace's formula) does not afford an acceptable explanation even of the facts of classical physics. No "demon" can have contingent knowledge without a slight margin of uncertainty. For example, if he knew the positions of all the molecules in this room with the unlikely error or one-millionth of one per cent at the present time, this knowledge would be completely wiped out by interactions in one microsecond. Hence this dynamically detailed manner of description is essentially illusory.

(2) Probability in science is not a subjective index of confidence in the outcome of an event: it is a regular and normal physical quantity, wholly on a par with other observables which serve as ultimate determinants of events in the physical universe. Their reduction to more elementary dynamic certainties is not always possible; an indeterminate world does not presuppose a crypto-determinism.

(3) The probabilities of quantum mechanics are of this irreducible kind.

(4) The uncertainty principle does not involve merely lack of knowledge, is not a proscription of certain kinds of measurement. It affects being as well as knowledge and is an expression of the latency of observables.

The numberless attempts made before radically stochastic theories like quantum mechanics were known cannot be reviewed here. Most widely accepted, perhaps, is a view set forth in the writings of Kant and Cassirer. Their philosophy (i.e., transcendental idealism) regards causality as a category of human understanding, a necessary form in which all knowledge of events must be cast. For things in themselves, which lie beyond our comprehension, causality and all other basic modes of thought are irrelevant. From this point of view universal causality or determinism, whether of the classical or the quantum-mechanical sort, must not be regarded as a metaphysical constraint upon all forms of being. It must be distinguished from what Cassirer calls a "dinglichen Zwang." Freedom, too, is a transcendental principle, but one regulating our actions, and it therefore controls another realm. If both were factual, descriptive attributes of the world, they would indeed collide; only their transcendental nature keeps them out of conflict and makes them "complementary," to use a currently cherished term.

Now it seems to me that classical determinism and freedom do collide—in a factual sense if both are taken as ultimate metaphysical principles, and in the form of logical irreconcilables if they are transcendental modes of explanation that regulate our understanding. Let me illustrate the meaning of this claim by reference to a trivial example.

Suppose I am asked to raise my hand. I can do this mechanically without thought and without engaging my will. In that case, habit acquired during my student days will probably cause me to raise my right hand. One may look upon this action as a causal one, whose result is predictable in terms of conditions existing in my brain, of associations acquired of neural pathways previously established, and so on. But notice: I took care to say that I would *probably* raise my right hand, thereby implying something less than strict predictability.

But if I am told: raise whichever hand you wish, the sequence of events is different. I am somehow challenged to think and then to make a choice. To believe that, during the moment of reflection preceding the decision to raise my left hand, the configuration of the molecules in my body, the currents in my brain cells, or even the psychological variables composing my mental state have already predetermined that I must raise my left hand is clearly false, for it contradicts the most elementary, reliable, self-declarative awareness of choice which accompanies this act. Thus a serious contradiction arises if strict causality is a metaphysical fact.

Nor can the situation be saved by saying, with Kant and Cassirer, that causality is merely a transcendental principle in terms of which we are required to conceive things. For in that case we should require *one* principle of understanding to comprehend the sequence of events that compose the objective course leading to the raising of my left hand, and a different, incompatible one to explain my feeling of freedom. Reason does not tolerate two incoherent principles where a single one will do. I shall now show that the loosening of causality required by quantum mechanics enlarges the scope of that principle sufficiently to allow removal of these difficulties and to cover both determination and freedom.

What we hope to accomplish needs careful statement. It might seem to be a proof that quantum mechanics has solved the problem of freedom. This is a vastly different task from showing that quantum mechanics has removed an essential obstacle from the road toward its solution, while the problem remains unsolved in its major details. The following analysis is directed toward this latter, much more modest aim. As we approach it, many of the difficulties, whose resolution constitutes the difference between the first and second tasks, will helpfully move into view.

Let it be noted again that we are not raising moral questions at this point. The decision which hand to raise is totally without ethical relevance; it merely illustrates the contrast between instinctive-reflexive, almost mechanical behavior and an action that involves thought and will, thereby engaging to a small extent the quality of freedom. The question of motivation, so essential in ethics, hardly enters at all. Or if it does, if for some conscious reason—perhaps the desire to surprise my partner—I have chosen to lift my left arm when he expected the right one to be raised, that reason is far from the concerns of ethics. The distance from here to choices that can be said to be *morally* good or bad, that conform or do not conform to ethical principles, is very great. Yet somehow it can be traveled by vehicles already at our disposal. Most theories of ethics achieve their end, the explanation of moral behavior, *once the possibility of freedom and motivation is established.* These qualities, however, are present at least in embryonic form in the example we chose for discussion; hence we return to it. Its relative simplicity is an important advantage.

Precisely what happened to me as a conscious person during that crucial interval in which I "made up my mind" to raise my left hand? Of the enormous variety of physical and chemical processes that took place in my body I am not aware. I do know, however, that the physical condition before the arm raising and that after the act were connected by a continuous series of objective physical happenings. And the entire series could have been different because of my decision, because of a choice of physical possibilities that were open to me.

The mental processes during the crucial interval are likewise difficult to record in detail. Nevertheless the following is perfectly clear. I was aware of having a choice; there was a moment of reflection, perhaps a brief recall of past occasions; then came a glimmer of rudimentary satisfaction in doing the unexpected, next a decision, and finally the act. The choice was enacted within consciousness, and it evidently was permitted, but merely permitted, by the physical processes that took place.

One thing, then, is utterly apparent: freedom is not wholly a problem of physical science but one involving biology, physiology, and psychology as well. Upon realizing this one immediately confronts the standard question of reducibility: Are the laws of psychophysiology merely elaborate versions of those encountered in the physicochemical world, or do they differ radically? The first alternative, which assumes the possibility of reducing all behavior to physicochemical bases, need not be tied to the naive supposition that all the laws of these basic sciences are now known, and it will not be construed in this narrow sense here. The second, which maintains a radical difference, takes two essential forms. First, one may interpret the difference as mere transcendence, secondly as outright violation of physicochemical laws.

To avoid circumlocutions, let us refer to the first alternative, that of reducibility, as I. The second will be labeled II, and we shall designate transcendence by IIa, violation by IIb. As already mentioned, acceptance of I does not commit us to the view that all basic laws of nature are already known.

The precise meaning of IIa involves a theory of levels of complexity among physical phenomena. It is most simply illustrated by recalling the relation between the mechanics of point masses and the statistical mechanics of gases which are here viewed as large assemblages of molecules, each in the form of a point mass. To describe the mechanical state of every individual molecule one needs to specify its position and its velocity, nothing more. The totality of molecules, the gas, however, exhibits measurable properties like pressure, temperature, and entropy which have no meaning whatever with respect to single molecules. In this sense they are radically different from the properties of point masses. Yet if the positions and velocities of all molecules were known, the aggregate observables, i.e., pressure, temperature, and entropy, could be calculated. These latter characterize a level of complexity above the mechanics of mass points. Explanation is *continuous from below*; the concepts of the lower level have meaning on the upper, but not the reverse.

It is seen, therefore, that thesis IIa asserts no incompatibility between concepts and principles on two different levels. The physicochemical and the physicopsychological can probably be regarded similarly as two different levels of complexity, even though the differences are so great that the full connection is not at present in evidence. The view, however, seems reasonable. If it is accepted, and the gap can some day be filled, the higher-level concepts can be reached from below and thus be "reduced."

The bearing of alternative IIa upon the problem of freedom, which as we have seen and encountered in the upper realm, is now apparent. Freedom cannot appear in the domains of physiology and psychology if indeterminacy is not already lodged in physics. Strict causality among the molecules, applied upward as a principle of nature to explain the behavior of aggregates, cannot entail freedom because of the requirement of continuity from below. It is equally impossible to engender freedom in the realm of psychology when strict determinism rules physics, so long as hypothesis IIa is maintained.

For our present purpose, therefore, IIa can be identified with I: neither permits freedom unless strict determinism is abandoned in physics.

Only alternative IIb provides the possibility of freedom in the face of unrelieved classical causality as it is understood in pre-quantum physics. That view cannot be rejected out of hand; indeed it is still occasionally held.

Since it is forced to assume the occurrence of violations of the normal order of nature, it is tantamount to a belief in miracles. As for myself, I refuse to regard freedom as a miracle so long as other avenues of explanation are open. This is the case if alternative I or IIa is adopted, *provided physical indeterminacy is taken seriously.*

I judge IIa to be the safest hypothesis, and propose to describe its consequences. This is a somewhat unpopular course; it forces us to part company with many distinguished moral philosophers who see the autonomy of ethics threatened when a relation of any sort is assumed to exist between that august discipline and science. For centuries, humanists have been impressed by the slogan that science deals with facts, ethics with values, and these two categories are so disparate that they must forever stand apart. If unanalyzed, this is a foolish and a dangerous dogma. Some feel that a view that finds a root of freedom in physical science desecrates and demeans the high estate of ethics, whose legitimate concerns should not seek refuge in the indeterminacies of natural events. Ethics, says Cassirer, should not be forced to build its nests in the gaps of physical causation; but he fails to tell where else it should build them, if at all.

Throughout this article one single physical law is continually called upon to do extremely heavy duty, namely Heisenberg's indeterminacy principle. It is unreasonable to suppose that this item of knowledge is absolute in its present understanding, forever immune to reformulation and refinement. Future discoveries will doubtless place it in a new light, but it is difficult to see how its essence, which is drawn upon in the present context, can ever be relinquished. No one, of course, can rule out this possibility. But it seems far more likely, and here is where I would place my bet, that further principles even more widely restrictive of Laplacian causality will enter science, in which case the position here taken will be reenforced.

Some of them are already on the horizon, vaguely visible but portentous. The theory of turbulence in hydrodynamics, meteorology, and several promising new approaches in the behavioral sciences seem radically and perhaps irreducibly stochastic.

I now turn to a few logical arguments which have been leveled against the possibility of freedom, quite apart from physical indeterminacy. J. J. C. Smart has attempted to dispose of freedom as an inconsistent concept by employing a simple and seemingly cogent logical argument. He constructs two theses, which he assumes to be exhaustive of all possibilities and also mutually exclusive: (1) one is Laplacian determinism as we have discussed it earlier; the other (2) is the view that "there are some events that even a superhuman calculator could not predict, however precise his knowledge of however wide a region of the universe at some previous time" (294). Freedom, he holds, violates both of them and is therefore ruled out.

Certainly the requirement of impeccable logic is to be imposed on every phase of scientific and philosophic reasoning; nevertheless logic, in spite of its merited vogue, is not the sole arbiter of truth. There are instances where the diffuseness of the meaning of terms makes its formal application impossible and its conclusions spurious in spite of all the reverence it commands. In the present instance, its use to settle the argument concerning freedom is as ineffectual as the application of arithmetic to clouds.

Smart's algorithm was challenged neatly by Errol Harris, who rightly insists that the two alternatives above are not mutually exclusive. If freedom were identical with Smart's second alternative we would call it erratic behavior or caprice. What makes Harris's point important is, first of all, the looseness that afflicts the term 'event' ('state' or 'observation' or 'measurement' would be more acceptable) and, second, the fact that physical indeterminacy is precisely the kind of intermediate alternative that is coincident neither with proposition 1 nor with proposition 2.

One of the most serious confusions about freedom arises in connection with the uniqueness of history. The course of events in the universe is a single flow; there is no ambiguity about the happenings at any given time, aside from our knowledge of them, and if a super-human intellect knew everything that happened up to a certain time *t*, he would perceive, in looking backward, not only clear determinism but a rigid, filled space-time structure of events. He could in fact, if he were a mathematician, write a formula—with a proper qualitative text defining the nature of all events—which would represent all history up to *t*. Where, then, can freedom enter in the presence of that timeless formula which, although it was unknown at times before *t*, nevertheless "existed" in a mathematical sense?

The answer involves recognition of the fact that retrodiction is not the same as prediction. Indeterminacy permits the former but not the latter. Only Laplacian determinism makes inferences along the time axis symmetric in both directions. In classical mechanics, full knowledge of the state of a physical system at time *t* allows in principle the calculation of its state at any time before or after *t*. Indeterminacy introduces a peculiar asymmetry into states with respect to their temporal implications: the past is certain, and the future is not. This causal irreversibility of time is, and must always be, asserted along with the affirmation of freedom. It is the agency which, in the phraseology of William James, "transforms an equivocal future into an inalterable and simple past."

Nor is this without consequences with respect to the nature of time. A recent controversy concerning emergence or becoming, and the sense in which time is a fourth dimension of space is strongly affected by it. Relativity theory, which is thus far an outgrowth of classical mechanics and does not incorporate indeterminacy, speaks in Laplace's voice and precludes creativity and emergence of features not already foreshadowed in the presence. In the controversy to which reference has just been made, Capek is right in arguing for emergence, not for any philosophic reason but for the simple scientific fact that ordinary four-dimensional relativity, the basis for the claim of frozen passage, is not applicable to the atomic domain.

Having reviewed the most common objections met by those who affirm human freedom and having attempted to expose their weaknesses, let me now summarize and state my case.

Classical determinism made freedom intrinsically impossible, unless its application to psychophysical phenomena is arbitrarily interdicted.

Historic arguments designed to reconcile freedom with classical causality were able merely to establish a subjective illusion, a personal feeling of freedom.

Modern physics, through Heisenberg's principle of indeterminacy, has loosened Laplacian determinism sufficiently to allow *uncaused* atomic events, creating in certain specifiable situations the occurrence of genuine chance.

The consequences of such microcosmic indeterminacies, while usually insignifcant in the molar world, do ingress into the macrocosm at least in several known instances. It is very likely that they play a role in delicate neurophysical and chemical processes.

Physics thus makes understandable the occurrence of *chance*, of true alternatives upon which the course of events can seize. Physics alone, in its present state, can account for unpredictable, erratic human behavior.

Human freedom involves more than chance: it joins chance with deliberate *choice*. But it needs the chance. In so far, and so long, as science can say nothing about this latter active, decisive, creative element it has not fully solved the problem of freedom.

But it has lifted it out of the wastebasket of illusions and paradoxes and re-established it as a challenging problem to be further resolved.

And now an afterthought. Suppose physical science, perhaps with the aid of sister disciplines like psychology, philosophy, and even theology, had solved the problem of choice supervening upon chance to explain freedom, would this fuller understanding not restore determinism? If we can *explain* how the agency effecting choice selects from the alternatives presented by physics a particular one, will the inclusion of that agency into the scheme of things not leave us where we started, i.e., with an amplified Laplacian formula?

The answer cannot be foreseen. It may be affirmative, but I strongly doubt it. For if that agency were one which looked into the future rather than into the past, were drawn by purposes rather than impelled by drives, partook of the liveliness of the incalculable human spirit—freedom in a unique sense would survive. At any rate causal closure of any conceivable sort, inasmch as it must accommodate probabilistic science, will have a scope far larger than any branch of knowledge now known as science or philosophy.

Quantum Mechanics and the End of Scientism

John Lukacs

. . . it has become, by and large, possible to say that "modern man has moved on beyond the classical, medieval and the modern world to a new physics and philosophy which combines consistently some of the basic causal and ontological assumptions of each," as F. S. C. Northrop put it in his attempt to introduce Heisenberg's Gifford Lectures to American readers in 1957. He added that this "coming together of this new philosophy of physics with the respective philosophies of the culture of mankind . . . is the major event in today's and tomorrow's world." He asked the question: "How is the philosophy of physics expounded by Heisenberg to be reconciled with moral, political and legal science and philosophy?" Let me attempt to answer: through historical consciousness; through historical thinking.

Heisenberg's Recognitions: The End of the Scientific World-View

Let me, therefore, insist that what follows is not the breathless attempt of an enthusiastic historian to hitch his wagon to Heisenberg's star, or to jump on Heisenberg's bandwagon, to use a more pedestrian metaphor. Rather, the contrary: my wagon is self-propelled, and a Heisenberg bandwagon does not exist (at least in the United States, among one hundred people who know the name of Einstein, not more than one may know of Heisenberg). It is the philosophical, rather than the experimental, part of Heisenberg's physics that I am qualified to discuss; my principal interest in this chapter springs from the condition that among the physicists of this century who have made excursions into philosophy I have found Heisenberg's philosophical exposition especially clear, meaningful and relevant to the general theme of this book; and I have drawn upon some of his writings in this chapter because I want to present some of his courageous epistemological recognitions in a form which every English-speaking historian may read and understand easily. I have arranged these matters in order to sum them up in the form of ten propositions, the phrasing, the selection, and the organization of which is entirely my own: it is but their illustrations which come from the sphere of physics, described as some of them were by Heisenberg, mostly

Source: John Lukacs, *Historical Consciousness: Or, the Remembered Past* (New York: Harper & Row, Publishers, 1968), Chapter 7, pp. 278-289.

in his Gifford Lectures. They are illustrations in the literal sense: they are intended to illustrate, to illuminate new recognitions, certain truths, in the assertion of which this writer, as indeed any historian in the twentieth century, is no longer alone.

First: there is no scientific certitude. Atomic physics found that the behavior of particles is considerably unpredictable: but, what is more important, this uncertainty is not "the outcome of defects in precision or measurement but a principle that could be demonstrated by experiment." Physicists have now found that while they can reasonably predict the average reactions of great numbers of electrons in an experiment, they cannot predict what a single electron will do, and not even when it will do it. The implications of this are, of course, the limitations of measurement; of accuracy; of scientific predictability—all fundamental shortcomings of "classical," or Newtonian, physics—they suggest the collapse of absolute determinism even in the world of matter.

Second: the illusory nature of the ideal of objectivity. In quantum mechanics the very act of observing alters the nature of the object, "especially when its quantum numbers are small." Quantum physics, Heisenberg says, "do not allow a completely objective description of nature." "As it really happened" (or "as it is really happening") is, therefore, an incomplete statement in the world of matter, too. We are ahead of Ranke. "In our century," Heisenberg wrote in *The Physicist's Conception of Nature*, "it has become clear that the desired objective reality of the elementary particle is too crude an oversimplification of what really happens. . . ." "We can no longer speak of the behaviour of the particle independently of the process of observation. As a final consequence, the natural laws formulated mathematically in quantum theory no longer deal with the elementary particles themselves but with our knowledge of them." In *Physics and Philosophy* he explained this further:

> We cannot completely objectify the result of an observation, we cannot describe what "happens" between [one] observation and the next . . . any statement about what has "actually happened" is a statement in terms of the [Newtonian] classical concepts and—because of the thermo-dynamics and of the uncertainty relations—by its very nature incomplete with respect of the details of the atomic events involved. The demand "to describe what happens" in the quantum-theoretical process between two successive observations is a contradiction *in adjecto*, since the word "describe" refers to the use of classical concepts, while these concepts cannot be applied in the space between the observations; they can only be applied at the points of observation.

In biology, too, "it may be important for a complete understanding that the questions are asked by the species man which itself belongs to the genus of living organisms, in other words, that we already know what life is even before we have defined it scientifically." The recognition of personal participation is inescapable.

Third: the illusory nature of definitions. It seems that the minds of most physicists during the present interregnum still clung to the old, "logical" order of things: they were always giving names to newly discovered atomic particles, to such elements of the atomic kernel that did not "fit." Yet the introduction of the name "wavicle" does preciously little to solve the problem of whether light consists of waves or of particles; and it may be that the continuing nominalistic habit of proposing new terms (sometimes rather silly-sounding ones, such as "neutrino") suggests that illusion of the modern mind which tends to substitute vocabulary for thought, tending to believe that once we name or define something we've "got it." Sometimes things may get darker through definitions, Dr. Johnson said: and Heisenberg seems to confirm the limited value of definitions even in the world of matter:

> Any concepts or words which have been formed in the past through the interplay between the world and ourselves are not really sharply defined with respect to their meaning; that is to say, we do not know exactly how far they will help us in finding our way in the world. We often know that they can be applied to a wide range of inner or outer experience but we practically never know precisely the limits of their applicability. This is true even of the simplest and most general concepts like "existence" and "space and time".... The words "position" and "velocity" of an electron, for instance, seemed perfectly well defined as to both their meaning and their possible connections, and in fact they were clearly defined concepts within the mathematical framework of Newtonian mechanics. But actually they were not well defined, as is seen from the relations of uncertainty. One may say that regarding their position in Newtonian mechanics they were well defined, but in their relation to nature they were not.

Fourth: the illusory nature of the absolute truthfulness of mathematics. The absoluteness of mathematical "truth" was disproven by Gödel's famous theorem in 1931, but even before that, in the 1920's, physicists were beginning to ask themselves this uneasy question; as Heisenberg put it:

> Is it true that only such experimental situations can arise in nature as can be expressed in the mathematical formalism? The assumption that this was actually true led to limitations in the use of those concepts that had been the basis of physics since Newton. One could speak of the position and of the velocity of an electron as in Newtonian mechanics and one could observe and measure these quantities. But one could not fix both these quantities simultaneously with an arbitrarily high accuracy.... One had learned that the old concepts fit nature only inaccurately.

Mathematical truth is neither complete nor infinite (the velocity of light added to the velocity of light may amount to the velocity of light; on the other end of the physical scale there can be no action smaller than the quantum of action; and under certain physical conditions two by two do not always amount to four). Quantum theory found, too, that certain mathematical statements depend on the time element: Heisenberg realized that p times q is not always the equivalent of q times p in physics (when, for example, p means momentum and q position). What this suggests is that certain basic mathematical operations are not independent of human concepts of time and perhaps not even of purpose. That certain quantities do not always obey arithmetical rules was suggested already in the 1830's by the Irish mathematical genius Hamilton; and the Englishman Dirac, still to some extent influenced by nominalism, tried in the 1920's to solve this problem by asserting the necessity to deal with a set of so-called "Q numbers" which do not always respond to the rules of multiplication. But perhaps the "problem" may be stated more simply: the order in which certain mathematical (and physical) operations are performed affects their results.

Fifth: the illusory nature of "factual" truth. Change is an essential component of all nature: this ancient principle reappears within quantum physics. We have seen that the physicist must reconcile himself to the condition that he cannot exactly determine both the position and the speed of the atomic particle. He must reconcile himself, too, to the consequent condition that in the static, or factual, sense a basic unit of matter does not exist. It is not measurable; it is not even ascertainable; it is, in a way, a less substantial concept than such "idealistic" concepts as "beauty" or "mind." We can never expect to see a static atom or electron, since they do not exist as "immutable facts"; at best, we may see the trace of their motions. Einstein's relativity theory stated that matter is transmutable, and that it is affected by time; but the full implications of this condition were not immediately recognized, since they mean, among other things, that the earlier watertight distinctions between "organic" and "inorganic" substances no longer hold. "A sharp distinction between animate and inanimate matter," writes Heisenberg, "cannot be made." "There is only one kind of matter, but it can exist in different discrete stationary conditions." Heisenberg doubts "whether physics and chemistry will, together with the concept of evolution, some day offer a complete description of the living organism."

Sixth: the breakdown of the mechanical concept of causality. We have seen how, for the historian, *causa* must be more than the *causa efficiens*, and that the necessarily narrow logic of mechanical causality led to deterministic systems that have harmed our understanding of history, since in reality, through life and in history this kind of causation almost always "leaks." But now not even in physics is this kind of causation universally applicable: it is inadequate, and moreover, "fundamentally and intrinsically undemonstrable." There is simply no satisfactory way of picturing the fundamental atomic processes of nature in categories of space and time and causality. The multiplicity and the complexity of causes reappears in the world of physical relationships, in the world of matter.

Seventh: the principal importance of potentialities and tendencies.

Quantum physics brought the concept of potentiality back into physical science—a rediscovery, springing from new evidence, of some of the earliest Greek physical and philosophical theories. Heraclitus was the first to emphasize this in the reality of the world: *panta rei*, his motto, "Everything Moves," "imperishable change that renovates the world"; he did not, in the Cartesian and Newtonian manner, distinguish between being and becoming; to him fire was *both* matter and force. Modern quantum theory comes close to this when it describes energy, according to Heisenberg, anything that moves: "it may be called the primary cause of all change, and energy can be transformed into matter or heat or light." To Aristotle, too, matter was not by itself a reality but a *potentia*, which existed by means of form: through the processes of nature the Aristotelian "essence" passed from mere possibility through form into actuality. When we speak of the temperature of the atom, says Heisenberg, we can only mean an expectation, "an objective tendency or possibility, a *potentia* in the sense of Aristotelian philosophy." An accurate description of the elementary particle is impossible: "the only thing which can be written down as description is a probability function"; the particle "exists" only as a possibility, "a possibility for being or a tendency for being." But this probability is not merely the addition of the element of "chance," and it is something quite different from mathematical formulas of probabilities:

> Probability in mathematics or in statistical mechanics [writes Heisenberg] means a statement about our degree of knowledge of the actual situation. In throwing dice we do not know the fine details of the motion of our hands which determine the fall of the dice and therefore we say the probability for throwing a special number is just one in six. The probability wave of Bohr, Kramers, Slater, however, meant more than that; it meant a tendency for something. It was a quantitative version of the old concept of *potentia* in Aristotelian philosophy.

We have already met Heisenberg's question: "What happens 'really' in an atomic event?" The mechanism of the results of the observation can always be stated in the terms of the Newtonian concepts: "but what one deduces from an observation is a probability function . . . [which] does not itself represent a course of events in the course of time. It represents a tendency for events and our knowledge of events."

Eighth: not the essence of "factors" but their relationship counts. Modern physics now admits, as we have seen, that important factors may not have clear definitions: but, on the other hand, these factors *may* be clearly defined, as Heisenberg puts it, "with regard to their connections." These relationships are of primary importance: just as no "fact" can stand alone, apart from its associations with other "facts" and other matters, modern physics now tends to divide its world not into "different groups of objects but into different groups of connections." In the concepts of modern mathematics, too, it is being increasingly recognized how the functions of dynamic connections may be more important than the static definitions of "factors." Euclid had said that a point is something which has no parts and which occupies no space. At the height of positivism, around 1890, it was generally believed that an even more perfect statement would consist in exact definitions of "parts" and of "space." But certain mathematicians have since learned that this tinkering with definitions tends to degenerate into the useless nominalism of semantics, and consequently they do not bother with definitions of "points" or "lines" or "connection"; their interest is directed, instead, to the axiom that two points can be always connected by a line, to the relationships of lines and points and connections.

Ninth: the principles of "classical" logic are no longer unconditional: new concepts of truths are recognized. "Men fail to imagine any relation between two opposing truths and so they assume that to state one is to deny the other," Pascal wrote. Three centuries later Heisenberg wrote about some of C. F. von Weizaecker's propositions:

> It is especially one fundamental principle of classical logic which seems to require a modification. In classical logic it is assumed that, if a statement has any meaning at all, either the statement or the negation of the statement must be correct. Of "here is a table" or "here is not a table" either the first or the second statement must be correct. "Tertium non datur," a third possibility does not exist. It may be that we do not know whether the statement or its negation is correct; but in "reality" one of the two is correct.
>
> In quantum theory this law "tertium non datur" is to be modified . . . Weizaecker points out that one may distinguish various levels of language. . . . In order to cope with [certain quantum situations] Weizaecker introduced the concept "degree of truth". . . . [By this] the term "not decided" is by no means equivalent to the term "not known." . . . There is still complete equivalence between the two levels of language with respect to the correctness of a statement, but not with respect to the incorrectness. . . .

Knowledge means not certainty, and a half-truth is not 50 percent truth; everyday language cannot be eliminated from any meaningful human statement of truth, including propositions dealing with matter; after all is said, logic is human logic, our own creation,

Tenth: at the end of the Modern Age the Cartesian partition falls away. Descartes's framework, his partition of the world into objects and subjects, no longer holds:

> The mechanics of Newton [Heisenberg writes] and all the other parts of classical physics constructed after its model started out from the assumption that one can describe the world without speaking about God or ourselves. This possibility seemed almost a necessary condition for natural science in general.
>
> But at this point the situation changed to some extent through quantum theory . . . we cannot disregard the fact [I would say: the condition] that sci-

> ence is formed by men. Natural science does not simply describe and explain nature; it is a part of the interplay between nature and ourselves; it describes nature as exposed to our method of questioning. This was a possibility of which Descartes could not have thought [?] but it makes the sharp separation between the world and the I impossible.
>
> If one follows the great difficulty which even eminent scientists like Einstein had in understanding and accepting the Copenhagen interpretation of quantum theory, one can trace the roots of this difficulty to the Cartesian partition. This partition has penetrated deeply into the human mind during the three centuries following Descartes and it will take a long time for it to be replaced by a really different attitude toward the problem of reality.

We cannot avoid the condition of our participation. Throughout this book I tried to draw attention to the personal and moral and historical implications of this recognition, that instead of the cold and falsely aseptic remoteness of observation we need the warmth and the penetration of personal interest: but this is no longer the solitary longing of a humanist, a poetic exhortation. For "even in science," as Heisenberg says in *The Physicist's Conception of Nature*, "the object of research is no longer nature itself, but man's investigation of nature. Here, again, man confronts himself alone." And the recognition of this marks the beginning of a revolution not only in physical and philosophical but also in biological (and, ultimately, medical) concepts, springing from the empirical realization that there is a closer connection between mind and matter than what we have been taught to believe. Still, because of our interregnum, decades and disasters may have to pass until this revolution will bring its widely recognizable results. Yet we may at least look back at what we have already begun to leave behind.

After three hundred years the principal tendency in our century is still to believe that life is a scientific proposition, and to demonstrate how all of our concepts are but the products of complex mechanical causes that may be ultimately determinable through scientific methods. Thus Science, in Heisenberg's words, produced "its own, inherently uncritical"—and, let me add, inherently unhistorical—philosophy. But now "the scientific method of analysing, [defining] and classifying has become conscious"—though, let me add, far from sufficiently conscious—"of its limitations, which rise out of the [condition] that by its intervention science alters and refashions the object of investigation. In other words, methods and object can no longer be separated. *The scientific world-view has ceased to be a scientific view in the true sense of the word.*"

These are Heisenberg's italics. They correspond with the arguments of this book, in which I have tried to propose the historicity of reality as something which is prior to its mathematicability. They represent a reversal of thinking after three hundred years: but, in any event, such recognitions involve not merely philosophical problems or problems of human perception but the entirety of human involvement in nature, a condition from which we, carriers of life in its highest complexity, cannot separate ourselves. The condition of this participation is the recognition of our limitations which is, as I wrote earlier, our gateway to knowledge. "There is no use in discussing," Heisenberg writes, "what could be done if we were other beings than what we are." We must even keep in mind that the introduction of the "Cartesian" instruments such as telescopes and microscopes, which were first developed in the seventeenth century, do not, in spite of their many practical applications, bring us ***always and necessarily*** closer to reality—since they are interpositions, *our* interpositions, between our senses and the "object." We may even ask ourselves whether *our* task is still to "see" more rather than to see better, since not only does our internal deepening of human understanding now lag behind our accumulation of external information, but too, this external information is becoming increasingly abstract and unreal. Hence the increasing breakdown of internal communications: for, in order to see better, we must understand our own limitations better and also trust ourselves better. At the very moment of history when enormous governments are getting ready to shoot selected men hermetically encased in plastic bubbles out of the earth onto the moon, the importance of certain aspects of the "expanding universe" has begun to decline, and not only for humanitarian reasons alone; we are, again, in the center of the universe—inescapably as well as hopefully so.

Our problems—all of our problems—concern primarily human nature. The human factor is the basic factor. These are humanistic platitudes. But they have now gained added meaning, through the unexpected support from physics. It is thus that the recognitions of the human condition of science, and of the historicity of science—let me repeat that Heisenberg's approach is also historical—may mark the way toward the next phase in the evolution of human consciousness, in the Western world at least.

10 Epilogue: The Contemporary Revolt against Science

Throughout the comments and readings of this collection, I have argued that the content and methods of science have had both a broad and a deep impact on Western culture during the past three and one-half centuries. Basic patterns for our most pervasive and important attempts to order and understand the social and spiritual as well as the material aspects of the world have been drawn from natural science. But this process has not been unchallenged. While science has tended to dominate our world-view, it has never completely overwhelmed its critics. The widespread Romantic movement of the early nineteenth century, for example, involved an explicit rejection of the Enlightenment drive to apply scientific methods of structural analysis, dissection, and quantification to all realms of experience. The agonized cries of Goethe, who argued in his *Die Farbenlehre* that he wanted to do away with numbers even in optics and to talk about colors as the *deeds* and *sufferings* of light, became the symbol of a strong cultural movement which saw the Newtonian world as "substituting for our world of quality and sense perception, the world in which we live and die, another world—the world of quantity; of reified geometry—a world in which though there is a place for every*thing*, there is no place for man."[1]

For the romantics, science erred in at least three important ways. First, it appealed exclusively to "reason" or "rationality" as the ultimate arbiter of all disputes and disregarded our feelings and aesthetic sensibilities. Thus science seemed to cut itself off from a wide range of important human experience. Secondly, science abstracted and generalized from concrete reality, i.e., it dealt with concepts rather than with the wholeness of experience. So science invariably failed to account for or respond to the uniqueness of situa-

---

[1]This description of the Newtonian world is by Alexandre Koyré, "The Significance of the Newtonian Synthesis," *Newtonian Studies* (Cambridge, Mass.: Harvard University Press, 1965), p. 23; emphasis mine.

tions, and it tended to be analytic and destructive rather than creative.[2] Thirdly, science—especially the physical and mathematical sciences which dominated the enlightenment—tended to be mechanistic and materialistic. Consequently, it seemed to place far too much value on the dead and inert as opposed to the vital and spiritual, and it seemed to support atheism against deeply held Christian beliefs about the primacy of the soul over the body.

Throughout the nineteenth and early twentieth centuries, anti-scientific feelings related to the romantic critique existed in constant tension with the dominant scientism of Western society. In spite of the important impact of Social Darwinism (with its virtual equation of men and lower animals, for example), Victorian England saw an immensely strong upper-class tradition which followed Matthew Arnold in placing human perfection "in an internal condition, in the growth and predominance of our humanity proper, as distinguished from our animality." This tradition, which defined "culture" to exclude science and which rejected the materialism fostered by the scientific revolution, has played an important, continuing role in British intellectual and political life.[3]

A much more intense reaction against science set in among literary and artistic intellectuals on the continent toward the end of the nineteenth century, as their optimistic hopes for the scientism of Comte and his followers failed to materialize. In 1880 Emile Zola could still claim to base his approach to literature on the model of Claude Bernard's positivistic *Introduction to the Study of Experimental Medicine*,[4] but all around him authors, artists, and philosophers were rejecting the objective and (to them) sterile guides of science. They were seeking a less rational and mechanized, more intensely emotional and spontaneous approach to the world. Friedrich Neitzsche published his first major work, *The Birth of Tragedy* in 1871, signaling a new wave of emotionalism or subjectivism which manifested itself powerfully in the *fin de siècle* Expressionist movement and in the post-war movements of Dadaism and Surrealism in art and Existentialism in philosophy.

During most of the nineteenth and early twentieth centuries the anti-rationalist, anti-mechanistic and, hence, anti-scientific movements centered around relatively small intellectual elites. Since the third decade of this century, however, antagonism to at least some aspects of science has become

---

[2]Goethe's justification for associating conceptualization and abstraction with a "destructive" nature of science is dealt with in Robert Bloch, "Goethe, Idealistic Morphology, and Science," *American Scientist* 40 (1952), 317-322.

[3]On the continuing estrangement between men of "culture" and scientists, see C. P. Snow's *The Two Cultures: And a Second Look*, an expanded version of *The Two Cultures and the Scientific Revolution* (Cambridge, England: Cambridge University Press, 1963).

[4]Emile Zola's *The Experimental Novel and Other Essays* (New York: Cassell Publishing Co., 1893) explains how the author of a novel must play the role of a scientific experimenter, observing his characters function according to the laws of nature in various circumstances which he can control.

an increasingly important element of popular Western culture. And since the mid-1960's the antagonism toward science and technology has become so powerful and widespread among young intellectuals that it appears destined for a dominant position.

This mid-twentieth century movement against science retains many of the traditional Romantic criticisms. In some senses it may even be seen as a democratization of the old British aristocratic tradition; for the increasing affluence of Western society leaves ever larger numbers of young people free of material worries and able to concentrate on the spiritual, aesthetic, and specifically human concerns which were once the province of small elites. But the major focus of the contemporary revolt has shifted from cultural to political concerns. This shift has been the result of important attempts to extend scientific techniques beyond the realm of material nature into the realm of social and political action and of the increasing interdependence between scientific expertise and political power.

From the beginning of the Enlightenment movement to study men and society scientifically, it was implicitly accepted that human behavior must be subsumed under deterministic laws just as natural events are—as Voltaire ironically put it in his *Ignorant Philosopher*, "it would be very singular that all nature, all the planets should obey eternal laws, and that there should be a little animal, five feet high, who in contempt of these laws could act as he pleased, solely according to his caprice." But until the early twentieth century there seemed to be little practical threat that man might be as predictable and hence as controllable as inanimate objects and the lower animals. Furthermore, to those few who believed in the possibility of human control through science during the early twentieth century, it seemed a hopeful possibility; psychology could help to reduce mental illness, genetic control could help to reduce physical deformity and hereditary defects, and sociology could provide a pattern for the ideal society.

In 1910, for example, the great American geneticist, Herman Muller, wrote: "Science, in the form especially of psychology and sociology, will discover what qualities are desirable for the most efficient cooperation and for the best enjoyment of life; and science, in the form especially of physiology and genetics . . . will discover what the elementary bases of these qualities are and how to procure them for man."[5]

It was not until 1924 that a fictional account seriously raised the negative side of the potential for scientific human control and thereby opened up the problem of the relation of science to authority. In that year, Eugene Zamiatin wrote *We*, a novel which showed how the technological

---

[5]Quoted from Garland E. Allen, "Science and Society in the Eugenic Thought of H. J. Muller," *Bio Science* 20 (1970), 349. Many of those who supported the American eugenics movement from 1900 to 1930 were certain that poverty and crime could be virtually abolished by controlling the reproduction of the genetically inferior lower classes. See Donald K. Pickens' *Eugenics and the Progressives* (Vanderbilt University Press, 1968) for a good discussion of the eugenics movement in America.

products of science might lead to a totalitarian repressive society and to a destruction of the individualism so highly prized by Western man. In *We*, human beings have no names, only numbers; they live in glass apartments so they can be constantly watched; and all "human" values are sacrificed to the goals of power and efficiency, goals closely associated with science and technology. Then in 1932, Aldous Huxley published ***Brave New World***, a prophetic novel which investigated the potentials of eugenics and behavioral psychology for creating a dystopia of horrifying dimensions. Biological science produces carbon copy test-tube people, and the reinforcement techniques of behaviorism keeps them gratified and happy in their robot-like existence; if boredom or malaise should set in, a Soma tablet sends them on a pleasant trip. Finally, in 1949, George Orwell presented what is probably the most terrifying of all dystopian visions, ***1984***, in which thought police, and telescreens provide the means for mass extermination of human individuality under the scientific dictatorship of Big Brother.

The revulsion toward technological society and toward its scientific foundations portrayed in the new dystopian fiction was strongly reinforced by the reality of German National Socialism in the 1930's. The Nazis demonstrated concretely that a biologically based ideology could support totalitarianism. And the complicity of German scientists in the development of efficient techniques to destroy a whole race of humans corroded the belief of many that science was basically "good," or at worst, value neutral. Similarly, the creation and immediate use of the atomic bomb with its unprecedented potential for destruction raised new questions about the benevolence of science.

It became particularly clear in the Nazi example and in the development of nuclear weaponry that science—especially modern science with its voracious appetite for monetary and social support—was closely bound up with the interests of politically dominant groups. Even if this connection is deplored by many scientists and by the orthodox vision of science as a disinterested search for truth, the practical political exigencies of modern society tend to make the connection ever more intimate. As Herbert Marcuse has written in a penetrating article:

> Science today is in a position of power that almost immediately translates pure scientific achievement into political and military weapons of global use and effectiveness. The fact that the organization and control of whole populations, in peace as well as in war, have become, in a strict sense, a *scientific* control and organization (from the most ordinary household gadgets to the highly sophisticated methods in public opinion formation, publicity, and propaganda) inexorably unites scientific research and experiment with the powers and plans of the economic, political, and military establishment.[6]

---

[6]Herbert Marcuse, "The Responsibility of Science" in *The Responsibility of Power*, edited by Leonard Krieger, (Garden City, N. Y.: Doubleday & Co., 1969), pp. 24-25.

The situation depicted by Marcuse has given rise to a bewildering variety of antagonistic responses. These join selected elements of the Romantic critique of science to a spectrum of political and social attitudes ranging from the social commitment and political hope of some radical activists to the total rejection of both society and politics by the deeply alienated hippie culture.[7]

For the most part, hippie complaints about science and technology differ from radical criticisms more in their intensity and pessimism than in the fundamental subjects of concern. Since radical criticisms have been better articulated and are more widely accepted, I will concentrate on them in what follows, referring only occasionally to the hippie extensions of radical arguments.

Radical opposition to science and technology can only be understood if we recognize the basic ordering of values which modern radicals share with the Romantic tradition. First of all, they prize direct, personal, I-thou encounters between unique human beings; thus any value, institution, or ideology which would diminish the frequency and importance of person-to-person relationships is suspect. Furthermore, relationships which involve exploitation, manipulation, control, or domination—i.e., relationships in which one human treats another as an object rather than as a unique individual—are abhorred. In addition, radical values include an openness not only to other human beings, but also to all aspects of one's own personality—all feelings, impulses, and fantasies as well as intellectualized thoughts. Thus no cause which urges the rejection or denial of part of one's self is completely acceptable.[8]

In light of these values one can understand why the close connection between science and political authority should seem threatening. The problem is particularly grave precisely because science is not seen as a mere tool providing manipulative power to political authorities whose legitimacy may be under strong attack for a variety of reasons—it is that, but it is much more. The very essence of science, claim the radicals, is manipulative; science seeks to predict the course of events in order to be able to control them. So authorities seem to be encouraged in their manipulative designs by the very *telos* of science. Furthermore, say the critics, it is from the sciences that our tendency to quantify and "objectify" men devolves. The social sciences in particular encourage us to view men as statistical entities and consequently to deal with them on impersonal terms. Theodore Roszak has eloquently presented the radical repugnance toward the detached observational procedures of social science in speaking of the approach of psychologists, anthropologists, and political scientists to their subjects:

---

[7]I follow Kenneth Keniston's valuable and sensitive typology of contemporary dissenters here. See "The Sources of Student Dissent," *The Journal of Social Issues* 23 (1967), 108-137.

[8]These values are discussed in Kenniston, *op. cit.*

> . . . in all such cases what the observer may very well be saying to the observed is the same: ". . . I shall observe this behavior of yours and record it. I shall not enter into your life, your task, your condition of existence. Do not turn to me or appeal to me or ask me to become involved with you. I am here only as a temporary observer whose role is to stand back and record and later make my own sense out of what you seem to be doing or intending. I assume I can adequately understand what you are doing or intending without entering wholly into your life. I am not particularly interested in what *you* uniquely are; I am interested only in the general pattern to which you conform. I assume that I have the right to use you to perform this process of classification. I assume that I have the right to reduce all that you are to an integer in my science."[9]

Radicals contend that if this attitude were found only among a few scientists it might be distressing but not dangerous. But all too often the impersonal approach of the social scientist is adopted by social and political authorities and made the cornerstone of vast bureaucratic systems. Men are classified, sorted, given social security numbers and selective service numbers. In short, it seems to some that their individuality is being eroded in order that they may be more easily handled by the mechanistic scientific system. It is almost as if scientific classification has become the paradigm for political organization.

It is not clear that the nature of science demands a reduction of human individuality. But on this point the criticisms of scientism and postivistic science cannot be easily brushed away. It is certainly true that even in its purest, most theoretical form science seeks uniformities among phenomena and things; it often ignores complicating factors—in a fundamental sense, then, science is insensitive to individuality and uniqueness.[10] If the scientist could remain aware that his theories involve a simplification of reality necessitated by his inability to grasp its full complexity, his uniformitarian demands would cause no trouble. But in the world of action the seductive tendency to simplify and pigeon-hole has dangerous and important manifestations.

The applied scientist exploits uniformities in nature to control nature and the technologist exploits them to produce uniform products—identical batches of chemicals, identical automobiles. He does this in part by taking natural products of varying constitution and by manipulating and refining them until the variations are eliminated or at least reduced to negligible proportions. As long as he deals with inanimate objects, these manipulations are usually unobjectionable (although we are becoming increasingly aware

---

[9]Theodore Roszak, *The Making of a Counter Culture: Reflections on the Technocratic Society and Its Youthful Opposition* (Garden City, N. Y.: Doubleday & Co., 1969), pp. 222-223. Roszak is more virulent in his attack on objectivity than most radicals, agreeing with what is more commonly the hippie position.

[10]It is clear here that the exact sciences rather than the biological sciences are taken as most representative of science. Natural history, and to a degree taxonomy, glory in variety and genetics and certain branches of psychology are interested in individual differences.

of the frequent and often unfortunate modifications of our environment which they may entail). But when the applied social scientist seeks to manipulate human beings to make them more uniform and to make society more easily exploited, men might rightly rebel and ask, "to what ends are individuals and societies to be molded into the likeness of machines?"[11]

This problem of the ends to which an increasingly scientific view of the world is to be directed brings us to the question of who will determine these ends (and once more to the critical problem of the relationship between science and political authority which was raised in its most drastic form by George Orwell in *1984*) which are at the bottom of much of the contemporary fear of science. This problem was beautifully illustrated in a 1967 debate between the anti-establishment poet, Allen Ginsberg, and James Fox, a representative of the Food and Drug Administration, over the makeup of a proposed committee to control research on marijuana and LSD:

> "Who is going to be on that committee?" asked Ginsberg, "is Timothy Leary going to be on that committee, is Alan Watts going to be on that committee . . . will I be on the committee?" In other words, who is going to be the selector with the controls, the controller, and here we have a *1984* problem, because if Goddard [F.D.A. administrator] is biased against aesthetic experience and against judgments of value which other people hold dear, then you are going to have a group composed of very hard-nosed doctors or hypocrites.[12]

One need not agree with all of Ginsberg's values to acknowledge that he makes a telling point. It is the natural tendency of those in power to share their authority primarily with technically trained experts, since the complexity of scientific and technological systems seems to make it impossible for any but a scientific-technological elite to understand and hence to make decisions about the use of the products of modern science. Thus more and more crucial societal decisions are being made by technocrats whose principal values are efficiency and order rather than spontaneity and variety, and who judge the successes and failures of social institutions in impersonal, objective, and quantitative rather than in human and qualitative terms.

---

[11]The threat to individuality and spontaneity seems particularly intense today because it is a central issue in a whole constellation of basic social problems only some of which are related directly and consciously to the issue of rationalizing society through science. For example, it is one of the central issues raised by intellectuals among ethnic minorities in America today. Many Blacks and Chicanos question the idea that they should conform to the social standards of the white majority and ask, "Why must we be willing to sacrifice our uniqueness and individuality to be accepted fully within society?" Similarly, one of the basic issues in the revolt of the youth culture against all forms of authority—parental, governmental, and academic—centers on the pressures to accept a very limited range of life styles—pressures which are seen as restricting spontaneity, creativity, and individuality in favor of a bland orderliness.

[12]Allen Ginsberg and James H. Fox, "Seminar on Marijuana and LSD Controls" in Charles Hollander, ed., ***Background Papers on Student Drug Involvement*** (Washington, D. C.: United States National Student Association, 1967).

Modern Western men (especially those with some humanistic leanings) with their emphasis on the importance of individual worth and self-determination rebel at placing themselves in the hands of the scientific experts. They have been well warned by Huxley and Orwell about what could happen, and they have seen our society move in ominous directions already, as decisions seem to be made on technical rather than on human grounds. Thus, today there is a large-scale and often very perceptive reaction against those aspects of science which militate against human freedoms.

Unfortunately, justifiable radical critiques of certain characteristics and distortions of science spill over among a few activists and many hippies into an uncritical or overcritical rejection of nearly all things scientific (mind-expanding drugs, electronic amplifiers, and recordings are often conspicuously excepted), just as in many of the cases I have presented in this collection the uncritical acceptance of scientific insights has spilled over into an unconcern for basically human values.

The new alienated Romantic (hippie) mood borrows from the valuable humanistic writings of the young Marx, the suggestive neo-Freudian writings of Erich Fromm, Herbert Marcuse, and Norman O. Brown, the long tradition of Eastern religious mysticism and the anti-intellectual element in the American democratic tradition in order to reject both the products and the modes of understanding of modern science.[13]

It would return to a primitivism which seeks a direct communion with nature, man, and God to replace the set of relations built upon the complex intellectual inheritance of Western civilization.

What are those of us, who see Western civilization and the scientific tradition within it as fundamentally compatible with the fullest realization of the human potential, to say to those who fear and reject the manifestations of modern science? Ultimately we can state only a series of hopes and beliefs which accept many of the current criticisms of modern institutional science and its impact on society but which argue that a fundamental *reformation* of our acceptance of science rather than a *revolution* against all of its foundations and manifestations is called for.[14]

---

[13]The scope of the present work does not allow a detailed analysis of the roots of the contemporary so-called "counterculture." Elements of humanistic Marxism can be seen especially in the writings of Herbert Marcuse and Erich Fromm, neither of whom would go as far as their younger readers in rejecting all elements of a scientific world view. Norman Brown's neo-Freudian *Life against Death* (Middletown, Conn.: Wesleyan University Press, 1959) and *Love's Body* (New York: Random House, 1966) go much further in rejecting the whole rationalistic bias in Western thought and in opting for mythical modes of perception. The role of transcendental meditation, yoga, and the Hari Krishna movement in the youth culture testifies to the Eastern element; the rhetorical cry of "power to the people" is closely related to the movement's demand for direct democracy in which the reign of all expertise and authority is overthrown. Though this call for direct democracy is by no means uniquely American and devolves in part from Marxist ideology, it nonetheless has clear historical roots in traditional American values.

[14]The notion that science and technology, valuable in themselves, have been corrupted in modern society and that a return to earlier, purer, attitudes within their history is needed—that

Let me deal first with the spectre raised by early and often naive interpretations of such systems as behavioral psychology that men do fit into a deterministic system and are thus ultimately controllable. It would be absurd to argue that scientific techniques are not already capable of asserting important influences on men—of modifying not only patterns of overt behavior but also of changing even certain objects of desire. But there are fundamental theoretical reasons for thinking that man's deterministic element is more limited and that our ability to manipulate men is less extensive than Huxley and Orwell feared. As Henry Margenau argued in his "Quantum Mechanics, Free Will, and Determinism" (pp. 283-292), modern physics provides a rationale for believing that there is an irreducible element of "freedom" in the world. And neurophysiologists are well on the way to demonstrating that any attempt to establish a form of psycho-physical determinism which can deal with significant aspects of human thought and behavior is in principle impossible.[15]

Secondly, let us consider the extent to which control and manipulation are necessary concomitants of a scientific tradition. On this topic I would like to distinguish between technological and scientific values. Radicals may be justified in arguing that the popular Anglo-American view of science, based on the Baconian dictum that knowledge is power, makes no distinction between the two. But this popular view may simply be misleading in fundamental ways. Most scientists throughout history would have argued that the goal of science is *understanding* while the goal of technology may be manipulation—and they could point to numerous historical examples of explicitly nonmanipulative sciences like traditional cosmology, historical geology, descriptive evolutionary biology, etc.[16]

To a tyrannized man it may make little immediate difference whether the tyrant is called a technocrat or scientist; but if we are interested in the justice of the contemporary rejection of scientific modes of perception, then the distinction is significant. Intellection is no less an element of human existence than are emotions, and the Romantic should be no more willing to reject his understanding than the scientist should be willing to renounce his emotions, unless there are very strong reasons for doing so. Insofar as the Radical or hippie rejects scientific aspects of our world view because science is assumed to be manipulative, he commits a gross injustice and limits his own personality.

What, then, are we to say of the scientific demand for objectivity? Is it true that men must diminish their humanity to function as scientists? It is no

---

something akin to the Reformation of the Church must take place—has been developed by Paul Goodman. See his "Can Technology Be Humane?" *The New York Review of Books* 13 (November 20, 1969), 27-34.

[15]See, for example, Donald O. Walter, "The Indeterminacies of the Brain," *Perspectives in Biology and Medicine* (Winter, 1968), 203-207.

[16]See Melvin Kranzberg, "The Disunity of Science-Technology," *American Scientist* 56 (1968), 21-33.

more true that scientific objectivity demands that we renounce our feelings and emotional involvements than it is that mystical experience demands that we suspend our rationality (which has an equal claim to being human). So the fact that scientific methodology requires a temporary emphasis on a limited range of human capacities provides no justification for seeing it as uniquely dehumanizing. Romantic tendencies which constrain human existence within the limits of emotion and immediate intuition seem almost as narrow as scientific tendencies to emphasize rationality and abstract speculation. Human beings are incapable of simultaneously exploring all elements of the vast range of their existence, so if we are to be fully open to all aspects of our personality (as the radicals profess), we must be open to objective modes as well as subjective modes of experience.

In spite of these comments, the radical and hippie criticisms of scientific objectivity are an important and valuable counter to what has become the dominant cultural attitude. Roszak is perceptive when he says that "Objectivity as a state of being fills the very air we breathe in a scientific culture; it grips us subliminally in all we say, feel, and do. . . . We seek to adapt our lives to the dictates of that mentality or, at the very least, we respond to it acquiescently in the myriad images and pronouncements in which it manifests itself during every waking hour."[17] Thus, if we are to retain some balance in our lives, we must frequently be reminded of the value of alternatives to "objective" ideals.

Let us turn now from the internal characteristics of science to some of its political ramifications. In spite of the limitations inherent in scientific systems to predict human behavior and provide subtle means for human control, it remains true that scientific developments have made possible increasing concentrations of power and information. And this power has generally been allied with established authorities to be used at their whim for their purposes. What is to be said of the relationship between science and authority? I have already mentioned the great temptations which draw them together. But is it inevitable that science serve those in power?[18]

There are both theoretical and empirical reasons for denying that this should be the case. Within the explicit value system adopted by the scientific community, an anti-authoritarian concern for free inquiry and for the intensely critical evaluation of scientific results based on objective criteria rather than on the prestige of the investigator or on social desirability has been conspicuously present. No doubt the prescriptions of scientific ideology are imperfectly applied, especially in connection with the internal working of

---

[17]Theodore Roszak, *The Making of a Counter Culture* . . . , op. cit., p. 216.

[18]In responding to this question, I want to deal exclusively with the contemporary situation. The very close historical relationship between anti-authoritarian movements and modern science is beautifully discussed in David Thomson, "Scientific Thought and Revolutionary Movements," *The Impact of Science on Society* 6 (1955), 3-29.

science,[19] so that the objectivity of science can easily be interpreted as myth rather than as reality. But the prescriptive myth—if it be that—does often provide justification for scientific challenges to social and political orthodoxy. A good example appears in the work of the South African anatomist, Phillip Tobias, who has challenged apartheid views on the inferiority of Negroes through his work on brain size. Tobias justifies his critical work by writing:

> Science requires . . . the constant reexamination and critical reappraisal of premises and assumptions, the elasticity of mind which permits, nay demands, old hypotheses to be modified when new facts emerge which cannot be adequately explained by them, resistance to the tendency to develop a vested interest in a particular viewpoint, avoidance of ascribing motives to scientists of opposite viewpoint in favour of the unbiased examination of the evidence they may advance. . . .[20]

Furthermore, there is a current and rapidly growing trend among scientists to question the uses to which their work is put and to challenge the present uncritical complicity between science and political authority. Without going into detail, I will mention only a few examples. The contemporary social concern of scientists first manifested itself in the scientists' movement at the end of World War II. Manhattan project scientists defied Defense Department directives in order to inform Congress and the public about the dramatic political implications of atomic energy and to urge the creation of a civilian- rather than a military-dominated Atomic Energy Commission.[21] During the 1950's, the Committee for Environmental Information (formerly the Committee on Nuclear Information), publicized the crucial issue of the effects of nuclear fallout from atomic weapons testing; and its continuing fight with the AEC and innumerable other commercial and governmental agencies to preserve the integrity of our environment has been more successful than its limited resources seem to warrant.[22] Finally, the recent opposi-

---

[19]The attack on Immanuel Velikovsky by the scientific establishment in response to the challenge presented by his *Worlds in Collision* (New York: Delta Paperback edition, 1965) to orthodox cosmology provides a graphic illustration of the violation of explicitly accepted canons of objectivity by modern scientific authorities. See Alfred De Grazia, *The Velikovsky Affair: The Warfare of Science and Scientists* (New Hyde Park, N. Y.: University Books, 1966).

[20]"Brain Size, Gray Matter, and Race—Fact or Fiction?" *American Journal of Physical Anthropology* 32 (1970), 22.

[21]The story of the scientists' movement is thoroughly told in Alice K. Smith, *A Peril and a Hope* (Chicago: University of Chicago Press, 1965). That the civilian AEC has not been effective in subordinating military to civilian priorities should not be used as evidence against the scientists' intent.

[22]*Environment* (formerly *Scientist and Citizen*), a publication of the Committee for Environmental Information and the Scientists' Institute for Public Information, is filled with sensitive articles about the control of science and with evidence of moderate successes, such as instituting

tion to ABM on the part of large numbers of scientists demonstrates their social concern.[23] These are only the beginnings of a spreading movement among the practitioners of science who see that it must be more carefully subordinated to human needs and desires, and that its products must be introduced only with the fullest possible awareness of their broad implications. The emphasis on prediction, power, and efficiency inherent in technology creates great difficulties for those who would humanize our society; but the evidence does not suggest that these difficulties must be insurmountable.

Finally, I believe that on balance, scientific and technological developments have increased rather than diminished men's opportunities to develop themselves both materially and intellectually. Surely the eradication of disease, the increase in agricultural productivity, the multiplication of means of communication and transportation, and the provision of a huge variety of materials and techniques for creating objects of use and of beauty must, in themselves, be counted as gains. Too often these gains have cost more than necessary in the spoilage of our environment and in the regimentation of our lives; but with our renewed awareness of nature and humanity we may find a better balance between the enjoyment of human and natural creations than we have in the recent past.

The uncritical celebration of scientific systems and technological artifacts and the ignorant cry for immediate and natural experience are equally restrictive and limiting—both are less than fully human. Science has provided a variety of things upon which man can act as well as a new range of experience for him to contemplate; and if we are vigilant, it need not destroy old modes of action or of thought while it opens up new possibilities for existence.

---

programs for eliminating lead poisoning in slums in spite of Establishment indifference (and, occasionally, hostility) and blocking the erection of atomic power installations where danger to the environment seems to be great.

[23]See A. Chayes and J. B. Wiesner, eds., *ABM: An Evaluation of the Decision To Deploy an Antiballistic Missile System* (New York: Harper & Row, Publishers, 1969).

11 Bibliographical Essay

Since this collection of readings is intended primarily for undergraduate students, I have included only a few of the most outstanding non-English sources. The source references of the selections included in this anthology may be extremely useful to interested students. In general, books from which selections have been presented are not included again below.

1 General Treatments of the Interaction between Science and Culture

Preeminent among the almost countless survey works dealing with the impact of science on culture is Alfred North Whitehead, *Science and the Modern World* (New York: The Macmillan Company, 1925). Other good general treatments are Everet W. Hall, *Modern Science and Human Values* (Princeton, N. J.: D. Van Nostrand Co., 1956), William Thomas Jones, *The Sciences and the Humanities: Conflict and Reconciliation* (Berkeley: University of California Press, 1965), Floyd W. Matson, *The Broken Image: Man, Science, and Society* (New York: Braziller, 1964), Herbert J. Muller, *Science and Criticism: The Humanistic Tradition in Contemporary Thought* (New Haven, Conn.: Yale University Press, 1943), and R. B. Lindsay, *The Role of Science in Civilization* (New York: Harper & Row, Publishers, 1963). John Herman Randall, *The Making of the Modern Mind*, rev. ed. (Boston: Houghton Mifflin Co., 1940), is one of the few general intellectual histories to deal in a concerted way with the scientific aspect of modern culture.

A number of first-rate collections of articles have dealt broadly with the subjects discussed in this book. Among them, Ruth Nanda Anshen, ed., *Science and Man* (New York: Harcourt, Brace & Co., 1942), though it deals largely with the material impact of science, is important for presenting a number of Marxist interpretations of the impact of science on social thought. Robert C. Stauffer, ed., *Science and Civilization* (Madison: University of Wisconsin Press, 1949), brings together the insights of a number of historians of science, and the two volumes edited by Gerald James Holton, *Science and the Modern Mind* (Boston: Houghton Mifflin Co., 1958) and *Science and Culture* (Boston: Houghton Mifflin Co., 1965), contain excellent interpretive essays which first appeared in *Daedalus*.

A huge number of books has been written on the interactions between science and religion. Andrew Dickson White's *A History of the Warfare of Science with Theology* (New York: D. Appleton and Company, 1896) is now available in paperback (New York: Dover, 1960); it is one of the great classics in this field as are Bertrand Russell's less scholarly *Religion and Science* (London: Oxford University Press, 1935), and Michael Polanyi's *Science, Faith, and Society* (London: Oxford University Press, 1946).

On the metaphysical and epistemological implications of science see Milič Čapek, *Philosophical Impact of Contemporary Physics* (Princeton: D. Van Nostrand Co., 1961), whose content is more inclusive than its title would suggest; Robin George Collingwood, *The Idea of Nature* (Oxford: Oxford University Press, 1945); and Alfred North Whitehead, *Essays in Science and Philosophy* (New York: Philosophical Library, 1947). Science and ethical concerns are dealt with by Hiram Bentley Glass in *Science and Ethical Values* (Chapel Hill: University of North Carolina Press, 1965) and by Henry Margenau in "Western Culture, Scientific Method, and the Problem of Ethics," *American Journal of Physics* 15 (1947), 218-228.

Harry Elmer Barnes' *Social Thought from Lore to Science*, 2 vols. (Boston: D. C. Heath & Co., 1938) is a good sympathetic account of the increasing scientific influence on social and political thought, while Sheldon S. Wolin's article, "Political Theory as a Vocation," *American Political Science Review* 63 (1969), 1062-1082, provides a carefully argued critique of the tendency to make all political thought scientific. Karl W. Deutsch's "Mechanism, Organization, and Society: Some Models in Natural and Social Science," *Philosophy of Science* 18 (1951), 230-252, Michael Walzer's "On the Role of Symbolism in Political Thought," *Political Science Quarterly* 82 (1967), 191-204, and H. E. Barnes, "Representative Biological Theories of Society," *Sociological Review* 17 (1925), 120-130, 182-194, 294-300, and 18 (1926), 100-105, 231-243, 306-314, all emphasize the use of scientific models in political thought.

The first chapter of David Joravsky's *Soviet Marxism and Natural Science: 1917-1932* (New York: Columbia University Press, 1961), provides a good introduction to the role of scientific ideas in various formulations of Marxist thought.

The relation of scientific thought to literature and art can best be approached by looking first at Marjorie Nicolson's superb short annotated bibliography, "Resource Letter SL-1 on Science and Literature," *American Journal of Physics* 33 (1965), 175-183.

The general point of view implicit in this book—that models and metaphors provide the basis for all important explanations—is developed in Max Black, *Models and Metaphors* (Ithaca, N. Y.: Cornell University Press, 1962) and by Douglas Berggren who concludes his "The Use and Abuse of Metaphor," *Reviews of Metaphysics* 16 (1962-1963), 237-258, 450-472, with the contention that "truly creative and nonmythic thought, whether in the arts, the sciences, religion, or metaphysics, must be invariably and irreducibly metaphorical."

2 and 3 The New Philosophy of the Seventeenth Century and the Mechanical Philosophy

Two of the best widely available introductions to the scientific revolution are Marie Boas, *The Scientific Renaissance: 1450-1630* (London: Collins, 1962) and Alfred Rupert Hall, *From Galileo to Newton: 1630-1720* (London: Collins, 1963). The general interaction between seventeenth-century science and culture is described in an elementary but very useful manner by John Frederick West in his *The Great Intellectual Revolution* (London: John Murray, Ltd., 1965) and by Preserved Smith in *Origins of Modern Culture: 1543-1687* (New York: Holt, Rinehart and Winston, 1934). More detailed analyses of the interactions between philosophy, theology, and science can be found in Alexandre Koyre, *From the Closed World to the Infinite Universe* (Baltimore: The Johns Hopkins Press, 1957) and Edwin Arthur Burtt, *The Metaphysical Foundations of Modern Physical Science* (New York: The Humanities Press, 1952), reprinted from 1925 original. Among the many more specialized monographs and articles on science and culture in the seventeenth century, the following stand out: Francis Rarick Johnson, *Astronomical Thought in Renaissance England* (Baltimore: Johns Hopkins Press, 1937); Thomas S. Kuhn, *The Copernican Revolution* (Cambridge, Mass.: Harvard University Press, 1957), especially Chapter 6; Richard Foster Jones, *Ancients and Moderns: A Study of the Rise of the Scientific Movement in Seventeenth-Century England*, 2nd ed. (St. Louis: The Washington University Press, 1961); Richard S. Westfall, *Science and Religion in Seventeenth-Century England* (New Haven, Conn.: Yale University Press, 1958); Leonard Marsak, "Bernard de Fontenelle: the Idea of Science in the French Enlightenment," *Transactions of the American Philosophical Society*, New Series 49 (1959), part 7; W. H. Greenleaf, *Order, Empiricism, and Politics* (London: Oxford University Press, 1964); James Schall, "Cartesianism and Political Theory," *Review of Politics* 24 (1962), 260-282; Henry G. Van Leeuwen, *The Problem of Certainty in English Thought, 1630-1690* (The Hague: Martinus Nijhoff, 1963); Hedley Howell Rhys, ed., *Seventeenth-Century Science and the Arts* (Princeton, N.

J.: Princeton University Press, 1961); and a series of works by Marjorie Hope Nicolson, including *Science and Imagination* (Ithaca: Cornell University Press, 1956), *The Breaking of the Circle: Studies in the Effect of the New Science upon Seventeenth-Century Poetry* (New York: Columbia University Press, 1960) and *Pepys' Diary and the New Science* (Charlottesville, N. C.: University Press of Virginia, 1965).

4 Newtonian Scientism and Other Aspects of Eighteenth- and Early Nineteenth-Century Interactions between Science and Culture

The general significance of scientific development for the Enlightenment has been well discussed in Alexandre Koyre, "The Significance of the Newtonian Synthesis," reprinted in *Newtonian Studies* (Cambridge, Mass.: Harvard University Press, 1965), in Ernst Cassirer, *The Philosophy of the Enlightenment* (Princeton, N. J.: Princeton University Press, 1951) and in Carl Becker's *The Heavenly City of the Eighteenth-Century Philosophers* (New Haven, Conn.: Yale University Press, 1932). Becker's simplistic emphasis on Newtonianism has been astutely criticized by Henry Guerlac in "Newton's Changing Reputation in the Eighteenth Century" in Raymond O. Rockwood, ed., *Carl Becker's Heavenly City Revisited* (Ithaca, N.Y.: Cornell University Press, 1958).

Hélène Metzger's *Attraction Universelle et Religion Naturelle chez Quelques Commentateurs Anglais de Newton* (Paris: Hermann et Cie., 1938), is the classic study of Newtonian Religion, and Robert Harris Hurlbutt's *Hume, Newton, and the Design Argument* (Lincoln: University of Nebraska Press, 1965), provides a clear analysis of the major religious implications of Newtonianism.

The response of social and political thinkers to scientific developments during the French Enlightenment is dealt with by Kingsley Martin in his *French Liberal Thought in the Eighteenth Century* (London: Ernest Benn, Ltd., 1929). Gladys Bryson, *Man and Society: The Scottish Inquiry of the Eighteenth Century* (Princeton, N. J.: Princeton University Press, 1945), does a similar job for the Scottish Enlightenment, and Frank Edward Manuel's *The Prophets of Paris* (Cambridge, Mass.: Harvard University Press, 1962), carries the analysis into the early nineteenth century. Thomas Neill's "Quesnay and Physiocracy," *Journal of the History of Ideas* 9 (1948), 153-173, discusses the underpinnings of the new science of economics and M. J. La Boulle, "La Mathématique Sociale: Condorcet et ses Prédécesseurs," *Revue d'Histoire Litteraire de la France* 46 (1939), 33-55, deals more generally with the attempt to quantify and objectify social thought. In addition to Frederick August von Hayek's *The Counter-Revolution of Science* (Glencoe, Ill.: The Free Press of Glencoe, 1952), a number of critiques and analyses of scientism in social theory have been written. Of these, Eric Voeglins' "The Origins of Scientism," *Social Research* 15 (1948), 462-494, is representative. David Thomson's "Scientific Thought and Revolutionary Movements," *Impact of Science on Society* 6 (1955), 3-29, is one of the few recent articles which see scientific influences on political ideology as revolutionary rather than reactionary during the early nineteenth century. Harry Hayden Clark's "The Influence of Science on American Ideas from 1775-1809," *Transactions of the Wisconsin Academy of Sciences, Arts, and Letters* 35 (1944), 305-349, provides an introduction to the religious, political, and social interpretations of Newtonianism in the early days of the United States.

The literary impact of eighteenth-century science is discussed in William P. Jones, *The Rhetoric of Science: A Study of Scientific Ideas and Imagery in Eighteenth-Century Poetry* (Berkeley: University of California Press, 1966) and in Marjorie Nicolson, *Newton Demands the Muse: Newton's Opticks and the Eighteenth-Century Poets* (Princeton, N. J.: Princeton University Press, 1946).

5 Darwinism

There are a number of first-rate general studies of Darwin and his cultural impact. A good short treatment is John C. Greene, *Darwin and the Modern World View* (Baton Rouge: Louisiana State University Press, 1961). Gertrude Himmelfarb, *Darwin and the Darwinian Revolution* (Garden City, N. Y.: Doubleday & Co., 1959), provides more information on Darwin, the man, than Greene, and also discusses his impact. Alvar Ellegård, *Darwin and the General Reader* (Göteborg, Sweden: Elanders Boktrycheri Aktiebolag, 1958), carefully analyzes the reaction to Darwin in the British periodical press. Two anthologies on Darwinism seem to be particularly outstanding: Bernard R. Kogan, ed., *Darwin and His Critics: The Darwinian Revolution* (San Francisco: Wadsworth, 1960), collects much information from the nineteenth-century reactions to Darwin; Phillip Appleman, ed., *Darwin: Texts, Backgrounds, Contemporary Opinion, Critical Essays* (New York: Norton Press, 1970), provides just what its title suggests.

Darwin's literary influence, while inseparable from his religious, social, and philosophical impact, is discussed extensively in Jacques Barzun's *Darwin, Marx, and Wagner: Critique of a Heritage* (Boston: Little, Brown and Co., 1941); in William Irvine's *Apes, Angels, and Victorians: Darwin, Huxley, and Evolution* (New York: McGraw-Hill Book Co., 1955), and "The Influence of Darwin on Literature," *Proceedings of the American Philosophical Society* 103 (1959), 616-628, in Leo Justin Henkin's *Darwinism in the English Novel, 1860-1910* (New York: Corporate Press, Inc.,

1940), and in Basil Willey's *Darwin and Butler: Two Versions of Evolution* (London and New York: Chatto and Windus, 1960).

The background to the religious impact of evolutionary ideas is discussed in Charles C. Gillespie, *Genesis and Geology, a Study in the Relations of Scientific Thought, Natural Theology, and Social Opinion in Great Britain, 1790-1850* (Cambridge, Mass.: Harvard University Press, 1951), and the specifically Darwinian impact on religion has been presented in Walter J. Ong, ed., *Darwin's Vision and Christian Perspectives* (New York: The Macmillan Company, 1960), and in John C. Greene, "Darwin and Religion," *Proceedings of the American Philosophical Society* 103 (1959), 716-725, as well as in more general works.

Darwin's importance for conservative social thought is emphasized in Gloria McConnaughey, "Darwin and Social Darwinism," *Osiris* 9 (1950), 397-412, and in Richard Hofstadter, *Social Darwinism in American Thought, 1816-1915* (Philadelphia: University of Pennsylvania Press, 1944), while the cooperative aspects of Social Darwinism are emphasized in Frederick Webb Headley, *Darwinism and Modern Socialism* (London: Methuen and Co., 1909) and in Ashley Montague, *Darwin; Competition and Cooperation* (New York: Henry Schuman, 1952). J. W. Burrow, *Evolution and Society: A Study in Victorian Social Theory* (London: Cambridge University Press, 1966), and George E. Simpson, "Darwinism and Social Darwinism," *Antioch Review* 19 (1959), 33-45, attempt more dispassionate analyses.

6 Energetics

One of the most ambitious attempts to place a scientific theory within its full cultural context is Stephen Brush, "Thermodynamics and History," *The Graduate Journal* 7 (1967), 477-565. N. F. Mott's "Physical Science and the Beliefs of the Victorians" in the BBC's *Ideas and Beliefs of the Victorians: An Historic Reevaluation of the Victorian Age*, (London: Sylvan Press, 1949), pp. 215-221, provides another interesting general view.

André Lalande's *La Dissolution Opposée à L'Évolution dans Les Sciences Physiques et Morales* (Paris: Alcan, 1899), deals with the moral implications of the second law of thermodynamics, and William Persehouse Delisle Wightman, *Science and Monism* (London: Allen and Unwin, 1934), and Donald Meyer, "Paul Carus and the Religion of Science," *American Quarterly* 14 (1962), 597-607, deal with the religious implications of the thermodynamic approach. William H. Jordy, *Henry Adams, Scientific Historian* (New Haven, Conn.: Yale University Press, 1952), is an excellent guide to the scientific element in Adams' work.

7 Freud

The best single short introduction to Freud's theories and their extensions is James Alexander Campbell Brown, *Freud and the Post-Freudians* (Baltimore: Penguin Books, 1961). Two good collections of essays that deal with many aspects of the Freudian impact on contemporary culture are Iago Galdston, ed., *Freud and Contemporary Culture* (New York: International Universities Press, 1957), and Benjamin N. Nelson, ed., *Freud and the Twentieth Century* (New York: Meridian Books, 1957).

The literary impact of Freud is emphasized by Lionel Trilling in *Freud and the Crisis of Our Culture* (Boston: Beacon Press, 1955), in Frederick John Hoffman, *Freudianism and the Literary Mind*, 2nd ed. (Baton Rouge: Louisiana State University Press, 1957), and in Lawrence Durrell's *A Key to Modern British Poetry* (Norman: University of Oklahoma Press, 1952).

Thomas E. Johnston's *Freud and Political Thought* (New York: The Citadel Press, 1965), is an uninspired but immensely useful book. Lewis Samuel Feuer's *Psychoanalysis and Ethics* (Springfield, Ill.: Charles C Thomas, 1955), while it is a polemical work, deals with the crucial issue of Freud's relationship to traditional human value structures.

8 Positivism and Behaviorism

Walter Simon's *European Positivism in the Nineteenth Century* (Ithaca, New York: Cornell University Press, 1963) provides a clear discussion of traditional Comtean positivism from a relatively dispassionate perspective. Lesyek Kalakowski's *The Alienation of Reason: A History of Positivist Thought* (Garden City, N. Y.: Doubleday & Co., 1968) covers the topic from medieval times into the twentieth century and combines history with criticism of positivist doctrines.

D. G. Charlton deals with the apex of French positivistic thought in his *Positive Thought in France during the Second Empire: 1852-1870* (Oxford, England: The Clarendon Press, 1959), and *Claude Bernard and His Place in the History of Ideas* (Lincoln: University of Nebraska Press, 1960) by Reino Virtanen discusses the central figure in French positivism after Comte. Frank E. Manuel's *The Prophets of Paris* (Cambridge, Mass.: Harvard University Press, 1962) provides a marvelous introduction to Comte and the early days of Positivism. No understanding of late nineteenth-century positivism would be complete without a look at Ernst Mach's *The Science of Mechanics: A Critical and Historical Account of Its Development*, 6th ed. (La Salle, Ill.: The Open Court Publishing Company, 1960). A. J. Ayers' introductory essay in his *Logical Positivism* (Glencoe, Ill.: The Free Press of

Glencoe, 1959) provides a good introduction to the Vienna circle and to modern logical positivism, as do the first two essays in Peter Achinstein and Stephen Barker, eds., *The Legacy of Logical Positivism* (Baltimore: Johns Hopkins Press, 1969).

The literature on behaviorism in psychology and social theory is so vast that I will mention only B. F. Skinner's *Walden Two* (New York: Macmillan, 1949), a Utopian novel which provides a thorough insight into the basic character of the behavioral approach.

9 Modern Physics

A number of prominent modern physicists have written sensitive books and articles on the philosophical and cultural implications of their work. Of these, Niels Henrik David Bohr's *Essays, 1958/1962 on Atomic Physics and Human Knowledge* (New York: John Wiley, 1963), P. W. Bridgman's "Philosophical Implications of Physics," *Bulletin of the American Academy of Arts and Sciences* 3 (1950) No. 5, Albert Einstein, *Ideas and Opinions* (New York: Crown Publishers, 1954), Werner Heisenberg, *Physics and Philosophy; The Revolution in Modern Science* (New York: Harper & Row, Publishers, 1962), and Erwin Schrödinger, *Science and Humanism* (Cambridge, England: Cambridge University Press, 1951), provide a representative sampling. Susan Stebbing's *Philosophy and the Physicists* (London: Methuen and Company, 1937), provides a semipopular account of the philosophical problems raised by modern physics.

The impact of relativity theory on modern literature is dealt with in Lawrence Durrell's *A Key to Modern British Poetry* (Norman: University of Oklahoma Press, 1952), and in numerous articles exemplified by Alfred Bork's "Durrell and Relativity," *The Centennial Review of Arts and Sciences* 7 (1963), 191-203, and Theodore Ziolkowski's "Herman Broch and Relativity in Fiction," *Wisconsin Studies in Contemporary Literature* 8 (1967), 365-376. Relativity's implications for theology are considered in Karl Heim, *The Transformation of the Scientific World View* (New York: Harper & Row, Publishers, 1953) and in M. Davidson, "Modern Cosmology and the Theologians," *Vistas in Astronomy* 1 (1955), 166-172.

The implications of Quantum mechanics have been most generally discussed by the physicists named above, but there is a growing semipopular periodical literature exemplified by Richard Lichtman, "Indeterminacy in the Social Sciences," *Inquiry* 10 (1967), 139-150, and Earl R. MacCormac, "Indeterminacy and Theology," *Religion in Life* 36 (1967), 355-370.

10 Epilogue—The Contemporary Revolt against Science

One of the most sympathetic, insightful, and broadly ranging analyses of the contemporary youth culture is Theodore Roszak, *The Making of a Counter-Culture: Reflections on the Technocratic Society and Its Youthful Opposition* (Garden City, N. Y.: Doubleday & Co., 1969). The *Harvard University Program on Technology, Fourth Annual Report: 1967-1968* (Cambridge, Mass.: Harvard University Program on Technology and Society, 1968), especially Emmanuel Mesthene's essay, "Technology and Values," provides one of the very few recent considered defenses of technology. John McDermott's "Technology: The Opiate of the Intellectuals," *New York Review of Books* 13 (July 31, 1969), 25-35, is a short, bitter, and cogent critique of the Harvard Program Report and all that it stands for. If you have only a few minutes to consider the modern radical critique of technological society, this is the place to look. The single most extensive analysis of the domination of technology in contemporary society is Jacques Ellul, *The Technological Society* (New York: Alfred Knopf, 1964); and one of the most probing examinations of the cultural implications of accepting the values of technology is Herbert Marcuse, *One-Dimensional Man* (Boston: Beacon Press, 1964). Among the many anthologies devoted to the problems arising out of our technological society, John Burke's *The New Technology and Human Values* (Belmont, Calif.: Wadsworth, 1966) stands out.